T0334392

LONDON MATHEMATICAL SOCIETY LECTURE NOTE SERIES

Managing Editor: Professor M. Reid, Mathematics Institute, University of Warwick, Coventry CV4 7AL, United Kingdom

The titles below are available from booksellers, or from Cambridge University Press at www.cambridge.org/mathematics

216 Stochastic partial differential equations, A. ETHERIDGE (ed)
217 Quadratic forms with applications to algebraic geometry and topology, A. PFISTER
218 Surveys in combinatorics, 1995, P. ROWLINSON (ed)
220 Algebraic set theory, A. JOYAL & I. MOERDIJK
221 Harmonic approximation, S.J. GARDINER
222 Advances in linear logic, J.-Y. GIRARD, Y. LAFONT & L. REGNIER (eds)
223 Analytic semigroups and semilinear initial boundary value problems, K. TAIRA
224 Computability, enumerability, unsolvability, S.B. COOPER, T.A. SLAMAN & S.S. WAINER (eds)
225 A mathematical introduction to string theory, S. ALBEVERIO *et al*
226 Novikov conjectures, index theorems and rigidity I, S.C. FERRY, A. RANICKI & J. ROSENBERG (eds)
227 Novikov conjectures, index theorems and rigidity II, S.C. FERRY, A. RANICKI & J. ROSENBERG (eds)
228 Ergodic theory of $\mathbf{Z}^d$ actions, M. POLLICOTT & K. SCHMIDT (eds)
229 Ergodicity for infinite dimensional systems, G. DA PRATO & J. ZABCZYK
230 Prolegomena to a middlebrow arithmetic of curves of genus 2, J.W.S. CASSELS & E.V. FLYNN
231 Semigroup theory and its applications, K.H. HOFMANN & M.W. MISLOVE (eds)
232 The descriptive set theory of Polish group actions, H. BECKER & A.S. KECHRIS
233 Finite fields and applications, S. COHEN & H. NIEDERREITER (eds)
234 Introduction to subfactors, V. JONES & V.S. SUNDER
235 Number theory: Séminaire de théorie des nombres de Paris 1993–94, S. DAVID (ed)
236 The James forest, H. FETTER & B.G. DE BUEN
237 Sieve methods, exponential sums, and their applications in number theory, G.R.H. GREAVES *et al* (eds)
238 Representation theory and algebraic geometry, A. MARTSINKOVSKY & G. TODOROV (eds)
240 Stable groups, F.O. WAGNER
241 Surveys in combinatorics, 1997, R.A. BAILEY (ed)
242 Geometric Galois actions I, L. SCHNEPS & P. LOCHAK (eds)
243 Geometric Galois actions II, L. SCHNEPS & P. LOCHAK (eds)
244 Model theory of groups and automorphism groups, D.M. EVANS (ed)
245 Geometry, combinatorial designs and related structures, J.W.P. HIRSCHFELD *et al* (eds)
246 p-Automorphisms of finite p-groups, E.I. KHUKHRO
247 Analytic number theory, Y. MOTOHASHI (ed)
248 Tame topology and O-minimal structures, L. VAN DEN DRIES
249 The atlas of finite groups - ten years on, R.T. CURTIS & R.A. WILSON (eds)
250 Characters and blocks of finite groups, G. NAVARRO
251 Gröbner bases and applications, B. BUCHBERGER & F. WINKLER (eds)
252 Geometry and cohomology in group theory, P.H. KROPHOLLER, G.A. NIBLO & R. STÖHR (eds)
253 The q-Schur algebra, S. DONKIN
254 Galois representations in arithmetic algebraic geometry, A.J. SCHOLL & R.L. TAYLOR (eds)
255 Symmetries and integrability of difference equations, P.A. CLARKSON & F.W. NIJHOFF (eds)
256 Aspects of Galois theory, H. VÖLKLEIN, J.G. THOMPSON, D. HARBATER & P. MÜLLER (eds)
257 An introduction to noncommutative differential geometry and its physical applications (2nd edition), J. MADORE
258 Sets and proofs, S.B. COOPER & J.K. TRUSS (eds)
259 Models and computability, S.B. COOPER & J. TRUSS (eds)
260 Groups St Andrews 1997 in Bath I, C.M. CAMPBELL *et al* (eds)
261 Groups St Andrews 1997 in Bath II, C.M. CAMPBELL *et al* (eds)
262 Analysis and logic, C.W. HENSON, J. IOVINO, A.S. KECHRIS & E. ODELL
263 Singularity theory, W. BRUCE & D. MOND (eds)
264 New trends in algebraic geometry, K. HULEK, F. CATANESE, C. PETERS & M. REID (eds)
265 Elliptic curves in cryptography, I. BLAKE, G. SEROUSSI & N. SMART
267 Surveys in combinatorics, 1999, J.D. LAMB & D.A. PREECE (eds)
268 Spectral asymptotics in the semi-classical limit, M. DIMASSI & J. SJÖSTRAND
269 Ergodic theory and topological dynamics of group actions on homogeneous spaces, M.B. BEKKA & M. MAYER
271 Singular perturbations of differential operators, S. ALBEVERIO & P. KURASOV
272 Character theory for the odd order theorem, T. PETERFALVI. Translated by R. SANDLING
273 Spectral theory and geometry, E.B. DAVIES & Y. SAFAROV (eds)
274 The Mandelbrot set, theme and variations, T. LEI (ed)
275 Descriptive set theory and dynamical systems, M. FOREMAN, A.S. KECHRIS, A. LOUVEAU & B. WEISS (eds)
276 Singularities of plane curves, E. CASAS-ALVERO
277 Computational and geometric aspects of modern algebra, M. ATKINSON *et al* (eds)
278 Global attractors in abstract parabolic problems, J.W. CHOLEWA & T. DLOTKO
279 Topics in symbolic dynamics and applications, F. BLANCHARD, A. MAASS & A. NOGUEIRA (eds)
280 Characters and automorphism groups of compact Riemann surfaces, T. BREUER
281 Explicit birational geometry of 3-folds, A. CORTI & M. REID (eds)
282 Auslander-Buchweitz approximations of equivariant modules, M. HASHIMOTO
283 Nonlinear elasticity, Y.B. FU & R.W. OGDEN (eds)
284 Foundations of computational mathematics, R. DEVORE, A. ISERLES & E. SÜLI (eds)
285 Rational points on curves over finite fields, H. NIEDERREITER & C. XING
286 Clifford algebras and spinors (2nd edition), P. LOUNESTO
287 Topics on Riemann surfaces and Fuchsian groups, E. BUJALANCE, A.F. COSTA & E. MARTÍNEZ (eds)
288 Surveys in combinatorics, 2001, J.W.P. HIRSCHFELD (ed)
289 Aspects of Sobolev-type inequalities, L. SALOFF-COSTE
290 Quantum groups and Lie theory, A. PRESSLEY (ed)
291 Tits buildings and the model theory of groups, K. TENT (ed)

292 A quantum groups primer, S. MAJID
293 Second order partial differential equations in Hilbert spaces, G. DA PRATO & J. ZABCZYK
294 Introduction to operator space theory, G. PISIER
295 Geometry and integrability, L. MASON & Y. NUTKU (eds)
296 Lectures on invariant theory, I. DOLGACHEV
297 The homotopy category of simply connected 4-manifolds, H.-J. BAUES
298 Higher operads, higher categories, T. LEINSTER (ed)
299 Kleinian groups and hyperbolic 3-manifolds, Y. KOMORI, V. MARKOVIC & C. SERIES (eds)
300 Introduction to Möbius differential geometry, U. HERTRICH-JEROMIN
301 Stable modules and the D(2)-problem, F.E.A. JOHNSON
302 Discrete and continuous nonlinear Schrödinger systems, M.J. ABLOWITZ, B. PRINARI & A.D. TRUBATCH
303 Number theory and algebraic geometry, M. REID & A. SKOROBOGATOV (eds)
304 Groups St Andrews 2001 in Oxford I, C.M. CAMPBELL, E.F. ROBERTSON & G.C. SMITH (eds)
305 Groups St Andrews 2001 in Oxford II, C.M. CAMPBELL, E.F. ROBERTSON & G.C. SMITH (eds)
306 Geometric mechanics and symmetry, J. MONTALDI & T. RATIU (eds)
307 Surveys in combinatorics 2003, C.D. WENSLEY (ed.)
308 Topology, geometry and quantum field theory, U.L. TILLMANN (ed)
309 Corings and comodules, T. BRZEZINSKI & R. WISBAUER
310 Topics in dynamics and ergodic theory, S. BEZUGLYI & S. KOLYADA (eds)
311 Groups: topological, combinatorial and arithmetic aspects, T.W. MÜLLER (ed)
312 Foundations of computational mathematics, Minneapolis 2002, F. CUCKER *et al* (eds)
313 Transcendental aspects of algebraic cycles, S. MÜLLER-STACH & C. PETERS (eds)
314 Spectral generalizations of line graphs, D. CVETKOVIĆ, P. ROWLINSON & S. SIMIĆ
315 Structured ring spectra, A. BAKER & B. RICHTER (eds)
316 Linear logic in computer science, T. EHRHARD, P. RUET, J.-Y. GIRARD & P. SCOTT (eds)
317 Advances in elliptic curve cryptography, I.F. BLAKE, G. SEROUSSI & N.P. SMART (eds)
318 Perturbation of the boundary in boundary-value problems of partial differential equations, D. HENRY
319 Double affine Hecke algebras, I. CHEREDNIK
320 L-functions and Galois representations, D. BURNS, K. BUZZARD & J. NEKOVÁŘ (eds)
321 Surveys in modern mathematics, V. PRASOLOV & Y. ILYASHENKO (eds)
322 Recent perspectives in random matrix theory and number theory, F. MEZZADRI & N.C. SNAITH (eds)
323 Poisson geometry, deformation quantisation and group representations, S. GUTT *et al* (eds)
324 Singularities and computer algebra, C. LOSSEN & G. PFISTER (eds)
325 Lectures on the Ricci flow, P. TOPPING
326 Modular representations of finite groups of Lie type, J.E. HUMPHREYS
327 Surveys in combinatorics 2005, B.S. WEBB (ed)
328 Fundamentals of hyperbolic manifolds, R. CANARY, D. EPSTEIN & A. MARDEN (eds)
329 Spaces of Kleinian groups, Y. MINSKY, M. SAKUMA & C. SERIES (eds)
330 Noncommutative localization in algebra and topology, A. RANICKI (ed)
331 Foundations of computational mathematics, Santander 2005, L.M PARDO, A. PINKUS, E. SÜLI & M.J. TODD (eds)
332 Handbook of tilting theory, L. ANGELERI HÜGEL, D. HAPPEL & H. KRAUSE (eds)
333 Synthetic differential geometry (2nd edition), A. KOCK
334 The Navier–Stokes equations, N. RILEY & P. DRAZIN
335 Lectures on the combinatorics of free probability, A. NICA & R. SPEICHER
336 Integral closure of ideals, rings, and modules, I. SWANSON & C. HUNEKE
337 Methods in Banach space theory, J.M.F. CASTILLO & W.B. JOHNSON (eds)
338 Surveys in geometry and number theory, N. YOUNG (ed)
339 Groups St Andrews 2005 I, C.M. CAMPBELL, M.R. QUICK, E.F. ROBERTSON & G.C. SMITH (eds)
340 Groups St Andrews 2005 II, C.M. CAMPBELL, M.R. QUICK, E.F. ROBERTSON & G.C. SMITH (eds)
341 Ranks of elliptic curves and random matrix theory, J.B. CONREY, D.W. FARMER, F. MEZZADRI & N.C. SNAITH (eds)
342 Elliptic cohomology, H.R. MILLER & D.C. RAVENEL (eds)
343 Algebraic cycles and motives I, J. NAGEL & C. PETERS (eds)
344 Algebraic cycles and motives II, J. NAGEL & C. PETERS (eds)
345 Algebraic and analytic geometry, A. NEEMAN
346 Surveys in combinatorics 2007, A. HILTON & J. TALBOT (eds)
347 Surveys in contemporary mathematics, N. YOUNG & Y. CHOI (eds)
348 Transcendental dynamics and complex analysis, P.J. RIPPON & G.M. STALLARD (eds)
349 Model theory with applications to algebra and analysis I, Z. CHATZIDAKIS, D. MACPHERSON, A. PILLAY & A. WILKIE (eds)
350 Model theory with applications to algebra and analysis II, Z. CHATZIDAKIS, D. MACPHERSON, A. PILLAY & A. WILKIE (eds)
351 Finite von Neumann algebras and masas, A.M. SINCLAIR & R.R. SMITH
352 Number theory and polynomials, J. MCKEE & C. SMYTH (eds)
353 Trends in stochastic analysis, J. BLATH, P. MÖRTERS & M. SCHEUTZOW (eds)
354 Groups and analysis, K. TENT (ed)
355 Non-equilibrium statistical mechanics and turbulence, J. CARDY, G. FALKOVICH & K. GAWEDZKI
356 Elliptic curves and big Galois representations, D. DELBOURGO
357 Algebraic theory of differential equations, M.A.H. MACCALLUM & A.V. MIKHAILOV (eds)
358 Geometric and cohomological methods in group theory, M.R. BRIDSON, P.H. KROPHOLLER & I.J. LEARY (eds)
359 Moduli spaces and vector bundles, L. BRAMBILA-PAZ, S.B. BRADLOW, O. GARCÍA-PRADA & S. RAMANAN (eds)
360 Zariski geometries, B. ZILBER
361 Words: Notes on verbal width in groups, D. SEGAL
362 Differential tensor algebras and their module categories, R. BAUTISTA, L. SALMERÓN & R. ZUAZUA
363 Foundations of computational mathematics, Hong Kong 2008, M.J. TODD, F. CUCKER & A. PINKUS (eds)
364 Partial differential equations and fluid mechanics, J.C. ROBINSON & J.L. RODRIGO (eds)
365 Surveys in combinatorics 2009, S. HUCZYNSKA, J.D. MITCHELL & C.M. RONEY-DOUGAL (eds)
366 Highly oscillatory problems, B. ENGQUIST, A. FOKAS, E. HAIRER & A. ISERLES (eds)
367 Random matrices, High dimensional phenomena, G. BLOWER

Geometric and Cohomological Methods in Group Theory

Edited by

MARTIN R. BRIDSON
University of Oxford

PETER H. KROPHOLLER
University of Glasgow

IAN J. LEARY
The Ohio State University

Shaftesbury Road, Cambridge CB2 8EA, United Kingdom

One Liberty Plaza, 20th Floor, New York, NY 10006, USA

477 Williamstown Road, Port Melbourne, VIC 3207, Australia

314–321, 3rd Floor, Plot 3, Splendor Forum, Jasola District Centre, New Delhi – 110025, India

103 Penang Road, #05–06/07, Visioncrest Commercial, Singapore 238467

Cambridge University Press is part of Cambridge University Press & Assessment, a department of the University of Cambridge.

We share the University's mission to contribute to society through the pursuit of education, learning and research at the highest international levels of excellence.

www.cambridge.org
Information on this title: www.cambridge.org/9780521757249

First published 2009

A catalogue record for this publication is available from the British Library

ISBN 978-0-521-75724-9 Paperback

Contents

Preface

More than eighty mathematicians from a variety of countries gathered in Durham in July 2003 for the London Mathematical Society's symposium on Geometry and Cohomology in Group Theory. This was the third symposium in an influential sequence of meetings that began with the meeting organised by Scott and Wall in 1976 and continued with the Kropholler–Stöhr meeting in 1994. As with these previous meetings, the 2003 Symposium attracted many of the world's leading researchers in this highly active field of mathematics.

The meeting came at an exciting time in the field, marked by a deepening of the fertile interactions with logic, analysis and large-scale geometry, as well as striking progress on classical problems at the heart of cohomological group theory. The symposium was built around six lecture courses exposing important aspects of these recent developments. The lecturers were A. Adem, W. Lück, J. McCammond, L. Mosher, R. Oliver, and Z. Sela.

The structure of this volume reflects that of the symposium: major survey articles form the backbone of the book, providing an extended tour through a selection of the most important trends in modern geometric group theory; these are supported by shorter research articles on diverse topics. All of the articles were refereed and we thank the referees for their hard work.

The articles corresponding to the minicourses are written in a style that researchers approaching the field for the first time should find inviting. In the first, Bestvina and Feighn present their own interpretation of Sela's theory of limit groups. (Important aspects of the theory are developed in the many exercises and an appendix by Wilton guides the reader through these.) Lück's essay on L^2-methods in geometry, topology and group theory is crafted specifically for an algebraically-minded audience. Mosher's article on the quasi-isometric rigidity of certain mapping class groups begins with a general introduction to quasi-isometric rigidity. McCammond's account of non-positive curvature in group theory focuses on the explicit construction of examples, emphasising the utility of combinatorics and computational group theory in this regard.

We thank all of the authors who contributed to this volume and apologise to them for our tardiness in gathering their work into final form. We thank the London Mathematical Society and the Engineering and Physical Sciences Research Council of the United Kingdom for their continuing support of the Durham Symposia series. We recall with particular fondness the contribution that the late Karl Gruenberg made to this symposium and its predecessors. We thank him and all of the participants of the 2003 symposium, each of whom made a contribution to its congenial atmosphere and mathematical success. We hope that some of the mathematical excitement generated at the meeting will be transmitted to the reader through these proceedings.

Martin R. Bridson, University of Oxford
Peter H. Kropholler, University of Glasgow
Ian J. Leary, The Ohio State University

List of Participants

A. Adem, Madison Wisconsin	adem@math.ubc.ca
M. Batty, Newcastle	Michael.Batty@ncl.ac.uk
M. Bestvina, Utah	bestvina@math.utah.edu
I. C. Borge, Oslo	ingerbo@math.uio.no
B. H. Bowditch, Southampton	B.H.Bowditch@warwick.ac.uk
T. Brady, Dublin	tom.brady@dcu.ie
M. R. Bridson, Imperial	Bridson@maths.ox.ac.uk
C. J. B. Brookes, Cambridge	C.J.B.Brookes@pmms.cam.ac.uk
J. Brookman, Edinburgh	brookman@maths.ed.ac.uk
R. Bryant, UMIST	bryant@umist.ac.uk
M. Burger, ETH Zürich	marc.burger@fim.math.ethz.ch
R. Charney, Ohio State	charney@brandeis.edu
I. L. Chatterji, Cornell	indira@math.ohio-state.edu
I. M. Chiswell, QM, London	I.M.Chiswell@qmul.ac.uk
D. Collins, QM, London	D.J.Collins@qmul.ac.uk
J. Crisp, Bourgogne	jcrisp@u-bourgogne.fr
M. W. Davis, Ohio State	mdavis@math.ohio-state.edu
A. Duncan, Newcastle	a.duncan@ncl.ac.uk
M. J. Dunwoody, Southampton	m.j.dunwoody@maths.soton.ac.uk
J. Dymara, Wroclaw	dymara@math.uni.wroc.pl
V. Easson, Oxford	V.Easson@dpmms.cam.ac.uk
B. Eckmann, ETH Zürich	beno.eckmann@math.ethz.ch
M. Feighn, Rutgers	feighn@rutgers.edu
K. Fujiwara, Tohoku	fujiwara@math.is.tohoku.ac.jp
I. Galvez, London Metropolitan	i.galvezicarrillo@londonmet.ac.uk

With affiliations at time of the Symposium and email addresses at time of writing.

R. Geoghegan, Binghamton	ross@math.binghamton.edu
K. Goda, Newcastle	k.m.goda@ncl.ac.uk
J. P. C. Greenlees	j.greenlees@sheffield.ac.uk
D. Groves, Oxford	groves@math.uic.edu
K. W. Gruenberg✠, QM, London	
I. Hambleton, McMaster	hambleton@mcmaster.ca
A. Harkins, Cambridge	a.harkins@pmms.cam.ac.uk
S. Hermiller, Nebraska	smh@math.unl.edu
J. Howie, Heriot-Watt	J.Howie@ma.hw.ac.uk
J. R. Hunton, Leicester	jrh7@mcs.le.ac.uk
A. Iozzi, ETH Zürich	iozzi@math.ethz.ch
T. Januszkiewicz, Wroclaw	tjan@math.ohio-state.edu
D. Juan-Pineda, UNAM Morelia	daniel@matmor.unam.mx
A. Kirk, QM, London	A.Kirk@qmul.ac.uk
N. Kopteva, Heriot-Watt	natasha@math.nsc.ru
A. Korzeniewski, Edinburgh	A.J.Korzeniewski@sms.ed.ac.uk
P. H. Kropholler, Glasgow	P.H.Kropholler@maths.gla.ac.uk
C. Leedham-Green, QM, London	C.R.Leedham-Green@qmul.ac.uk
R. Levi, Aberdeen	ran@maths.abdn.ac.uk
I. J. Leary, Southampton	leary@math.ohio-state.edu
P. Linnell, Virginia Tech	linnell@math.vt.edu
W. Lück, Münster	wolfgang.lueck@math.uni-muenster.de
Z. Lykova, Newcastle	z.a.lykova@ncl.ac.uk
A. MacIntyre, Edinburgh	A.Macintyre@qmul.ac.uk
J. McCammond, UC Santa Barbara	jon.mccammond@math.ucsb.edu
J. Meier, Lafayette	meierj@lafayette.edu
G. Mislin, ETH Zürich	guido.mislin@math.ethz.ch
N. Monod, Chicago	nicolas.monod@epfl.ch
L. Mosher, Rutgers	mosher@andromeda.rutgers.edu
Th. Müller, QM, London	T.W.Muller@qmul.ac.uk
G. A. Niblo, Southampton	g.a.niblo@soton.ac.uk
B. E. A. Nucinkis, ETH Zürich	B.E.A.Nucinkis@soton.ac.uk
R. Oliver, Paris 13	bobol@math.univ-paris13.fr
A. Piggott, Oxford	adam.piggott@tufts.edu
S. J. Pride, Glasgow	sjp@maths.gla.ac.uk
F. Quinn, Virginia Tech	quinn@math.vt.edu
A. Ranicki, Edinburgh	a.ranicki@ed.ac.uk
S. Rees, Newcastle	Sarah.Rees@ncl.ac.uk
H. Reich, Münster	reich@math.uni-duesseldorf.de
A. Rhemtulla, Alberta	akbar@malindi.ualberta.ca
J. Rickard, Bristol	j.rickard@bristol.ac.uk
T. Riley, Yale	tim.riley@bris.ac.uk
C. Röver, Newcastle	claas.roever@nuigalway.ie
P. Rowley, UMIST	peter.rowley@umist.ac.uk

M Saadetoğlu, Southampton muge.saadetoglu@emu.edu.tr
M. Sageev, Technion sageevm@tx.technion.ac.il
R. Sánchez-García, Southampton sanchez@math.uni-duesseldorf.de
M. Sapir, Vanderbilt m.sapir@vanderbilt.edu
R. Sauer, Münster romansauer@member.ams.org
Th. Schick, Göttingen schick@uni-math.gwdg.de
Z. Sela, Jerusalem zlil@math.huji.ac.il
K. Shackleton, Southampton kjs2006@alumni.soton.ac.uk
V. P. Snaith, Southampton V.Snaith@sheffield.ac.uk
R. Stöhr, UMIST Ralph.Stohr@umist.ac.uk
J. Świątkowski, Wroclaw Jacek.Swiatkowski@math.uni.wroc.pl
O. Talelli, Athens otalelli@math.uoa.gr
A. Tonks, London Metropolitan a.tonks@londonmet.ac.uk
M. Tweedale, Imperial m.tweedale@bristol.ac.uk
A. Valette, Neuchâtel alain.valette@unine.ch
K. Vogtmann, Cornell vogtmann@math.cornell.edu
C. T. C. Wall, Liverpool ctcw@liv.ac.uk
Th. Weigel, Milan - Bicocca thomas.weigel@unimib.it
E. Wharton, Birmingham whartone@for.mat.bham.ac.uk
G. Williams, UC Dublin gwill@essex.ac.uk
J. S. Wilson, Birmingham wilsonjs@maths.ox.ac.uk
R. Wilson, Birmingham R.A.Wilson@qmul.ac.uk
H. Wilton, Imperial henry.wilton@math.utexas.edu

Notes on Sela's work: Limit groups and Makanin-Razborov diagrams

Mladen Bestvina* Mark Feighn*

Abstract

This is the first in a planned series of papers giving an alternate approach to Zlil Sela's work on the Tarski problems. The present paper is an exposition of work of Kharlampovich-Myasnikov and Sela giving a parametrization of $Hom(G, \mathbb{F})$ where G is a finitely generated group and $\mathbb{F}$ is a non-abelian free group.

Contents

1 The Main Theorem

1.1 Introduction

This is the first of a planned series of papers giving an alternative approach to Zlil Sela's work on the Tarski problems [35, 34, 36, 38, 37, 39, 40, 41, 31, 32].

*The authors gratefully acknowledge support of the National Science Foundation. This material is based upon work supported by the National Science Foundation under Grants Nos. DMS-0502441 and DMS-0805440.

The present paper is an exposition of the following result of Kharlampovich-Myasnikov [14, 15] and Sela [34]:

Theorem. *Let G be a finitely generated non-free group. There is a finite collection $\{q_i : G \to \Gamma_i\}$ of proper epimorphisms of G such that, for any homomorphism f from G to a free group F, there is $\alpha \in Aut(G)$ such that $f\alpha$ factors through some q_i.*

A more refined statement is given in the Main Theorem on page 7. Our approach, though similar to Sela's, differs in several aspects: notably a different measure of complexity and a more geometric proof which avoids the use of the full Rips theory for finitely generated groups acting on $\mathbb{R}$-trees; see Section 7. We attempted to include enough background material to make the paper self-contained. See Paulin [24] and Champetier-Guirardel [5] for accounts of some of Sela's work on the Tarski problems.

The first version of these notes was circulated in 2003. In the meantime Henry Wilton [45] made available solutions to the exercises in the notes. We also thank Wilton for making numerous comments that led to many improvements.

Remark 1.1. In the theorem above, since G is finitely generated we may assume that F is also finitely generated. If F is abelian, then any f factors through the abelianization of G mod its torsion subgroup and we are in the situation of Example 1.4 below. Finally, if F_1 and F_2 are finitely generated non-abelian free groups then there is an injection $F_1 \to F_2$. So, if $\{q_i\}$ is a set of epimorphisms that satisfies the conclusion of the theorem for maps to F_2, then $\{q_i\}$ also works for maps to F_1. Therefore, throughout the paper we work with a fixed finitely generated non-abelian free group $\mathbb{F}$.

Notation 1.2. Finitely generated (finitely presented) is abbreviated fg (respectively fp).

The main goal of [34] is to give an answer to the following:

Question 1. *Let G be an fg group. Describe the set of all homomorphisms from G to $\mathbb{F}$.*

Example 1.3. When G is a free group, we can identify $Hom(G, \mathbb{F})$ with the cartesian product $\mathbb{F}^n$ where $n = rank(G)$.

Example 1.4. If $G = \mathbb{Z}^n$, let $\mu : \mathbb{Z}^n \to \mathbb{Z}$ be the projection to one of the coordinates. If $h : \mathbb{Z}^n \to \mathbb{F}$ is a homomorphism, there is an automorphism $\alpha : \mathbb{Z}^n \to \mathbb{Z}^n$ such that $h\alpha$ factors through μ. This provides an explicit (although not 1-1) parametrization of $Hom(G, \mathbb{F})$ by $Aut(\mathbb{Z}^n) \times Hom(\mathbb{Z}, \mathbb{F}) \cong GL_n(\mathbb{Z}) \times \mathbb{F}$.

Example 1.5. When G is the fundamental group of a closed genus g orientable surface, let $\mu : G \to F_g$ denote the homomorphism to a free group of rank g induced by the (obvious) retraction of the surface to the rank g graph. It is a folk theorem[1] that for every homomorphism $f : G \to \mathbb{F}$ there is an automorphism $\alpha : G \to G$ (induced by a homeomorphism of the surface) so that $f\alpha$ factors through μ. The theorem was generalized to the case when G is the fundamental group of a non-orientable closed surface by Grigorchuk and Kurchanov [9]. Interestingly, in this generality the single map μ is replaced by a finite collection $\{\mu_1, \cdots, \mu_k\}$ of maps from G to a free group F. In other words, for all $f \in Hom(G, \mathbb{F})$ there is $\alpha \in Aut(G)$ induced by a homeomorphism of the surface such that $f\alpha$ factors through some μ_i.

1.2 Basic properties of limit groups

Another goal is to understand the class of groups that naturally appear in the answer to the above question, these are called limit groups.

Definition 1.6. Let G be an fg group. A sequence $\{f_i\}$ in $Hom(G, \mathbb{F})$ is *stable* if, for all $g \in G$, the sequence $\{f_i(g)\}$ is eventually always 1 or eventually never 1. The *stable kernel* of $\{f_i\}$, denoted $\underrightarrow{Ker}\ f_i$, is

$$\{g \in G \mid f_i(g) = 1 \text{ for almost all } i\}.$$

An fg group Γ is a *limit group* if there is an fg group G and a stable sequence $\{f_i\}$ in $Hom(G, \mathbb{F})$ so that $\Gamma \cong G/\underrightarrow{Ker}\ f_i$.

Remark 1.7. One can view each f_i as inducing an action of G on the Cayley graph of $\mathbb{F}$, and then can pass to a limiting $\mathbb{R}$-tree action (after a subsequence). If the limiting tree is not a line, then $\underrightarrow{Ker}\ f_i$ is precisely the kernel of this action and so Γ acts faithfully. This explains the name.

Definition 1.8. An fg group Γ is *residually free* if for every element $\gamma \in \Gamma$ there is $f \in Hom(\Gamma, \mathbb{F})$ such that $f(\gamma) \neq 1$. It is *ω-residually free* if for every finite subset $X \subset \Gamma$ there is $f \in Hom(\Gamma, \mathbb{F})$ such that $f|X$ is injective.

Exercise 2. *Residually free groups are torsion free.*

Exercise 3. *Free groups and free abelian groups are ω-residually free.*

Exercise 4. *The fundamental group of $n\mathbb{P}^2$ for n = 1, 2, or 3 is not ω-residually free, see [18].*

Exercise 5. *Every ω-residually free group is a limit group.*

Exercise 6. *An fg subgroup of an ω-residually free group is ω-residually free.*

Exercise 7. *Every non-trivial abelian subgroup of an ω-residually free group is contained in a unique maximal abelian subgroup. For example, $F \times \mathbb{Z}$ is not ω-residually free for any non-abelian F.*

[1] see Zieschang [46] and Stallings [43]

Lemma 1.9. *Let $G_1 \to G_2 \to \cdots$ be an infinite sequence of epimorphisms between fg groups. Then the sequence*

$$Hom(G_1, \mathbb{F}) \leftarrow Hom(G_2, \mathbb{F}) \leftarrow \cdots$$

eventually stabilizes (consists of bijections).

Proof. Embed $\mathbb{F}$ as a subgroup of $SL_2(\mathbb{R})$. That the corresponding sequence of varieties $Hom(G_i, SL_2(\mathbb{R}))$ stabilizes follows from algebraic geometry, and this proves the lemma. □

Corollary 1.10. *A sequence of epimorphisms between ($\omega-$)residually free groups eventually stabilizes.* □

Lemma 1.11. *Every limit group is ω-residually free.*

Proof. Let Γ be a limit group, and let G and $\{f_i\}$ be as in the definition. Without loss, G is fp. Now consider the sequence of quotients

$$G \to G_1 \to G_2 \to \cdots \to \Gamma$$

obtained by adjoining one relation at a time. If Γ is fp the sequence terminates, and in general it is infinite. Let $G' = G_j$ be such that $Hom(G', \mathbb{F}) = Hom(\Gamma, \mathbb{F})$. All but finitely many f_i factor through G' since each added relation is sent to 1 by almost all f_i. It follows that these f_i factor through Γ and each non-trivial element of Γ is sent to 1 by only finitely many f_i. By definition, Γ is ω-residually free. □

The next two exercises will not be used in this paper but are included for their independent interest.

Exercise 8. *Every ω-residually free group Γ embeds into $PSL_2(\mathbb{R})$, and also into $SO(3)$.*

Exercise 9. *Let Γ be ω-residually free. For any finite collection of nontrivial elements $g_1, \cdots, g_k \in \Gamma$ there is an embedding $\Gamma \to PSL_2(\mathbb{R})$ whose image has no parabolic elements and so that $g_1, \cdots, g_k$ go to hyperbolic elements.*

1.3 Modular groups and the statement of the main theorem

Only certain automorphisms, called *modular automorphisms*, are needed in the theorem on page 2. This section contains a definition of these automorphisms.

Definition 1.12. Free products with amalgamations and HNN-decompositions of a group G give rise to *Dehn twist automorphisms of* G. Specifically, if $G = A *_C B$ and if z is in the centralizer $Z_B(C)$ of C in B, then the automorphism α_z of G, called the *Dehn twist in* z, is determined as follows.

$$\alpha_z(g) = \begin{cases} g, & \text{if } g \in A; \\ zgz^{-1}, & \text{if } g \in B. \end{cases}$$

If $C \subset A$, $\phi : C \to A$ is a monomorphism, $G = A*_C = \langle A, t \mid tat^{-1} = \phi(a), a \in A\rangle$,[2] and $z \in Z_A(C)$, then α_z is determined as follows.

$$\alpha_z(g) = \begin{cases} g, & \text{if } g \in A; \\ gz, & \text{if } g = t. \end{cases}$$

Definition 1.13. A GAD[3] of a group G is a finite graph of groups decomposition[4] of G with abelian edge groups in which some of the vertices are designated QH[5] and some others are designated *abelian*, and the following holds.

- A QH-vertex group is the fundamental group of a compact surface S with boundary and the boundary components correspond to the incident edge groups (they are all infinite cyclic). Further, S carries a pseudoAnosov homeomorphism (so S is a torus with 1 boundary component or $\chi(S) \leq -2$).

- An abelian vertex group A is non-cyclic abelian. Denote by $P(A)$ the subgroup of A generated by incident edge groups. The *peripheral subgroup of* A, denoted $\overline{P}(A)$, is the subgroup of A that dies under every homomorphism from A to $\mathbb{Z}$ that kills $P(A)$, i.e.

$$\overline{P}(A) = \cap\{Ker(f) \mid f \in Hom(A, \mathbb{Z}), P(A) \subset Ker(f)\}.$$

The non-abelian non-QH vertices are *rigid.*

Remark 1.14. We allow the possibility that edge and vertex groups of GAD's are not fg.

Remark 1.15. If Δ is a GAD for a fg group G, and if A is an abelian vertex group of Δ, then there are epimorphisms $G \to A/P(A) \to A/\overline{P}(A)$. Hence, $A/P(A)$ and $A/\overline{P}(A)$ are fg. Since $A/\overline{P}(A)$ is also torsion free, $A/\overline{P}(A)$ is free,

[2] t is called a *stable letter.*

[3] Generalized Abelian Decomposition

[4] We will use the terms *graph of groups decomposition* and *splitting* interchangeably. Without further notice, splittings are always *minimal*, i.e. the associated G-tree has no proper invariant subtrees.

[5] Quadratically Hanging

and so $A = A_0 \oplus \overline{P}(A)$ with $A_0 \cong A/\overline{P}(A)$ a retract of G. Similarly, $A/\overline{P}(A)$ is a direct summand of $A/P(A)$. A summand complementary to $A/\overline{P}(A)$ in $A/P(A)$ must be a torsion group by the definition of $\overline{P}(A)$. In particular, $P(A)$ has finite index in $\overline{P}(A)$. It also follows from the definition of $\overline{P}(A)$ that any automorphism leaving $P(A)$ invariant must leave $\overline{P}(A)$ invariant as well. It follows that if A is torsion free, then any automorphism of A that is the identity when restricted to $P(A)$ is also the identity when restricted to $\overline{P}(A)$.

Definition 1.16. The *modular group* $Mod(\Delta)$ associated to a GAD Δ of G is the subgroup of $Aut(G)$ generated by

- inner automorphisms of G,
- Dehn twists in elements of G that centralize an edge group of Δ,
- unimodular[6] automorphisms of an abelian vertex group that are the identity on its peripheral subgroup and all other vertex groups, and
- automorphisms induced by homeomorphisms of surfaces S underlying QH-vertices that fix all boundary components. If S is closed and orientable, we require the homeomorphisms to be orientation-preserving[7].

The *modular group of* G, denoted $Mod(G)$, is the subgroup of $Aut(G)$ generated by $Mod(\Delta)$ for all GAD's Δ of G. At times it will be convenient to view $Mod(G)$ as a subgroup of $Out(G)$. In particular, we will say that an element of $Mod(G)$ is *trivial* if it is an inner automorphism.

Definition 1.17. A *generalized Dehn twist* is a Dehn twist or an automorphism α of $G = A *_C B$ or $G = A*_C$ where in each case A is abelian, α restricted to $\overline{P}(A)$ and B is the identity, and α induces a unimodular automorphism of $A/\overline{P}(A)$. Here $\overline{P}(A)$ is the peripheral subgroup of A when we view $A *_C B$ or $G = A$ as a GAD with one or zero edges and abelian vertex A. If C is an edge groups of a GAD for G and if $z \in Z_G(C)$, then C determines a splitting of G as above and so also a Dehn twist in z. Similarly, an abelian vertex A of a GAD determines[8] a splitting $A *_C B$ and so also generalized Dehn twists.

Exercise 10. *$Mod(G)$ is generated by inner automorphisms together with generalized Dehn twists.*

Definition 1.18. A *factor set* for a group G is a finite collection of proper epimorphisms $\{q_i : G \to G_i\}$ such that if $f \in Hom(G, \mathbb{F})$ then there is $\alpha \in Mod(G)$ such that $f\alpha$ factors through some q_i.

---

[6] The induced automorphism of $A/\overline{P}(A)$ has determinant 1.

[7] We will want our homeomorphisms to be products of Dehn twists.

[8] by folding together the edges incident to A

Main Theorem ([14, 15, 35]). *Let G be an fg group that is not free. Then, G has a factor set $\{q_i : G \to \Gamma_i\}$ with each Γ_i a limit group. If G is not a limit group, we can always take α to be the identity.*

We will give two proofs–one in Section 4 and the second, which uses less in the way of technical machinery, in Section 7. In the remainder of this section, we explore some consequences of the Main Theorem and then give another description of limit groups.

1.4 Makanin-Razborov diagrams

Corollary 1.19. *Iterating the construction of the Main Theorem (for Γ_i's etc.) yields a finite tree of groups terminating in groups that are free.*

Proof. If $\Gamma \to \Gamma'$ is a proper epimorphism between limit groups, then since limit groups are residually free, $Hom(\Gamma', \mathbb{F}) \subsetneq Hom(\Gamma, \mathbb{F})$. We are done by Lemma 1.9. □

Definition 1.20. The tree of groups and epimorphisms provided by Corollary 1.19 is called an *MR-diagram*[9] for G (with respect to $\mathbb{F}$). If

$$G \xrightarrow{q} \Gamma_1 \xrightarrow{q_1} \Gamma_2 \xrightarrow{q_2} \cdots \xrightarrow{q_{m-1}} \Gamma_m$$

is a branch of an MR-diagram and if $f \in Hom(G, \mathbb{F})$ then we say that f *MR-factors* through this branch if there are $\alpha \in Mod(G)$ (which is the identity if G is not a limit group), $\alpha_i \in Mod(\Gamma_i)$, for $1 \le i < m$, and $f' \in Hom(\Gamma_m, \mathbb{F})$ (recall Γ_m is free) such that $f = f' q_{m-1}\alpha_{m-1} \cdots q_1 \alpha_1 q \alpha$.

Remark 1.21. The key property of an MR-diagram for G is that, for $f \in Hom(G, \mathbb{F})$, there is a branch of the diagram through which f MR-factors. This provides an answer to Question 1 in that $Hom(G, \mathbb{F})$ is parametrized by branches of an MR-diagram and, for each branch as above, $Mod(G) \times Mod(\Gamma_1) \times \cdots \times Mod(\Gamma_{m-1}) \times Hom(\Gamma_m, \mathbb{F})$. Note that if Γ_m has rank n, then $Hom(\Gamma_m, \mathbb{F}) \cong \mathbb{F}^n$.

In [32], Sela constructed MR-diagrams with respect to hyperbolic groups. In her thesis [1], Emina Alibegović constructed MR-diagrams with respect to limit groups. More recently, Daniel Groves [10, 11] constructed MR-diagrams with respect to torsion-free groups that are hyperbolic relative to a collection of free abelian subgroups.

1.5 Abelian subgroups of limit groups

Corollary 1.22. *Abelian subgroups of limit groups are fg and free.*

[9] for Makanin-Razborov, cf. [19, 20, 25].

Along with the Main Theorem, the proof of Corollary 1.22 will depend on an exercise and two lemmas.

Exercise 11 ([34, Lemma 2.3]). *Let M be a non-cyclic maximal abelian subgroup of the limit group Γ.*

1. *If $\Gamma = A *_C B$ with C abelian, then M is conjugate into A or B.*

2. *If $\Gamma = A*_C$ with C abelian, then either M is conjugate into A or there is a stable letter t such that M is conjugate to $M' = \langle C, t\rangle$ and $\Gamma = A *_C M'$.*

As a consequence, if $\alpha \in Mod(\Gamma)$ is a generalized Dehn twist and $\alpha|M$ is non-trivial, then there is an element $\gamma \in \Gamma$ and a GAD $\Delta = M *_C B$ or $\Delta = M$ for Γ such that, up to conjugation by γ, α is induced by a unimodular automorphism of $M/\overline{P}(M)$ (as in Definition 1.17). (Hint: Use Exercise 7.)

Lemma 1.23. *Suppose that Γ is a limit group with factor set $\{q_i : \Gamma \to G_i\}$. If H is a (not necessarily fg) subgroup of Γ such that, for every homomorphism $f : \Gamma \to \mathbb{F}$, $f|H$ factors through some $q_i|H$ (pre-compositions by automorphisms of Γ not needed) then, for some i, $q_i|H$ is injective.*

Proof. Suppose not and let $1 \neq h_i \in Ker(q_i|H)$. Since Γ is a limit group, there is $f \in Hom(\Gamma, \mathbb{F})$ that is injective on $\{1, h_1, \cdots, h_n\}$. On the other hand, $f|H$ factors through some $q_i|H$ and so $h_i = 1$, a contradiction. □

Lemma 1.24. *Let M be a non-cyclic maximal abelian subgroup of the limit group Γ. There is an epimorphism $r : \Gamma \to A$ where A is free abelian and every modular automorphism of Γ is trivial*[10] *when restricted to $M \cap Ker(r)$.*

Proof. By Exercise 10, it is enough to find r such that $\alpha|M \cap Ker(r)$ is trivial for every generalized Dehn twist $\alpha \in Mod(\Gamma)$. By Exercise 11 and Remark 1.15, there is a fg free abelian subgroup M_α of M and a retraction $r_\alpha : \Gamma \to M_\alpha$ such that $\alpha|M \cap Ker(r_\alpha)$ is trivial. Let $r = \Pi_\alpha r_\alpha : \Gamma \to \Pi_\alpha M_\alpha$ and let A be the image of r. Since Γ is fg, so is A. Hence A is free abelian. □

Proof of Corollary 1.22. Let M be a maximal abelian subgroup of a limit group Γ. We may assume that M is not cyclic. Since Γ is torsion free, it is enough to show that M is fg. By restricting the map r of Lemma 1.24 to M, we see that $M = A \oplus A'$ where A is fg and each $\alpha|A'$ is trivial. Let $\{q_i : \Gamma \to \Gamma_i\}$ be a factor set for Γ given by Theorem 1.3. By Lemma 1.23, A' injects into some Γ_i. Since $Hom(\Gamma_i, \mathbb{F}) \subsetneq Hom(\Gamma, \mathbb{F})$, we may conclude by induction that A' and hence M is fg. □

[10] agrees with the restriction of an inner automorphism of Γ.

1.6 Constructible limit groups

It will turn out that limit groups can be built up inductively from simpler limit groups. In this section, we give this description and list some properties that follow.

Definition 1.25. We define a hierarchy of fg groups – if a group belongs to this hierarchy it is called a CLG[11].

Level 0 of the hierarchy consists of fg free groups.

A group Γ belongs to level $\leq n+1$ iff either it has a free product decomposition $\Gamma = \Gamma_1 * \Gamma_2$ with Γ_1 and Γ_2 of level $\leq n$ or it has a homomorphism $\rho : \Gamma \to \Gamma'$ with Γ' of level $\leq n$ and it has a GAD such that

- ρ is injective on the peripheral subgroup of each abelian vertex group.
- ρ is injective on each edge group E and at least one of the images of E in a vertex group of the one-edged splitting induced by E is a maximal abelian subgroup.
- The image of each QH-vertex group is a non-abelian subgroup of Γ'.
- For every rigid vertex group B, ρ is injective on the *envelope* $\tilde{B}$ *of* B, defined by first replacing each abelian vertex with the peripheral subgroup and then letting $\tilde{B}$ be the subgroup of the resulting group generated by B and by the centralizers of incident edge-groups.

Example 1.26. A fg free abelian group is a CLG of level one (consider a one-point GAD for $\mathbb{Z}^n$ and $\rho : \mathbb{Z}^n \to \langle 0 \rangle$). The fundamental group of a closed surface S with $\chi(S) \leq -2$ is a CLG of level one. For example, an orientable genus 2 surface is a union of 2 punctured tori and the retraction to one of them determines ρ. Similarly, a non-orientable genus 2 surface is the union of 2 punctured Klein bottles.

Example 1.27. Start with the circle and attach to it 3 surfaces with one boundary component, with genera 1, 2, and 3 say. There is a retraction to the surface of genus 3 that is the union of the attached surfaces of genus 1 and 2. This retraction sends the genus 3 attached surface say to the genus 2 attached surface by "pinching a handle". The GAD has a central vertex labeled $\mathbb{Z}$ and there are 3 edges that emanate from it, also labeled $\mathbb{Z}$. Their other endpoints are QH-vertex groups. The map induced by retraction satisfies the requirements so the fundamental group of the 2-complex built is a CLG.

Example 1.28. Choose a primitive[12] w in the fg free group F and form $\Gamma = F *_{\mathbb{Z}} F$, the *double of* F *along* $\langle w \rangle$ (so $1 \in \mathbb{Z}$ is identified with w on both sides). There is a retraction $\Gamma \to F$ that satisfies the requirements (both vertices are rigid), so Γ is a CLG.

[11] Constructible Limit Group
[12] no proper root

The following can be proved by induction on levels.

Exercise 12. *Every CLG is fp, in fact coherent. Every fg subgroup of a CLG is a CLG. (Hint: a graph of coherent groups over fg abelian groups is coherent.)*

Exercise 13. *Every abelian subgroup of a CLG Γ is fg and free, and there is a uniform bound to the rank. There is a finite $K(\Gamma, 1)$.*

Exercise 14. *Every non-abelian, freely indecomposable CLG admits a* principal splitting *over $\mathbb{Z}$: $A *_{\mathbb{Z}} B$ or $A *_{\mathbb{Z}}$ with A, B non-cyclic, and in the latter case $\mathbb{Z}$ is maximal abelian in the whole group.*

Exercise 15. *Every CLG is ω-residually free.*

The last exercise is more difficult than the others. It explains where the conditions in the definition of CLG come from. The idea is to construct homomorphisms $G \to \mathbb{F}$ by choosing complicated modular automorphisms of G, composing with ρ and then with a homomorphism to $\mathbb{F}$ that comes from the inductive assumption.

Example 1.29. Consider an index 2 subgroup H of an fg free group F and choose $g \in F \setminus H$. Suppose that $G := H *_{\langle g^2 \rangle} \langle g \rangle$ is freely indecomposable and admits no principal cyclic splitting. There is the obvious map $G \to F$, but G is not a limit group (Exercise 14 and Theorem 1.30). This shows the necessity of the last condition in the definition of CLG's. [13]

In Section 6, we will show:

Theorem 1.30. *For an fg group G, the following are equivalent.*

1. *G is a CLG.*
2. *G is ω-residually free.*
3. *G is a limit group.*

The fact that ω-residually free groups are CLG's is due to O. Kharlampovich and A. Myasnikov [16]. Limit groups act freely on $\mathbb{R}^n$-trees; see Remeslennikov [27] and Guirardel [13]. Kharlampovich-Myasnikov [15] prove that limit groups act freely on $\mathbb{Z}^n$-trees where $\mathbb{Z}^n$ is lexicographically ordered. Remeslennikov [26] also demonstrated that 2-residually free groups are ω-residually free.

[13] The element $g := a^2b^2a^{-2}b^{-1} \notin H := \langle a, b^2, bab^{-1} \rangle \subset F := \langle a, b \rangle$ is such an example. This can be seen from the fact that if $\langle x, y, z \rangle$ denotes the displayed basis for H, then $g^2 = x^2yx^{-2}y^{-1}z^2yz^{-2}$ is Whitehead reduced and each basis element occurs at least 3 times.

2 The Main Proposition

Definition 2.1. An fg group is *generic* if it is torsion free, freely indecomposable, non-abelian, and not a closed surface group.

The Main Theorem will follow from the next proposition.

Main Proposition. *Generic limit groups have factor sets.*

Before proving this proposition, we show how it implies the Main Theorem.

Definition 2.2. Let G and G' be fg groups. The minimal number of generators for G is denoted $\mu(G)$. We say that G is *simpler* than G' if there is an epimorphism $G' \to G$ and either $\mu(G) < \mu(G')$ or $\mu(G) = \mu(G')$ and $Hom(G, \mathbb{F}) \subsetneq Hom(G', \mathbb{F})$.

Remark 2.3. It follows from Lemma 1.9 that every sequence $\{G_i\}$ with G_{i+1} simpler than G_i is finite.

Definition 2.4. If G is an fg group, then by $RF(G)$ denote the *universal residually free quotient of* G, i.e. the quotient of G by the (normal) subgroup consisting of elements killed by every homomorphism $G \to \mathbb{F}$.

Remark 2.5. $Hom(G, \mathbb{F}) = Hom(RF(G), \mathbb{F})$ and for every proper quotient G' of $RF(G)$, $Hom(G', \mathbb{F}) \subsetneq Hom(G, \mathbb{F})$.

The Main Proposition implies the Main Theorem. Suppose that G is an fg group that is not free. By Remark 2.3, we may assume that the Main Theorem holds for groups that are simpler than G. By Remark 2.5, we may assume that G is residually free, and so also torsion free. Examples 1.4 and 1.5 show that the Main Theorem is true for abelian and closed surface groups. If $G = U * V$ with U non-free and freely indecomposable and with V nontrivial, then U is simpler than G. So, U has a factor set $\{q_i : U \to L_i\}$, and $\{q_i * Id_V : U * V \to L_i * V\}$ is a factor set for G.

If G is not a limit group, then there is a non-empty finite subset $\{g_i\}$ of G such that any homomorphism $G \to \mathbb{F}$ kills one of the g_i. We then have a factor set $\{G \to H_i := G/\langle\langle g_i \rangle\rangle\}$. Since $Hom(H_i, \mathbb{F}) \subsetneq Hom(G, \mathbb{F})$, by induction the Main Theorem holds for H_i and so for G.

If G is generic and a limit group, then the Main Proposition gives a factor set $\{q_i : G \to G_i\}$ for G. Since G is residually free, each G_i is simpler than G. We are assuming that the Main Theorem then holds for each G_i and this implies the result for G. □

3 Review: Measured laminations and $\mathbb{R}$-trees

The proof of the Main Proposition will use a theorem of Sela describing the structure of certain real trees. This in turn depends on the structure of measured laminations. In Section 7, we will give an alternate approach that only uses the lamination results. First these concepts are reviewed. A more leisurely review with references is [2].

3.1 Laminations

Definition 3.1. A *measured lamination* Λ on a simplicial 2-complex K consists of a closed subset $|\Lambda| \subset |K|$ and a *transverse measure* μ. $|\Lambda|$ is disjoint from the vertex set, intersects each edge in a Cantor set or empty set, and intersects each 2-simplex in 0, 1, 2, or 3 families of straight line segments spanning distinct sides. The measure μ assigns a non-negative number $\int_I \mu$ to every interval I in an edge whose endpoints are outside $|\Lambda|$. There are two conditions:

1. **(compatibility)** If two intervals I, J in two sides of the same triangle Δ intersect the same components of $|\Lambda| \cap \Delta$ then $\int_I \mu = \int_J \mu$.

2. **(regularity)** μ restricted to an edge is equivalent under a "Cantor function" to the Lebesgue measure on an interval in $\mathbb{R}$.

A path component of $|\Lambda|$ is a *leaf*.

Two measured laminations on K are considered equivalent if they assign the same value to each edge.

Proposition 3.2 (Morgan-Shalen [21]). *Let Λ be a measured lamination on compact K. Then*

$$\Lambda = \Lambda_1 \sqcup \cdots \sqcup \Lambda_k$$

so that each Λ_i is either minimal *(each leaf is dense in $|\Lambda_i|$) or* simplicial *(each leaf is compact, a regular neighborhood of $|\Lambda_i|$ is an I-bundle over a leaf and $|\Lambda_i|$ is a Cantor set subbundle).*

There is a theory, called the *Rips machine*, for analyzing minimal measured laminations. It turns out that there are only 3 qualities.

Example 3.3 (Surface type). Let S be a compact hyperbolic surface (possibly with totally geodesic boundary). If S admits a pseudoAnosov homeomorphism then it also admits *filling measured geodesic laminations* – these are measured laminations Λ (with respect to an appropriate triangulation) such that each leaf is a biinfinite geodesic and all complementary components are ideal polygons or crowns. Now to get the model for a general surface type lamination attach finitely many annuli $S^1 \times I$ with lamination $S^1 \times$ (Cantor set) to the surface along arcs transverse to the geodesic lamination. If these additional annuli do not appear then the lamination is of *pure surface type*. See Figure 1.

Example 3.4 (Toral type). Fix a closed interval $I \subset \mathbb{R}$, a finite collection of pairs (J_i, J_i') of closed intervals in I, and isometries $\gamma_i : J_i \to J_i'$ so that:

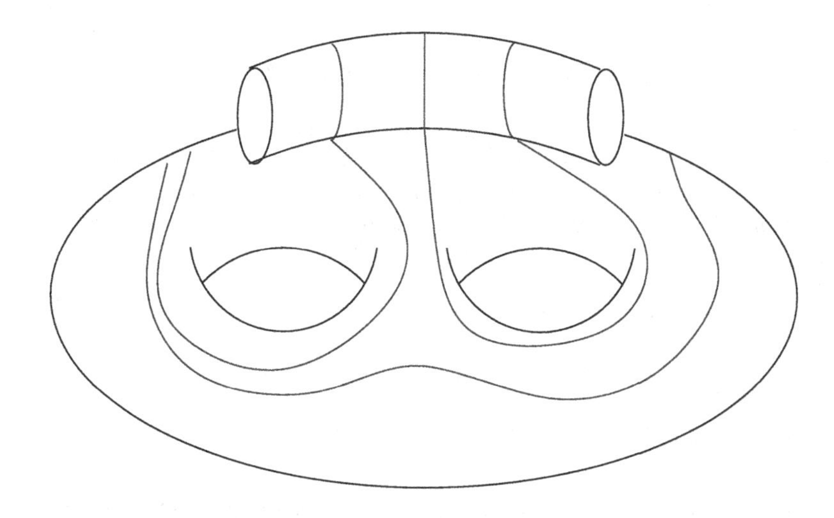

Figure 1: A surface with an additional annulus and some pieces of leaves.

1. If γ_i is orientation reversing then $J_i = J_i'$ and the midpoint is fixed by γ_i.

2. The length of the intersection of all J_i and J_i' (over all i) is more than twice the translation length of each orientation preserving γ_i and the fixed points of all orientation reversing γ_i are in the middle third of the intersection.

Now glue a foliated band for each pair (J_i, J_i') so that following the band maps J_i to J_i' via γ_i. Finally, using Cantor functions blow up the foliation to a lamination. There is no need to explicitly allow adding annuli as in the surface case since they correspond to $\gamma_i = Id$. The subgroup of $Isom(\mathbb{R})$ generated by the extensions of the γ_i's is the *Bass group*. The lamination is minimal iff its Bass group is not discrete.

Example 3.5 (Thin type). This is the most mysterious type of all. It was discovered by Gilbert Levitt, see [17]. In the *pure* case (no annuli attached) the leaves are 1-ended trees (so this type naturally lives on a 2-complex, not on a manifold). By performing certain moves (sliding, collapsing) that don't change the homotopy type (respecting the lamination) of the complex one can transform it to one that contains a (thin) band. This band induces a non-trivial free product decomposition of $\pi_1(K)$, assuming that the component is a part of a resolution of a tree (what's needed is that loops that follow leaves until they come close to the starting point and then they close up are non-trivial in π_1).

In the general case we allow additional annuli to be glued, just like in the surface case. Leaves are then 1-ended trees with circles attached.

Theorem 3.6 ("Rips machine"). *Let Λ be a measured lamination on a finite 2-complex K, and let Λ_i be a minimal component of Λ. There is a neighborhood N (we refer to it as a* standard *neighborhood) of $|\Lambda_i|$, a finite 2-complex N' with measured lamination Λ' as in one of 3 model examples, and there is a π_1-isomorphism $f : N \to N'$ such that $f^*(\Lambda') = \Lambda$.*

We refer to Λ_i as being of *surface*, *toral*, or *thin* type.

3.2 Dual trees

Let G be an fg group and let $\hat{K}$ be a simply connected 2-dimensional simplicial G-complex so that, for each simplex Δ of $\hat{K}$, $Stab(\Delta) = Fix(\Delta)$.[14] Let $\hat{\Lambda}$ be a G-invariant measured lamination in $\hat{K}$. There is an associated real G-tree $T(\hat{\Lambda})$ constructed as follows. Consider the pseudo-metric on $\hat{K}$ obtained by minimizing the $\hat{\Lambda}$-length of paths between points. The real tree $T(\hat{\Lambda})$ is the associated metric space[15]. There is a natural map $\hat{K} \to T(\hat{\Lambda})$ and we say that $(\hat{K}, \hat{\Lambda})$ is a *model* for $T(\hat{\Lambda})$ if

- for each edge $\hat{e}$ of $\hat{K}$, $T(\hat{\Lambda} \mid \hat{e}) \to T(\hat{\Lambda})$ is an isometry (onto its image) and
- the quotient $\hat{K}/G$ is compact.

If a tree T admits a model $(\hat{K}, \hat{\Lambda})$, then we say that T is *dual* to $(\hat{K}, \hat{\Lambda})$. This is denoted $T = Dual(\hat{K}, \hat{\Lambda})$. We will use the quotient $(K, \Lambda) := (\hat{K}, \hat{\Lambda})/G$ with simplices decorated (or labeled) with stabilizers to present a model and sometimes abuse notation by calling (K, Λ) a model for T.

Remark 3.7. Often the G-action on $\hat{K}$ is required to be free. We have relaxed this condition in order to be able to consider actions of fg groups. For example, if T is a minimal[16], simplicial G-tree (with the metric where edges have length one[17]) then there is a lamination $\hat{\Lambda}$ in T such that $Dual(T, \hat{\Lambda}) = T$.[18]

If S and T are real G-trees, then an equivariant map $f : S \to T$ is a *morphism* if every compact segment of S has a finite partition such that the restriction of f to each element is an isometry or trivial[19].

---

[14] $Stab(\Delta) := \{g \in G \mid g\Delta = \Delta\}$ and $Fix(\Delta) := \{g \in G \mid gx = x, x \in \Delta\}$
[15] identify points of pseudo-distance 0
[16] no proper invariant subtrees
[17] This is called the *simplicial metric* on T.
[18] The metric and simplicial topologies on T don't agree unless T is locally finite. But, the action of G is by isomorphisms in each structure. So, we will be sloppy and ignore this distinction.
[19] has image a point

If S is a real G-tree with G fp, then there is a real G-tree T with a model and a morphism $f : T \to S$. The map f is obtained by constructing an equivariant map to S from the universal cover of a 2-complex with fundamental group G. In general, if $(\hat{K}, \hat{\Lambda})$ is a model for T and if $T \to S$ is a morphism then the composition $\hat{K} \to T \to S$ is a *resolution* of S.

3.3 The structure theorem

Here we discuss a structure theorem (see Theorem 3.13) of Sela for certain actions of an fg torsion free group G on real trees. The actions we consider will usually be super stable[20], have primitive[21] abelian (non-degenerate) arc stabilizers, and have trivial tripod[22] stabilizers. There is a short list of basic examples.

Example 3.8 (Pure surface type). A real G-tree T is *of pure surface type* if it is dual to the universal cover of (K, Λ) where K is a compact surface and Λ is of pure surface type. We will usually use the alternate model where boundary components are crushed to points and are labeled $\mathbb{Z}$.

Example 3.9 (Linear). The tree T is *linear* if G is abelian, T is a line and there an epimorphism $G \to \mathbb{Z}^n$ such that G acts on T via a free $\mathbb{Z}^n$-action on T. In particular, T is dual to $(\hat{K}, \hat{\Lambda})$ where $\hat{K}$ is the universal cover of the 2-skeleton of an n-torus K. For simplicity, we often complete K with its lamination to the whole torus. This is a special case of a toral lamination.

Example 3.10 (Pure thin). The tree T is *pure thin* if it is dual to the universal cover of a finite 2-complex K with a pure thin lamination Λ. If T is pure thin then $G \cong F * V_1 * \cdots * V_m$ where F is non-trivial and fg free and $\{V_1, \cdots, V_m\}$ represents the conjugacy classes of non-trivial point stabilizers in T.

Example 3.11 (Simplicial). The tree T is *simplicial* if it is dual to $(\hat{K}, \hat{\Lambda})$ where all leaves of $\Lambda := \hat{\Lambda}/G$ are compact. If T is simplicial it is convenient to crush the leaves and complementary components to points in which case $\hat{K}$ becomes a tree isomorphic to T.

If $\mathcal{K}$ is a graph of 2-complexes with underlying graph of groups $\mathcal{G}$[23] then there is a simplicial $\pi_1(\mathcal{G})$-space $\hat{K}(\mathcal{K})$ obtained by gluing copies of $\hat{K}_e \times I$ and $\hat{K}_v$'s equipped with a simplicial $\pi_1(\mathcal{G})$-map $\hat{K}(\mathcal{K}) \to T(\mathcal{G})$ that crushes to points copies of $\hat{K}_e \times \{point\}$ as well as the $\hat{K}_v$'s.

Definition 3.12. A real G-tree is *very small* if the action is non-trivial[24], minimal, the stabilizers of non-degenerate arcs are primitive abelian, and the stabilizers of non-degenerate tripods are trivial.

---

[20] If $J \subset I$ are (non-degenerate) arcs in T and if $Fix_T(I)$ is non-trivial, then $Fix_T(J) = Fix_T(I)$

[21] root-closed

[22] a cone on 3 points

[23] for each bonding map $\phi_e : G_e \to G_v$ there are simplicial G_e- and G_v-complexes $\hat{K}_e$ and $\hat{K}_v$ together with a ϕ_e-equivariant simplicial map $\Phi_e : \hat{K}_e \to \hat{K}_v$

[24] no point is fixed by G

Theorem 3.13 ([33, special case of Section 3] See also [12]). *Let T be a real G-tree. Suppose that G is generic and that T is very small and super stable. Then, T has a model.*

Moreover, there is a model for T that is a graph of spaces such that each edge space is a point with non-trivial abelian stabilizer and each vertex space with restricted lamination is either

- (point) *a point with non-trivial stabilizer,*
- (linear) *a non-faithful action of an abelian group on the (2-skeleton of the) universal cover of a torus with an irrational* [25] *lamination, or*
- (surface) *a faithful action of a free group on the universal cover of a surface with non-empty boundary (represented by points with $\mathbb{Z}$-stabilizer) with a lamination of pure surface type.*

Remark 3.14. For an edge space $\{point\}$, the restriction of the lamination to $\{point\} \times I$ may or may not be empty. It can be checked that between any two points in models as in Theorem 3.13 there are Λ-length minimizing paths. Thin pieces do not arise because we are assuming our group is freely indecomposable.

Remark 3.15. Theorem 3.13 holds more generally if the assumption that G is freely indecomposable is replaced by the assumption that G is freely indecomposable rel point stabilizers, i.e. if $\mathcal{V}$ is the subset of G of elements acting elliptically[26] on T, then G cannot be expressed non-trivially as $A * B$ with all $g \in \mathcal{V}$ conjugate into $A \cup B$.

We can summarize Theorem 3.13 by saying that T is a non-trivial finite graph of simplicial trees, linear trees, and trees of pure surface type (over trivial trees). See Figure 2.

Corollary 3.16. *If G and T satisfy the hypotheses of Theorem 3.13, then G admits a non-trivial GAD Δ. Specifically, Δ may be taken to be the GAD induced by the boundary components of the surface vertex spaces and the simplicial edges of the model. The surface vertex spaces give rise to the QH-vertices of Δ and the linear vertex spaces give rise to the abelian vertices of Δ.*

3.4 Spaces of trees

Let G be a fg group and let $\mathcal{A}(G)$ be the set of minimal, non-trivial, real G-trees endowed with the Gromov topology[27]. Recall, see [22, 23, 4], that in the Gromov topology $\lim\{(T_n, d_n)\} = (T, d)$ if and only if: for any finite

[25] no essential loops in leaves
[26] fixing a point
[27] The second author thanks Gilbert Levitt for a helpful discussion on the Gromov and length topologies.

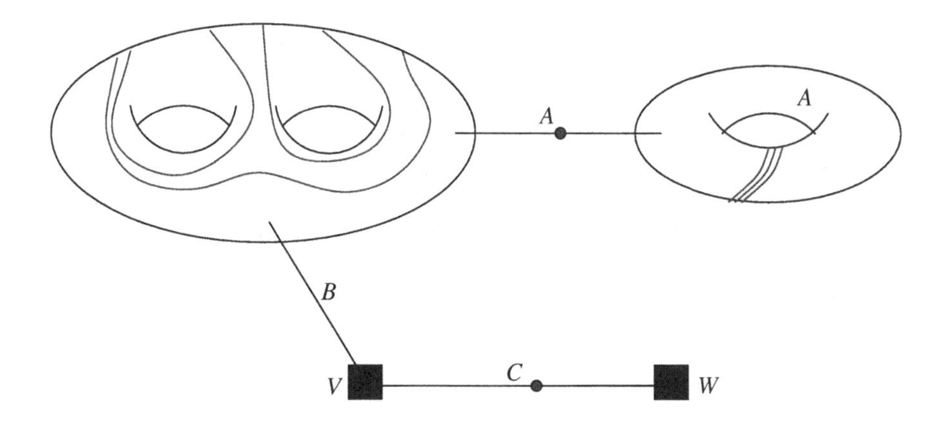

Figure 2: A model with a surface vertex space, a linear vertex space, and 2 rigid vertex spaces (the black boxes). The groups A, B and C are abelian with A and B infinite cyclic. Pieces of some leaves are also indicated by wavy lines and dots. For example, the dot on the edge labeled C is one leaf in a Cantor set of leaves.

subset K of T, any $\epsilon > 0$, and any finite subset P of G, for sufficiently large n, there are subsets K_n of T_n and bijections $f_n : K_n \to K$ such that

$$|d(gf_n(s_n), f_n(t_n)) - d_n(gs_n, t_n)| < \epsilon$$

for all $s_n, t_n \in K_n$ and all $g \in P$. Intuitively, larger and larger pieces of the limit tree with their restricted actions appear in nearby trees.

Let $\mathcal{PA}(G)$ be the set of non-trivial real G-trees modulo homothety, i.e. $(T, d) \sim (T, \lambda d)$ for $\lambda > 0$. Fix a basis for $\mathbb{F}$ and let $T_{\mathbb{F}}$ be the corresponding Cayley graph. Give $T_{\mathbb{F}}$ the simplicial metric. So, a non-trivial homomorphism $f : G \to \mathbb{F}$ determines $T_f \in \mathcal{PA}(G)$. Let X be the subset of $Hom(G, \mathbb{F})$ consisting of those homomorphisms with non-cyclic image. The space of interest is the closure $\mathcal{T}(G)$ of (the image of) $\{T_f \mid f \in X\}$ in $\mathcal{PA}(G)$.

Proposition 3.17 ([34]). *Every sequence of homomorphisms in X has a subsequence $\{f_n\}$ such that $\lim T_{f_n} = T$ in $\mathcal{T}(G)$. Further,*

1. *T is irreducible*[28].

2. *$\underrightarrow{Ker}\, f_n$ is precisely the kernel $Ker(T)$ of the action of G on T.*

3. *The action of $G/Ker(T)$ on T is very small and super stable.*

4. *For $g \in G$, $U(g) := \{T \in \mathcal{T}(G) \mid g \in Ker(T)\}$ is clopen*[29].

[28] T is not a line and doesn't have a fixed end

[29] both open and closed

Proof. The first statement follows from Paulin's Convergence Theorem [22].[30] The three numbered statements are exercises in Gromov convergence. □

Caution. Sela goes on to claim that stabilizers of minimal components of the limit tree are trivial (see Lemma 1.6 of [34]). However, it is possible to construct limit actions on the amalgam of a rank 2 free group F_2 and $\mathbb{Z}^3$ over $\mathbb{Z}$ where one of the generators of $\mathbb{Z}^3$ is glued to the commutator c of basis elements of F_2 and where the $\mathbb{Z}^3$ acts non-simplicially on a linear subtree with c acting trivially on the subtree but not in the kernel of the action. As a result, some of his arguments, though easily completed, are not fully complete.

Remark 3.18. There is another common topology on $\mathcal{A}(G)$, the length topology. For $T \in \mathcal{A}(G)$ and $g \in G$, let $\|g\|_T$ denote the minimum distance that g translates a point of T. The length topology is induced by the map $\mathcal{A}(G) \to [0,\infty)^G$, $T \mapsto (\|g\|_T)_{g\in G}$. Since the trees in $\mathcal{A}(G)$ are non-trivial, it follows from [42, page 64] that $\{0\}$ is not in the image[31]. Since $\mathcal{T}(G)$ consists of irreducible trees, it follows from [23] that the two topologies agree when restricted to $\mathcal{T}(G)$ and from [6] that $\mathcal{T}(G)$ injects into $\big([0,\infty)^G \setminus \{0\}\big)/(0,\infty)$.

Corollary 3.19. *$\mathcal{T}(G)$ is metrizable and compact.*

Proof. $[0,\infty)^G \setminus \{0\} \to \big([0,\infty)^G \setminus \{0\}\big)/(0,\infty)$ has a section over $\mathcal{T}(G)$ (e.g. referring to Footnote 31, normalize so that the sum of the translation lengths of words in $\mathcal{B}_G$ of length at most two is one). Therefore, $\mathcal{T}(G)$ embeds in the metrizable space $[0,\infty)^G$. In light of this, the main statement of Proposition 3.17 implies that $\mathcal{T}(G)$ is compact. □

Remark 3.20. Culler and Morgan [6] show that, if G is fg, then $\mathcal{PA}(G)$ with the length topology is compact. This can be used instead of Paulin's convergence theorem to show that $\mathcal{T}(G)$ is compact. The main lemma to prove is that, in the length topology, the closure in $\mathcal{PA}(G)$ of $\{T_f \mid f \in X\}$ consists of irreducible trees.

4 Proof of the Main Proposition

To warm up, we first prove the Main Proposition under the additional assumption that Γ has only trivial abelian splittings, i.e. every simplicial Γ-tree with abelian edge stabilizers has a fixed point. This proof is then modified to apply to the general case.

[30] Paulin's proof assumes the existence of convex hulls and so does not apply in the generality stated in his theorem. His proof does however apply in our situation since convex hulls do exist in simplicial trees.

[31] In fact, if $\mathcal{B}_G$ is a finite generating set for G then $\|g\|_T \neq 0$ for some word g that is a product of at most two elements of $\mathcal{B}_G$.

Proposition 4.1. *Suppose that Γ is a generic limit group and has only trivial abelian splittings*[32]. *Then, Γ has a factor set.*

Proof. Let $T \in \mathcal{T}(\Gamma)$. By Proposition 3.17, either $Ker(T)$ is non-trivial or T satisfies the hypotheses of Theorem 3.13. The latter case doesn't occur or else, by Corollary 3.16, $\Gamma/Ker(T)$ admits a non-trivial abelian splitting. In particular, $Ker(T)$ is non-trivial. Choose non-trivial $k_T \in Ker(T)$. By Item 4 of Proposition 3.17, $\{U(k_T) \mid T \in \mathcal{T}(\Gamma)\}$ is an open cover of $\mathcal{T}(\Gamma)$. Let $\{U(k_i)\}$ be a finite subcover. By definition, $\{\Gamma \to Ab(\Gamma)\} \cup \{q_i : \Gamma \to \Gamma/\langle\langle k_i \rangle\rangle\}$ is a factor set. □

The key to the proof of the general case is Sela's notion of a *short* homomorphism, a concept which we now define.

Definition 4.2. Let G be an fg group. Two elements f and f' in $Hom(G, \mathbb{F})$ are *equivalent*, denoted $f \sim f'$, if there is $\alpha \in Mod(G)$ and an element $c \in \mathbb{F}$ such that $f' = i_c \circ f \circ \alpha$.[33] Fix a set $\mathcal{B}$ of generators for G and by $|f|$ denote $\max_{g \in \mathcal{B}} |f(a)|$ where, for elements of $\mathbb{F}$, $|\cdot|$ indicates word length. We say that f is *short* if, for all $f' \sim f$, $|f| \leq |f'|$.

Note that if $f \in X$ and $f' \sim f$, then $f' \in X$. Here is another exercise in Gromov convergence. See [34, Claim 5.3] and also [28] and [2, Theorem 7.4].

Exercise 16. *Suppose that G is generic, $\{f_i\}$ is a sequence in $Hom(G, \mathbb{F})$, and $\lim T_{f_i} = T$ in $\mathcal{T}(G)$. Then, either*

- *$Ker(T)$ is non-trivial, or*
- *eventually f_i is not short.*

The idea is that if the first bullet does not hold, then the GAD of G given by Corollary 3.16 can be used to find elements of $Mod(G)$ that shorten f_i for i large.

Let Y be the subset of X consisting of short homomorphisms and let $\mathcal{T}'(G)$ be the closure in $\mathcal{T}(G)$ of $\{T_f \mid f \in Y\}$. By Corollary 3.19, $\mathcal{T}'(G)$ is compact.

Proof of the Main Proposition. Let $T \in \mathcal{T}'(\Gamma)$. By Exercise 16, $Ker(T)$ is non-trivial. Choose non-trivial $k_T \in Ker(T)$. By Corollary 3.19, $\{U(k_T) \mid T \in \mathcal{T}'(\Gamma)\}$ is an open cover of $\mathcal{T}'(\Gamma)$. Let $\{U(k_i)\}$ be a finite subcover. By definition, $\{\Gamma \to Ab(\Gamma)\} \cup \{q_i : \Gamma \to \Gamma/\langle\langle k_i \rangle\rangle\}$ is a factor set. □

Remark 4.3. Cornelius Reinfeldt and Richard Weidmann point out that a factor set for a generic limit group Γ can be found without appealing to Corollary 3.19 as follows. Let $\{\gamma_1, \dots\}$ enumerate the non-trivial elements

[32] By Proposition 3.17 and Corollary 3.16, generic limit groups have non-trivial abelian splittings. The purpose of this proposition is to illustrate the method in this simpler (vacuous) setting.

[33] i_c is conjugation by c

of Γ. Let $\mathcal{Q}_i := \{\Gamma/\langle\langle\gamma_1\rangle\rangle, \ldots, \Gamma/\langle\langle\gamma_i\rangle\rangle\}$. If $\mathcal{Q}_i$ is not a factor set, then there is $f_i \in Y$ that is injective on $\{\gamma_1, \ldots, \gamma_i\}$. By Paulin's convergence theorem, a subsequence of $\{T_{f_i}\}$ converges to a faithful Γ-tree in contradiction to Exercise 16.

JSJ-decompositions will be used to prove Theorem 1.30, so we digress.

5 Review: JSJ-theory

Some familiarity with JSJ-theory is assumed. The reader is referred to Rips-Sela [29], Dunwoody-Sageev [7], Fujiwara-Papasoglou [8]. For any freely indecomposable fg group G consider the class GAD's with at most one edge such that:

(JSJ) every non-cyclic abelian subgroup $A \subset G$ is elliptic.

We observe that

- Any two such GAD's are hyperbolic-hyperbolic[34] or elliptic-elliptic[35] (a hyperbolic-elliptic pair implies that one splitting can be used to refine the other. Since the hyperbolic edge group is necessarily cyclic by (JSJ), this refinement gives a free product decomposition of G).
- A hyperbolic-hyperbolic pair has both edge groups cyclic and yields a GAD of G with a QH-vertex group.
- An elliptic-elliptic pair has a common refinement that satisfies (JSJ) and whose set of elliptics is the intersection of the sets of elliptics in the given splittings.

Given a GAD Δ of G, we say that $g \in G$ is *Δ-elliptic*[36] if there is a vertex group V of Δ such that either:

- V is QH and g is conjugate to a multiple of a boundary component;
- V is abelian and g is conjugate into $\overline{P}(V)$; or
- V is rigid and g is conjugate into the envelope of V.

The idea is that Δ gives rise to a family of splittings[37] with at most one edge that come from edges of the decomposition, from simple closed curves in QH-vertex groups, and from subgroups A' of an abelian vertex A that contain

[34] each edge group of corresponding trees contains an element not fixing a point of the other tree

[35] each edge group of corresponding trees fixes a point of the other tree

[36] We thank Richard Weidmann for pointing out an error in a previous version of this definition and also for suggesting the correction.

[37] not necessarily satisfying (JSJ).

$\overline{P}(A)$ (equivalently $P(A)$) and with $A/A' \cong \mathbb{Z}$. For example, a non-peripheral element of A is hyperbolic in some 1-edge splitting obtained by blowing up the vertex A to an edge and then collapsing the original edges of Δ. An element is Δ-elliptic iff it is elliptic with respect to all these splittings with at most one edge. Conversely, any finite collection of GAD's with at most one edge and that satisfy (JSJ) gives rise to a GAD whose set of elliptics is precisely the intersection of the set of elliptics in the collection.

Definition 5.1. An *abelian JSJ-decomposition of G* is a GAD whose elliptic set is the intersection of elliptics in the family of *all* GAD's with at most one edge and that satisfy (JSJ).

Example 5.2. The group $G = F \times \mathbb{Z}$ has no 1-edge GAD's satisfying (JSJ) so the abelian JSJ-decomposition Δ of G is a single point labeled G. Of course, G does have (many) abelian splittings. If F is non-abelian, then every element of G is Δ-elliptic. If F is abelian, then only the torsion elements of G are Δ-elliptic.

To show that a group G admits an abelian JSJ-decomposition it is necessary to show that there is a bound to the complexity of the GAD's arising from finite collections of 1-edge splittings satisfying (JSJ). If G were fp the results of [3] would suffice. Since we don't know yet that limit groups are fp, another technique is needed. Following Sela, we use acylindrical accessibility.

Definition 5.3. A simplicial G-tree T is *n-acylindrical* if, for non-trivial $g \in G$, the diameter in the simplicial metric of the sets $Fix(g)$ is bounded by n. It is *acylindrical* if it is n-acylindrical for some n.

Theorem 5.4 (Acylindrical Accessibility: Sela [33], Weidmann [44]). *Let G be a non-cyclic freely indecomposable fg group and let T be a minimal k-acylindrical simplicial G-tree. Then, T/G has at most $1 + 2k(rank\ G - 1)$ vertices.*

The explicit bound in Theorem 5.4 is due to Richard Weidmann. For limit groups, 1-edge splittings satisfying (JSJ) are 2-acylindrical and finitely many such splittings give rise to GAD's that can be arranged to be 2-acylindrical. Theorem 5.4 can then be applied to show that abelian JSJ-decompositions exist.

Theorem 5.5 ([34]). *Limit groups admit abelian JSJ-decompositions.*

Exercise 17 (cf. Exercises 10 and 11). *If Γ is a generic limit group, then $Mod(\Gamma)$ is generated by inner automorphisms together with the generalized Dehn twists associated to 1-edge splittings of Γ that satisfy (JSJ); see [34, Lemma 2.1]. In fact, the only generalized Dehn twists that are not Dehn twists can be taken to be with respect to a splitting of the form $A *_C B$ where $A = C \oplus \mathbb{Z}$.*

Remark 5.6. Suppose that Δ is an abelian JSJ-decomposition for a limit group G. If B is a rigid vertex group of Δ or the peripheral subgroup of an abelian vertex of Δ and if $\alpha \in Mod(G)$, then $\alpha|B$ is trivial[38]. Indeed, B is Δ'-elliptic in any 1-edge GAD Δ' of G satisfying (JSJ) and so the statement is true for a generating set of $Mod(G)$.

6 Limit groups are CLG's

In this section, we show that limit groups are CLG's and complete the proof of Theorem 1.30.

Lemma 6.1. *Limit groups are CLG's*

Proof. Let Γ be a limit group, which we may assume is generic. Let $\{f_i\}$ be a sequence in $Hom(\Gamma, \mathbb{F})$ such that f_i is injective on elements of length at most i (with respect to some finite generating set for Γ). Define $\hat{f}_i$ to be a short map equivalent to f_i. According to Exercise 16, $q : \Gamma \to \Gamma' := \Gamma/\underrightarrow{Ker}\ \hat{f}_i$ is a proper epimorphism, and so by induction we may assume that Γ' is a CLG.

Let Δ be an abelian JSJ-decomposition of Γ. We will show that q and Δ satisfy the conditions in Definition 1.25. The key observations are these.

- Elements of $Mod(\Gamma)$ when restricted to the peripheral subgroup $\overline{P}(A)$ of an abelian vertex A of Δ are trivial (Remark 5.6). Since $\underrightarrow{Ker}\ f_i$ is trivial, $q|\overline{P}(A)$ is injective. Similarly, the restriction of q to the envelope of a rigid vertex group of Δ is injective.

- Elements of $Mod(\Gamma)$ when restricted to edge groups of Δ are trivial. Since Γ is a limit group, each edge group is a maximal abelian subgroup in at least one of the two adjacent vertex groups. See Exercise 7.

- The q-image of a QH-vertex group Q of Δ is non-abelian. Indeed, suppose that Q is a QH-vertex group of Δ and that $q(Q)$ is abelian. Then, eventually $\hat{f}_i(Q)$ is abelian. QH-vertex groups of abelian JSJ-decompositions are canonical, and so every element of $Mod(\Gamma)$ preserves Q up to conjugacy. Hence, eventually $f_i(Q)$ is abelian, contradicting the triviality of $\underrightarrow{Ker}\ f_i$.

□

Proof of Theorem 1.30. (1) $\Longrightarrow$ (2) $\Longrightarrow$ (3) were exercises. (3) $\Longrightarrow$ (1) is the content of Lemma 6.1. □

[38] Recall our convention that *trivial* means *agrees with the restriction of an inner automorphism.*

7 A more geometric approach

In this section, we show how to derive the Main Proposition using Rips theory for fp groups in place of the structure theory of actions of fg groups on real trees.

Definition 7.1. Let K be a finite 2-complex with a measured lamination (Λ, μ). The *length of* Λ, denoted $\|\Lambda\|$, is the sum $\Sigma_e \int_e \mu$ over the edges e of K.

If $\phi : \tilde{K} \to T$ is a resolution, then $\|\phi\|_K$ is the length of the induced lamination Λ_ϕ. Suppose that K is a 2-complex for G.[39] Recall that $T_{\mathbb{F}}$ is a Cayley graph for $\mathbb{F}$ with respect to a fixed basis and that from a homomorphism $f : G \to \mathbb{F}$ a resolution $\phi : (\tilde{K}, \tilde{K}^{(0)}) \to (T_{\mathbb{F}}, T_{\mathbb{F}}^{(0)})$ can be constructed, see [3]. The resolution ϕ depends on a choice of images of a set of orbit representatives of vertices in $\tilde{K}$. If ϕ minimizes $\|\cdot\|_K$ over this set of choices, then we define $\|f\|_K := \|\phi\|_K$.

Lemma 7.2. *Let K_1 and K_2 be finite 2-complexes for G. There is a number $B = B(K_1, K_2)$ such that, for all $f \in Hom(G, \mathbb{F})$,*

$$B^{-1} \cdot \|f\|_{K_1} \leq \|f\|_{K_2} \leq B \cdot \|f\|_{K_1}.$$

Proof. Let $\phi_1 : \tilde{K}_1 \to T_{\mathbb{F}}$ be a resolution such that $\|\phi_1\|_{K_1} = \|f\|_{K_1}$. Choose an equivariant map $\psi^{(0)} : \tilde{K}_2^{(0)} \to \tilde{K}_1^{(0)}$ between 0-skeleta. Then, $\phi_1\psi^{(0)}$ determines a resolution $\phi_2 : \tilde{K}_2 \to T_{\mathbb{F}}$. Extend $\psi^{(0)}$ to a cellular map $\psi^{(1)} : \tilde{K}_2^{(1)} \to \tilde{K}_1^{(1)}$ between 1-skeleta. Let B_2 be the maximum over the edges e of the simplicial length of the path $\psi^{(1)}(e)$ and let E_2 be the number of edges in K_2. Then,

$$\|f\|_{K_2} \leq \|\phi_2\|_{K_2} \leq B_2 N_2 \|\phi_1\|_{K_1} = B_2 N_2 \|f\|_{K_1}.$$

The other inequality is similar. □

Recall that in Definition 4.2, we defined another length $|\cdot|$ for elements of $Hom(G, \mathbb{F})$.

Corollary 7.3. *Let K be a finite 2-complex for G. Then, there is a number $B = B(K)$ such that for all $f \in Hom(G, \mathbb{F})$*

$$B^{-1} \cdot |f| \leq \|f\|_K \leq B \cdot |f|.$$

Proof. If $\mathcal{B}$ is the fixed finite generating set for G and if $R_{\mathcal{B}}$ is the wedge of circles with fundamental group identified with the free group on $\mathcal{B}$, then complete $R_{\mathcal{B}}$ to a 2-complex for G by adding finitely many 2-cells and apply Lemma 7.2. □

[39] i.e. the fundamental group of K is identified with the group G

Remark 7.4. Lemma 7.2 and its corollary allow us to be somewhat cavalier with our choices of generating sets and 2-complexes.

Exercise 18. *The space of (nonempty) measured laminations on K can be identified with the closed cone without 0 in $\mathbb{R}_+^E$, where E is the set of edges of K, given by the triangle inequalities for each triangle of K. The projectivized space $\mathcal{PML}(K)$ is compact.*

Definition 7.5. Two sequences $\{m_i\}$ and $\{n_i\}$ in $\mathbb{N}$ are *comparable* if there is a number $C > 0$ such that $C^{-1} \cdot m_i \leq n_i \leq C \cdot m_i$ for all i.

Exercise 19. *Suppose K is a finite 2-complex for G, $\{f_i\}$ is a sequence in $Hom(G, \mathbb{F})$, $\phi_i : \tilde{K} \to T_{\mathbb{F}}$ is an f_i-equivariant resolution, $\lim T_{f_i} = T$, and $\lim \Lambda_{\phi_i} = \Lambda$. If $\{|f_i|\}$ and $\{\|\phi_i\|_K\}$ are comparable, then, there is a resolution $\tilde{K} \to T$ that sends lifts of leaves of Λ to points of T and is monotonic (Cantor function) on edges of $\tilde{K}$.*

Definition 7.6. An element f of $Hom(G, \mathbb{F})$ is K*-short* if $\|f\|_K \leq \|f'\|_K$ for all $f' \sim f$.

Corollary 7.7. *Let $\{f_i\}$ be a sequence in $Hom(G, \mathbb{F})$. Suppose that $f_i' \sim f_i \sim f_i''$ where f_i' is short and f_i'' is K-short. Then, the sequences $\{|f_i'|\}$ and $\{\|f_i''\|_K\}$ are comparable.* □

Definition 7.8. If ℓ is a leaf of a measured lamination Λ on a finite 2-complex K, then (conjugacy classes of) elements in the image of $\pi_1(\ell \subset K)$ are *carried by* ℓ. Suppose that Λ_i is a component of Λ. If Λ_i is simplicial (consists of a parallel family of compact leaves ℓ), then elements in the image of $\pi_1(\ell \subset K)$ are *carried by* Λ_i. If Λ_i is minimal and if N is a standard neighborhood[40] of Λ_i, then elements in the image of $\pi_1(N \subset K)$ are *carried by* Λ_i.

Definition 7.9. Let K be a finite 2-complex for G. Let $\{f_i\}$ be a sequence of short elements in $Hom(G, \mathbb{F})$ and let $\phi_i : \tilde{K} \to T_{\mathbb{F}}$ be an f_i-equivariant resolution. We say that the sequence $\{\phi_i\}$ is *short* if $\{\|\phi_i\|_K\}$ and $\{|f_i|\}$ are comparable.

Exercise 20. *Let G be freely indecomposable. In the setting of Definition 7.9, if $\{\phi_i\}$ is short, $\Lambda = \lim \Lambda_{\phi_i}$, and $T = \lim T_{f_i}$, then Λ has a leaf carrying non-trivial elements of $Ker(T)$.*

The idea is again that, if not, the induced GAD could be used to shorten. The next exercise, along the lines of Exercise 15, will be needed in the following lemma.[41]

Exercise 21. *Let Δ be a 1-edge GAD of a group G with a homomorphism q to a limit group Γ. Suppose:*

[40] see Theorem 3.6

[41] It is a consequence of Theorem 1.30, but since we are giving an alternate proof we cannot use this.

- *the vertex groups of* Δ *are non-abelian,*
- *the edge group of* Δ *is maximal abelian in each vertex group, and*
- q *is injective on vertex groups of* Δ.

Then, G *is a limit group.*

Lemma 7.10. *Let* Γ *be a limit group and let* $q : G \to \Gamma$ *be an epimorphism such that* $Hom(G, \mathbb{F}) = Hom(\Gamma, \mathbb{F})$. *If* $\alpha \in Mod(G)$ *then* α *induces an automorphism* α' *of* Γ *and* α' *is in* $Mod(\Gamma)$.

Proof. Since $\Gamma = RF(G)$, automorphisms of G induce automorphisms of Γ. Let Δ be a 1-edge splitting of G such that $\alpha \in Mod(\Delta)$. It is enough to check the lemma for α. We will check the case that $\Delta = A *_C B$ and that α is a Dehn twist by an element $c \in C$ and leave the other (similar) cases as exercises. We may assume that $q(A)$ and $q(B)$ are non-abelian for otherwise α' is trivial. Our goal is to successively modify q until it satisfies the conditions of Exercise 21.

First replace all edge and vertex groups by their q-images so that the third condition of the exercise holds. Always rename the result G. If the second condition does not hold, pull[42] the centralizers $Z_A(c)$ and $Z_B(c)$ across the edge. Iterate. It is not hard to show that the limiting GAD satisfies the conditions of the exercise. So, the modified G is a limit group. Since $Hom(G, \mathbb{F}) = Hom(\Gamma, \mathbb{F})$, we have that $G = \Gamma$ and $\alpha = \alpha'$. □

Alternate proof of the Main Proposition. Suppose that Γ is a generic limit group, $T \in \mathcal{T}'(\Gamma)$, and $\{f_i\}$ is a sequence of short elements of $Hom(\Gamma, \mathbb{F})$ such that $\lim T_{f_i} = T$. As before, our goal is to show that $Ker(T)$ is non-trivial, so suppose it is trivial. Recall that the action of Γ on T satisfies all the conclusions of Proposition 3.17.

Let $q : G \to \Gamma$ be an epimorphism such that G is fp and $Hom(G, \mathbb{F}) = Hom(\Gamma, \mathbb{F})$. By Lemma 7.10, elements of the sequence $\{f_i q\}$ are short. We may assume that all intermediate quotients $G \to G' \to \Gamma$ are freely indecomposable[43].

Choose a 2-complex K for G and a subsequence so that $\Lambda = \lim \Lambda_{\phi_i}$ exists where $\phi_i : \tilde{K} \to T_{\mathbb{F}}$ is an $f_i q$-equivariant resolution and $\{\phi_i\}$ is short. For each component Λ_0 of Λ, perform one of the following moves to obtain a new finite laminated 2-complex for an fp quotient of G (that we will immediately rename (K, Λ) and G). Let G_0 denote the subgroup of G carried by Λ_0.

1. If Λ_0 is minimal and if G_0 stabilizes a linear subtree of T, then enlarge $N(\Lambda_0)$ to a model for the action of $q(G_0)$ on T.

[42] If A_0 is a subgroup of A, then the result of *pulling* A_0 across the edge is $A *_{\langle A_0, C\rangle} \langle A_0, B\rangle$, cf. moves of type IIA in [3].

[43] see [30]

2. If Λ_0 is minimal and if G_0 does not stabilize a linear subtree of T, then collapse all added annuli to their bases.

3. If Λ_0 is simplicial and G_0 stabilizes an arc of T, then attach 2-cells to leaves to replace G_0 by $q(G_0)$.

In each case, also modify the resolutions to obtain a short sequence on the new complex with induced laminations converging to Λ. The modified complex and resolutions contradict Exercise 20. Hence, $Ker(T)$ is non-trivial.

To finish, choose non-trivial $k_T \in Ker(T)$. As before, if $\{U(k_{T_i})\}$ is a finite cover for $\mathcal{T}'(\Gamma)$, then $\{\Gamma \to Ab(\Gamma)\} \cup \{\Gamma \to \Gamma/\langle\langle k_{T_i} \rangle\rangle\}$ is a factor set. □

References

[1] Emina Alibegović. Makanin-Razborov diagrams for limit groups. *Geom. Topol.*, 11:643–666, 2007.

[2] Mladen Bestvina. $\mathbb{R}$-trees in topology, geometry, and group theory. In *Handbook of geometric topology*, pages 55–91. North-Holland, Amsterdam, 2002.

[3] Mladen Bestvina and Mark Feighn. Bounding the complexity of simplicial group actions on trees. *Invent. Math.*, 103(3):449–469, 1991.

[4] M. R. Bridson and G. A. Swarup. On Hausdorff-Gromov convergence and a theorem of Paulin. *Enseign. Math. (2)*, 40(3-4):267–289, 1994.

[5] Christophe Champetier and Vincent Guirardel. Limit groups as limits of free groups. *Israel J. Math.*, 146:1–75, 2005.

[6] Marc Culler and John W. Morgan. Group actions on **R**-trees. *Proc. London Math. Soc. (3)*, 55(3):571–604, 1987.

[7] M. J. Dunwoody and M. E. Sageev. JSJ-splittings for finitely presented groups over slender groups. *Invent. Math.*, 135(1):25–44, 1999.

[8] K. Fujiwara and P. Papasoglu. JSJ-decompositions of finitely presented groups and complexes of groups. *Geom. Funct. Anal.*, 16(1):70–125, 2006.

[9] R. I. Grigorchuk and P. F. Kurchanov. On quadratic equations in free groups. In *Proceedings of the International Conference on Algebra, Part 1 (Novosibirsk, 1989)*, volume 131 of *Contemp. Math.*, pages 159–171, Providence, RI, 1992. Amer. Math. Soc.

[10] Daniel Groves. Limit groups for relatively hyperbolic group, I: the basic tools. preprint.

[11] Daniel Groves. Limit groups for relatively hyperbolic groups. II. Makanin-Razborov diagrams. *Geom. Topol.*, 9:2319–2358 (electronic), 2005.

[12] Vincent Guirardel. Actions of finitely generated groups on **R**-tree. to appear in Annales de l'Institut Fourier.

[13] Vincent Guirardel. Limit groups and groups acting freely on $\mathbb{R}^n$-trees. *Geom. Topol.*, 8:1427–1470 (electronic), 2004.

[14] O. Kharlampovich and A. Myasnikov. Irreducible affine varieties over a free group. I. Irreducibility of quadratic equations and Nullstellensatz. *J. Algebra*, 200(2):472–516, 1998.

[15] O. Kharlampovich and A. Myasnikov. Irreducible affine varieties over a free group. II. Systems in triangular quasi-quadratic form and description of residually free groups. *J. Algebra*, 200(2):517–570, 1998.

[16] O. Kharlampovich and A. Myasnikov. Description of fully residually free groups and irreducible affine varieties over a free group. In *Summer School in Group Theory in Banff, 1996*, volume 17 of *CRM Proc. Lecture Notes*, pages 71–80. Amer. Math. Soc., Providence, RI, 1999.

[17] Gilbert Levitt. La dynamique des pseudogroupes de rotations. *Invent. Math.*, 113(3):633–670, 1993.

[18] R. C. Lyndon. The equation $a^2b^2 = c^2$ in free groups. *Michigan Math. J*, 6:89–95, 1959.

[19] G. S. Makanin. Equations in a free group. *Izv. Akad. Nauk SSSR Ser. Mat.*, 46(6):1199–1273, 1344, 1982.

[20] G. S. Makanin. Decidability of the universal and positive theories of a free group. *Izv. Akad. Nauk SSSR Ser. Mat.*, 48(4):735–749, 1984.

[21] J. Morgan and P. Shalen. Valuations, trees, and degenerations of hyperbolic structures, I. *Ann. of Math.(2)*, 120:401–476, 1984.

[22] Frédéric Paulin. Topologie de Gromov équivariante, structures hyperboliques et arbres réels. *Invent. Math.*, 94(1):53–80, 1988.

[23] Frédéric Paulin. The Gromov topology on **R**-trees. *Topology Appl.*, 32(3):197–221, 1989.

[24] Frédéric Paulin. Sur la théorie élémentaire des groupes libres (d'après Sela). *Astérisque*, (294):ix, 363–402, 2004.

[25] A. A. Razborov. Systems of equations in a free group. *Izv. Akad. Nauk SSSR Ser. Mat.*, 48(4):779–832, 1984.

[26] V. N. Remeslennikov. ∃-free groups. *Sibirsk. Mat. Zh.*, 30(6):193–197, 1989.

[27] V. N. Remeslennikov. ∃-free groups and groups with length function. In *Second International Conference on Algebra (Barnaul, 1991)*, volume 184 of *Contemp. Math.*, pages 369–376. Amer. Math. Soc., Providence, RI, 1995.

[28] E. Rips and Z. Sela. Structure and rigidity in hyperbolic groups. I. *Geom. Funct. Anal.*, 4(3):337–371, 1994.

[29] E. Rips and Z. Sela. Cyclic splittings of finitely presented groups and the canonical JSJ decomposition. *Ann. of Math. (2)*, 146(1):53–109, 1997.

[30] G. P. Scott. Finitely generated 3-manifold groups are finitely presented. *J. London Math. Soc. (2)*, 6:437–440, 1973.

[31] Z. Sela. Diophantine geometry over groups VII: The elementary theory of a hyperbolic group. preprint.

[32] Z. Sela. Diophantine geometry over groups VIII: Stability. preprint.

[33] Z. Sela. Acylindrical accessibility for groups. *Invent. Math.*, 129(3):527–565, 1997.

[34] Z. Sela. Diophantine geometry over groups I. Makanin-Razborov diagrams. *Publ. Math. Inst. Hautes Études Sci.*, (93):31–105, 2001.

[35] Z. Sela. Diophantine geometry over groups and the elementary theory of free and hyperbolic groups. In *Proceedings of the International Congress of Mathematicians, Vol. II (Beijing, 2002)*, pages 87–92, Beijing, 2002. Higher Ed. Press.

[36] Z. Sela. Diophantine geometry over groups II. Completions, closures and formal solutions. *Israel J. Math.*, 134:173–254, 2003.

[37] Z. Sela. Diophantine geometry over groups IV. An iterative procedure for validation of a sentence. *Israel J. Math.*, 143:1–130, 2004.

[38] Z. Sela. Diophantine geometry over groups III. Rigid and solid solutions. *Israel J. Math.*, 147:1–73, 2005.

[39] Z. Sela. Diophantine geometry over groups V_1. Quantifier elimination. I. *Israel J. Math.*, 150:1–197, 2005.

[40] Z. Sela. Diophantine geometry over groups V_2. Quantifier elimination. II. *Geom. Funct. Anal.*, 16(3):537–706, 2006.

[41] Z. Sela. Diophantine geometry over groups VI. The elementary theory of a free group. *Geom. Funct. Anal.*, 16(3):707–730, 2006.

[42] Jean-Pierre Serre. *Trees.* Springer-Verlag, Berlin, 1980. Translated from the French by John Stillwell.

[43] John Stallings. How not to prove the Poincaré conjecture. In *Topology Seminar, Wisconsin, 1965*, Edited by R. H. Bing and R. J. Bean. Annals of Mathematics Studies, No. 60, pages ix+246. Princeton University Press, 1966.

[44] Richard Weidmann. The Nielsen method for groups acting on trees. *Proc. London Math. Soc. (3)*, 85(1):93–118, 2002.

[45] Henry Wilton. Solutions to Bestvina and Feighn's Exercises on limit groups. math.GR/0604137.

[46] Heiner Zieschang. Alternierende Produkte in freien Gruppen. *Abh. Math. Sem. Univ. Hamburg*, 27:13–31, 1964.

Solutions to Bestvina & Feighn's exercises on limit groups

Henry Wilton

Abstract

This article gives solutions to the exercises in Bestvina and Feighn's paper [2] on Sela's work on limit groups. We prove that all constructible limit groups are limit groups and give an account of the shortening argument of Rips and Sela.

Mladen Bestvina and Mark Feighn's beautiful first set of notes [2] on Zlil Sela's work on the Tarski problems (see [10] *et seq.*) provides a very useful introduction to the subject. It gives a clear description of the construction of Makanin–Razborov diagrams, and precisely codifies the structure theory for limit groups in terms of *constructible limit groups (CLGs)*. Furthermore, the reader is given a practical initiation in the subject with exercises that illustrate the key arguments. This article is intended as a supplement to [2], to provide solutions to these exercises. Although we do give some definitions in order not to interrupt the flow, we refer the reader to [2] for all the longer definitions and background ideas and references.

1 Definitions and elementary properties

In this section we present solutions to exercises 2, 3, 4, 5, 6 and 7, which give some of the simpler properties and the first examples and non-examples of limit groups.

1.1 ω-residually free groups

Fix $\mathbb{F}$ a free group of rank $r > 1$.

Definition 1.1 *A finitely generated group G is* ω-residually free *if, for any finite subset $X \subset G$, there exists a homomorphism $h : G \to \mathbb{F}$ whose restriction to X is injective. (Equivalently, whenever $1 \notin X$ there exists a homomorphism $h : G \to \mathbb{F}$ so that $1 \notin h(X)$.)*

Residually free groups inherit many of the properties of free groups; the first and most obvious property is being torsion-free.

Lemma 1.2 (Exercise 2 of [2]) *Any residually free group is torsion-free.*

Proof. Let G be ω-residually free (indeed G can be thought of as merely residually free). Then for any $g \in G$, there exists a homomorphism $h : G \to \mathbb{F}$ with $h(g) \neq 1$; so $h(g^k) \neq 1$ for all integers k, and $g^k \neq 1$. □

It is immediate that any subgroup of an ω-residually free group is ω-residually free (exercise 6 of [2]).

That the choice of $\mathbb{F}$ does not matter follows from the observation that all finitely generated free groups are ω-residually free.

Example 1.3 (Free groups) *Let F be a finitely generated free group. Realize $\mathbb{F}$ as the fundamental group of a rose Γ with r petals; that is, the wedge of r circles. Then Γ has an infinite-sheeted cover that corresponds to a subgroup F' of $\mathbb{F}$ of countably infinite rank. The group F can be realized as a free factor of F'; this exhibits an injection $F \hookrightarrow \mathbb{F}$. In particular, every free group is ω-residually free.*

Example 1.4 (Free abelian groups) *Let A be a finitely generated free abelian group, and let $a_1, \ldots, a_n \in A$ be non-trivial. Fix a basis for A, and consider the corresponding inner product. Let $z \in A$ be such that $\langle z, a_i \rangle \neq 0$ for all i. Then inner product with z defines a homomorphism $A \to \mathbb{Z}$ so that the image of every a_i is non-trivial, as required.*

Examples 1.3 and 1.4 give exercise 3 of [2].

1.2 Limit groups

Groups that are ω-residually free are natural examples of limit groups.

Definition 1.5 *Let $\mathbb{F}$ be as above, and Γ a finitely generated group. A sequence of homomorphisms $(f_n : \Gamma \to \mathbb{F})$ is* stable *if, for every $g \in G$, $f_n(g)$ is either eventually* 1 *or eventually not* 1. *The* stable kernel *of a stable sequence of homomorphisms (f_n) consists of all $g \in G$ with $f_n(g)$ eventually trivial; it is denoted $\underrightarrow{\ker} f_n$.*

A limit group *is a group arising as a quotient $\Gamma / \underrightarrow{\ker} f_n$ for (f_n) a stable sequence.*

Lemma 1.6 (Exercise 5 of [2]) *Every group that is ω-residually free group is a limit group.*

Proof. Let G be an ω-residually free group. Fix a generating set, and let $X_n \subset G$ be the ball of radius n about the identity in the word metric. Let $f_n : G \to \mathbb{F}$ be a homomorphism that is injective on X_n. Now f_n is a stable sequence and the stable kernel is trivial, so G is a limit group. □

In fact every limit group is ω-residually free (lemma 1.11 of [2]). Henceforth, we shall use the terms interchangeably.

1.3 Negative examples

Let's see some examples of groups that aren't limit groups. The first three examples are surface groups that aren't even residually free. It follows from lemma 1.2 that the fundamental group of the real projective plane is not a limit group. A slightly finer analysis yields some other negative examples.

Lemma 1.7 *The only 2-generator residually free groups are the free group of rank 2 and the free abelian group of rank 2.*

Proof. Let G be a residually free group generated by x and y. If G is non-abelian then $[x, y] \neq 1$ so there exists a homomorphism $f : G \to \mathbb{F}$ with $f([x, y]) \neq 1$. So $f(x)$ and $f(y)$ generate a rank 2 free subgroup of $\mathbb{F}$. Therefore G is free. □

In particular, the fundamental group of the Klein bottle is not a limit group. Only one other surface group fails to be ω-residually free. This was first shown by R. S. Lyndon in [5].

Lemma 1.8 (The surface of Euler characteristic -1) *Let Σ be a closed surface of Euler characteristic -1. Then any homomorphism $f_* : \pi_1(\Sigma) \to \mathbb{F}$ has abelian image. In particular, since $\pi_1(\Sigma)$ is not abelian, it is not residually free.*

Proof. Let Γ be a bouquet of circles so $\mathbb{F} = \pi_1(\Gamma)$. Realize the homomorphism f_* as a map from Σ to Γ, which we denote by f. Our first aim is to find an essential simple closed curve in the kernel of f_*.

Consider x the mid-point of an edge of Γ. Altering f by a homotopy, it can be assumed that f is transverse at x; in this case, $f^{-1}(x)$ is a collection of simple closed curves. Let γ be such a curve. If γ is null-homotopic in Σ then γ can be removed from $f^{-1}(x)$ by a homotopy. If all components of the pre-images of all midpoints x can be removed in this way then f_* was the trivial homomorphism. Otherwise, any remaining such component γ lies in the kernel of f as required.

We proceed with a case-by-case analysis of the components of $\Sigma \smallsetminus \gamma$.

1. If γ is 2-sided and separating then, by examining Euler characteristic, the components of $\Sigma \smallsetminus \gamma$ are a punctured torus or a Klein bottle, together with a Möbius band. So f factors through the one-point union $T \vee \mathbb{R}P^2$ or $K \vee \mathbb{R}P^2$. In either case, it follows that the image is abelian.

2. If γ is 2-sided and non-separating then $\Sigma - \gamma$ is the non-orientable surface with Euler characteristic -1 and two boundary components, so f_* factors through $\mathbb{Z} * \mathbb{Z}/2\mathbb{Z}$ and hence through $\mathbb{Z}$.

3. If γ is 1-sided then γ^2 is 2-sided and separating, and case 1 applies.

This finishes the proof. □

Lemmas 1.2, 1.7 and 1.8 give exercise 4 of [2]. Here is a more interesting obstruction to being a limit group. A group G is *commutative transitive* if every non-trivial element has abelian centralizer; equivalently, if $[x, y] = [y, z] = 1$ then $[x, z] = 1$. Note that $\mathbb{F}$ is commutative transitive, since every non-trivial element has cyclic centralizer.

Lemma 1.9 (Exercise 7 of [2]) *Limit groups are commutative transitive.*

Proof. Let G be ω-residually free, let $g \in G$, and suppose $a, b \in G$ commute with g. Then there exists a homomorphism

$$f : G \to \mathbb{F}$$

injective on the set $\{1, g, [a, b]\}$. Then $f([g, a]) = f([g, b]) = 1$; since $\mathbb{F}$ is commutative transitive, it follows that

$$f([a, b]) = 1.$$

So $[a, b] = 1$, as required. □

A stronger property also holds. A subgroup $H \subset G$ is *malnormal* if, whenever $g \notin H$, $gHg^{-1} \cap H = 1$. The group G is *CSA* if every maximal abelian subgroup is malnormal.

Remark 1.10 *If G is CSA then G is commutative transitive. For, let $g \in G$ with centralizer $Z(g)$. Consider maximal abelian $A \subset Z(g)$ and $h \in Z(g)$. Then $g \in hAh^{-1} \cap A$, so $h \in A$. Therefore $Z(g) = A$.*

Lemma 1.11 *Limit groups are CSA.*

Proof. Let $H \subset G$ be a maximal abelian subgroup, consider $g \in G$, and suppose there exists non-trivial $h \in gHg^{-1} \cap H$. Let $f : G \to \mathbb{F}$ be injective on the set

$$\{1, g, h, [g, h]\}.$$

Then $f([h, ghg^{-1}]) = 1$, which implies that $f(h)$ and $f(ghg^{-1})$ lie in the same cyclic subgroup. But in a free group, this is only possible if $f(g)$ also lies in that cyclic subgroup; so $f([g, h]) = 1$, and hence $[g, h] = 1$. By lemma 1.9 it follows that g commutes with every element of H, so $g \in H$. □

2 Embeddings in real algebraic groups

In this section we provide solutions to exercises 8 and 9 of [2], which show how to embed limit groups in real algebraic groups and also $PSL_2(\mathbb{R})$, and furthermore give some control over the nature of the embeddings. First, we need a little real algebraic geometry.

By, for example, proposition 3.3.13 of [3], every real algebraic variety V has an open dense subset $V_{\mathrm{reg}} \subset V$ with finitely many connected components, so that every component $V' \subset V_{\mathrm{reg}}$ is a manifold.

Lemma 2.1 *Consider a countable collection $V_1, V_2, \ldots \subset V$ of closed subvarieties. Then for any component V' of V_{reg} as above, either there exists k so that $V' \subset V_k$, or*

$$V' \cap \bigcup_{i=1}^{\infty} V_i$$

has empty interior.

Proof. Suppose $V' \cap \bigcup_i V_i$ doesn't have empty interior. Then, by Baire's Category Theorem, there exists k such that $V' \cap V_k$ doesn't have empty interior. Consider x in the closure of the interior of $V' \cap V_k$ and let f be an algebraic function on V' that vanishes on V_k. Then f has zero Taylor expansion at x, so f vanishes on an open neighbourhood of x. In particular, x lies in the interior of $V' \cap V_k$. So the interior of $V' \cap V_k$ is both open and closed, and $V' \subset V_k$ since V' is connected. □

Lemma 2.2 (Exercise 8 of [2]) *Let $\mathcal{G}$ be an algebraic group over $\mathbb{R}$ into which $\mathbb{F}$ embeds. Then for any limit group G there exists an embedding*

$$G \hookrightarrow \mathcal{G}.$$

In particular, G embeds into $SL_2(\mathbb{R})$ and $SO(3)$.

Proof. Consider the variety $V = \mathrm{Hom}(G, \mathcal{G})$. (If G is of rank r, then V is a subvariety of $\mathcal{G}^r$, cut out by the relations of G. By Hilbert's Basis Theorem, finitely many relations suffice.) For each $g \in G$, consider the subvariety

$$V_g = \{f \in V | f(g) = 1\}.$$

If G does not embed into $\mathcal{G}$ then V is covered by the subvarieties V_g for $g \neq 1$. By lemma 2.1 every component of V_{reg} is contained in some V_g, so

$$V = V_{g_1} \cup \ldots \cup V_{g_n}$$

for some non-trivial $g_1, \ldots, g_n \in G$.

So every homomorphism from G to $\mathcal{G}$ kills one of the g_i. But $\mathbb{F}$ embeds in $\mathcal{G}$, so this contradicts the assumption that G is ω-residually free. □

Remark 2.3 *Given a limit group G and an embedding $f : G \hookrightarrow SL_2(\mathbb{R})$ we have a natural map $G \to PSL_2(\mathbb{R})$. This is also an embedding since any element in its kernel satisfies $f(g)^2 = 1$ and G is torsion-free.*

We can gain more control over embeddings into $PSL_2(\mathbb{R})$ by considering the trace function.

Lemma 2.4 (Exercise 9 of [2]) *If G is a limit group and $g_1, \ldots, g_n \in G$ are non-trivial then there is an embedding $G \hookrightarrow PSL_2(\mathbb{R})$ whose image has no non-trivial parabolic elements, and so that the images of $g_1, \ldots, g_n$ are all hyperbolic.*

Proof. Abusing notation, we identify each element of the variety $V = \mathrm{Hom}(G, SL_2(\mathbb{R}))$ with the corresponding element of $\mathrm{Hom}(G, PSL_2(\mathbb{R}))$, and call it elliptic, hyperbolic or parabolic accordingly. For each $g \in G$, consider the closed subvariety U_g of homomorphisms that map g to a parabolic, and the open set W_g of homomorphisms that map g to a hyperbolic. (Note that $\gamma \in SL_2(\mathbb{R})$ is parabolic if $|\mathrm{tr}\gamma| = 2$ and hyperbolic if $|\mathrm{tr}\gamma| < 2$.) Fix an embedding $\mathbb{F} \hookrightarrow SL_2(\mathbb{R})$ whose image in $PSL_2(\mathbb{R})$ is the fundamental group of a sphere with open discs removed; such a subgroup is called a *Schottky group*, and every non-trivial element is hyperbolic. Call a component V' of V_{reg} *essential* if its closure contains a homomorphism $G \to SL_2(\mathbb{R})$ that factors through $\mathbb{F} \hookrightarrow SL_2(\mathbb{R})$ and maps the g_i non-trivially.

Suppose every essential component V' of V_{reg} is contained in some $U_{g'}$ for non-trivial g'. Then, since there are only finitely many components, for certain non-trivial $g'_1, \ldots, g'_m$, every homomorphism $G \to \mathbb{F}$ kills one of the g_i or one of the g'_j, contradicting the assumption that G is ω-residually free. Therefore, by lemma 2.1, there exists an essential component V' so that $V' \cap \bigcup_{g \neq 1} U_g$ has empty interior. In particular,

$$\bigcap_i W_{g_i} \smallsetminus \bigcup_{g \neq 1} U_g$$

is non-empty, as required. □

3 GADs for limit groups

In this section we provide solutions to exercises 10 and 11, and also the related exercise 17. For the definitions of the modular group $\mathrm{Mod}(G)$, and of generalized Dehn twists, see definitions 1.6 and 1.17 respectively in [2].

Lemma 3.1 (Exercise 10 of [2]) $\mathrm{Mod}(G)$ *is generated by inner automorphisms and generalized Dehn twists.*

Proof. Since the mapping class group of a surface is generated by Dehn twists (see, for example, [4]), it only remains to show that unimodular automorphisms of abelian vertices are generated by generalized Dehn twists. Given such a vertex A, we can write

$$G = A *_{\bar{P}(A)} B$$

for some subgroup B of G. Any unimodular automorphism of A is a generalized Dehn twist of this splitting. □

Lemma 3.2 (Exercise 11 of [2]) *Let M be a non-cyclic maximal abelian subgroup of a limit group G.*

1. *If $G = A *_C B$ for C abelian then M is conjugate into either A or B.*
2. *If $G = A*_C$ with C abelian then either M is conjugate into A or there is a stable letter t so that M is conjugate to $M' = \langle C, t\rangle$ and $G = A*_C M'$.*

Proof. We first prove 1. Suppose M is not elliptic in the splitting $G = A*_C B$. (Note that we don't yet know that M is finitely generated.) Non-cyclic abelian groups have no free splittings, so C is non-trivial. Let T be the Bass–Serre tree of the splitting. Either M fixes an axis in T, or it fixes a point on the boundary. In the latter case, there is an increasing chain of edge groups

$$C_1 \subset C_2 \subset \ldots M.$$

But every C_i is a conjugate of C_1, and since M is malnormal it follows that $C_i = C_1$. So $M = C_i$, contradicting the assumption that M is not elliptic.

If M fixes a line in T then M can be conjugated to M' fixing a line L so that A stabilizes a vertex v of L and M' is of the form $M' = C \oplus \mathbb{Z}$; C fixes L pointwise, and $\mathbb{Z}$ acts as translations of L. Consider the edges of L incident at v, corresponding to the cosets C and aC, for some $a \in A \setminus C$. Since $aCa^{-1} = C$ and C is non-trivial, it follows from lemma 1.11 that $a \in M'$. But a is elliptic so $a \in C$, a contradiction.

In the HNN-extension case, assuming M is not elliptic in the splitting we have as before that M preserves a line in the Bass–Serre tree T. Conjugating M to M', we may assume C fixes an edge in the preserved line L, so $C \subset M'$. The stabilizer of an adjacent edge is of the form $(ta)C(ta)^{-1}$, where $a \in A$ and t is the stable letter of the HNN-extension. Therefore $C = (ta)C(ta)^{-1}$, so since G is CSA it follows that $M' = C \oplus \langle ta\rangle$ and

$$G = A *_C M'$$

as required. □

Remark 3.3 *Note that, in fact, the proof of lemma 1.11 only used that the vertex groups are CSA and that the edge group was maximal abelian on one side.*

A one-edge splitting of G is said to satisfy *condition JSJ* if every non-cyclic abelian group is elliptic in it. Recall that a limit group is *generic* if it is freely indecomposable, non-abelian and not a surface group.

Lemma 3.4 (Exercise 17 of [2]) *If G is a generic limit group then the group* $\mathrm{Mod}(G)$ *is generated by inner automorphisms and generalized Dehn twists in one-edge splittings satisfying JSJ.*

Furthermore, the only generalized Dehn twists that are not Dehn twists can be taken to be with respect to a splitting of the from $G = A *_C B$ with $A = C \oplus \mathbb{Z}$.

Proof. By lemma 3.1, $\mathrm{Mod}(G)$ is generated by inner automorphisms and generalized Dehn twists. By lemma 3.2, any splitting of G as an amalgamated product satisfies JSJ. Consider, therefore, the splitting

$$G = A *_C .$$

A (generalized) Dehn twist δ_z in this splitting fixes A and maps the stable letter $t \mapsto tz$, for some $z \in Z_G(C)$.

Suppose that this HNN-extension doesn't satisfy JSJ, so there exists some (without loss, maximal) abelian subgroup M that is not elliptic in the splitting. By lemma 3.2, after conjugating M to M', we have that

$$G = A *_C M'$$

and $M' = C \oplus \langle t \rangle$ where t is the stable letter. Since $M' = Z_G(C)$, a Dehn twist δ_z along z (for $c \in C$ and $n \in \mathbb{Z}$) fixes A and maps

$$t \mapsto z + t.$$

But this is a generalized Dehn twist in the amalgamated product. So $\mathrm{Mod}(G)$ is, indeed, generated by generalized Dehn twists in one-edge splittings satisfying JSJ.

Any generalized Dehn twist δ that is not a Dehn twist is in a splitting of the form

$$G = A *_C B$$

with A abelian, and acts as a unimodular automorphism on A that preserves $\bar{P}(A)$. Recall that $A/\bar{P}(A)$ is finitely generated, by remark 1.15 of [2]. To show that it is enough to use splittings in which $A = C \oplus \mathbb{Z}$, we work by induction on the rank of $A/\bar{P}(A)$. Write $A = A' \oplus \mathbb{Z}$, where $\bar{P}(A) \subset A'$. Then there is a modular automorphism α, agreeing with δ on A', generated by generalized Dehn twists of the required form, by induction. Now δ and α differ by a generalized Dehn twist in the splitting

$$G = A *_{A'} (A' *_C B)$$

which is of the required form. □

4 Constructible Limit Groups

For the definition of CLGs see [2]. The definition lends itself to the technique of proving results by a nested induction, first on level and then on the number of edges of the GAD Δ. To prove that CLGs have a certain property, this technique often reduces the proof to the cases where G has a one-edge splitting over groups for which the property can be assumed. In this section we provide solutions to exercises 12, 13, 14 and 15, which give the first properties of CLGs, culminating in the result that all CLGs are ω-residually free.

4.1 CLGs are CSA

We have seen that limit groups are CSA. This section is devoted to proving that CLGs are also CSA. Knowing this will prove extremely useful in deducing the other properties of CLGs. Note that the property of being CSA passes to subgroups.

Lemma 4.1 *CLGs are CSA.*

By induction on the number of edges in the graph of groups Δ, it suffices to consider a CLG G such that

$$G = A *_C B$$

or

$$G = A*_C$$

where each vertex group is assumed to be CSA and the edge group is taken to be maximal abelian on one side. (In the first case, we will always assume that C is maximal abelian in A.) First, we have an analogue of lemma 3.2.

Lemma 4.2 *Let G be as above. Then G decomposes as an amalgamated product or HNN-extension in such a way that all non-cyclic maximal abelian subgroups are conjugate into a vertex group. Furthermore, in the HNN-extension case we have that $C \cap C^t = 1$ where C is the edge group and t is the stable letter.*

Proof. By induction, we can assume that the vertex groups are CSA. Note that the proof of the first assertion of lemma 3.2 only relies on the facts that the vertex groups are CSA and the edge group is maximal abelian in one vertex group. So the amalgamated product case follows.

Now consider the case of an HNN-extension. Suppose that, for some stable letter t, $C \cap C^t$ is non-trivial. Then since the vertex group A is commutative transitive, it follows that $C \subset C^t$ (or $C^t \subset C$, in which case replace t by t^{-1}). If $\rho : G \to G'$ is the retraction to a lower level then, since G' can be taken to be CSA, $\rho(C^t) = \rho(C)$. But ρ is injective on edge groups so $C = C^t$ and, furthermore, t commutes with C.

Otherwise, for every choice of stable letter t, $C \cap C^t$ is trivial. In either case, the result now follows as in the proof of the second assertion of lemma 3.2. □

Recall that a simplicial G-tree is *k-acylindrical* if the fixed point set of every non-trivial element of G has diameter at most k.

Lemma 4.3 *In the graph-of-groups decomposition given by lemma 4.2, the Bass–Serre tree is 2-acylindrical.*

Proof. In the amalgamated product case this is because, for any $a \in A \setminus C$, $C \cap C^a = 1$. Likewise, in the HNN-extension case this is because $C \cap C^t = 1$ where t is the stable letter. □

Proof of lemma 4.1. Let $M \subset G$ be a maximal abelian subgroup and suppose

$$1 \neq m \subset M^g \cap M.$$

Let T be the Bass–Serre tree of the splitting. If M is cyclic then it might act as translations on a line L in T. Then g also maps L to itself. But it follows from acylindricality that any element that preserves L lies in M; so $g \in M$ as required.

We can therefore assume that M fixes a vertex v of T. If g also fixes v then, since the vertex stabilizers are CSA, $g \in M$. Consider the case when

$$G = A *_C B;$$

the case of an HNN-extension is similar. Then without loss of generality $M \subset A$ and $g = ba$ for some $a \in A$ and $b \in B$ so M^g fixes a vertex stabilized by A^b for some $b \in B$. Since C is maximal abelian in A we have $C = M$ and since B is CSA and $m \in C \cap C^b$ it follows that b commutes with C so $b \in M$, hence $g \in M$. □

4.2 Abelian subgroups

It is no surprise that CLGs share the most elementary property of limit groups.

Lemma 4.4 *CLGs are torsion-free.*

Proof. The freely decomposable case is immediate by induction. Therefore assume G is a freely indecomposable CLG of level n, with Δ and $\rho : G \to G'$ as in the definition. Suppose $g \in G$ is of finite order. Then g acts elliptically on the Bass–Serre tree of Δ, so g lies in a vertex group. Clearly if the vertex is QH then $g = 1$, and by induction if the vertex is rigid then $g = 1$. Suppose therefore that the vertex is abelian. Then g lies in the peripheral subgroup. But ρ is assumed to inject on the peripheral subgroup, so by induction g is trivial. □

However, it is far from obvious that limit groups have abelian subgroups of bounded rank; indeed, it is not obvious that all abelian subgroups of limit groups are finitely generated. But this is true of CLGs.

Lemma 4.5 (Exercise 13 in [2]) *Abelian subgroups of CLGs are free, and there is a uniform (finite) bound on their rank.*

Proof. The proof starts by induction on the level of G. Let G be a CLG. Since non-cyclic abelian subgroups have no free splittings we can assume G is freely indecomposable. Let Δ be a generalized abelian decomposition. Let T be the Bass–Serre tree of Δ. If A fixes a vertex of T then the result follows by induction on level. Otherwise, A fixes a line T_A in T, on which it acts by translations. The quotient $\Delta' = T_A/A$ is topologically a circle; after some collapses Δ' is an HNN-extension; so the rank of A is bounded by the maximum rank of abelian subgroups of the vertex groups plus 1. So by induction the rank of A is uniformly bounded. That A is free follows from lemma 4.4. □

4.3 Heredity

Let Σ be a (not necessarily compact) surface with boundary. Then a boundary component δ is a circle or a line, and defines up to conjugacy a cyclic subgroup $\pi_1(\delta) \subset \pi_1(\Sigma)$. These are called the *peripheral subgroups* of Σ.

Remark 4.6 *Let Σ be a non-compact surface with non-abelian fundamental group. Then there exists a non-trivial free splitting of $\pi_1(\Sigma)$, with respect to which all peripheral subgroups are elliptic.*

Lemma 4.7 (Exercise 12 in [2]) *Let G be a CLG of level n and let H be a finitely generated subgroup. Then H is a free product of finitely many CLGs of level at most n.*

Proof. The subgroup H can be assumed to be freely indecomposable by Grushko's theorem, so we can also assume that G is freely indecomposable.

Let Δ and $\rho : G \to G'$ be as in the definition. The subgroup H inherits a graph-of-groups decomposition from Δ, namely the quotient of the Bass–Serre tree T by H. Since H is finitely generated, it is the fundamental group of some finite core $\Delta' \subset T/H$. Every vertex of Δ' covers a vertex of Δ, from which it inherits its designation as QH, abelian or rigid.

The edge groups of Δ' are subgroups of the edge groups of Δ, so they are abelian and ρ is injective on them. Furthermore, it follows from lemma 4.1 that H is commutative transitive, so each edge group of Δ' is maximal abelian on one side of the associated one-edge splitting.

Let V' be a vertex group of Δ', a subgroup of the vertex group V of Δ. There are three case to consider.

1. $V' \subset V$ are abelian. Since every map $f' : V' \to \mathbb{Z}$ with $f'(P(V')) = 0$ extends to a map $f : V \to \mathbb{Z}$ with $f(P(V)) = 0$ we have that
$$\bar{P}(V') \subset \bar{P}(V)$$
so ρ is injective on $\bar{P}(V')$.

2. $V' \subset V$ are QH. If V' is of infinite index in V then V' is the fundamental group of a non-compact surface, so by remark 4.6 H is freely decomposable. Therefore it can be assumed that V' is of finite degree m in V. In particular, V is the fundamental group of a compact surface that admits a pseudo-Anosov automorphism. Furthermore, let $g, h \in V$ be such that $\rho([g, h]) \neq 1$. Then because CLGs are commutative transitive,
$$\rho([g^m, h^m]) \neq 1.$$
But $g^m, h^m \in V'$, so $\rho(V')$ is non-abelian.

3. $V' \subset V$ are rigid. Then $\tilde{V}' \subset \tilde{V}$ because CLGs are commutative transitive, so $\rho|_{V'}$ is injective.

Therefore $\rho|_H : H \to G'$ and Δ' satisfy the properties for H to be a CLG. □

4.4 Coherence

A group is *coherent* if every finitely generated subgroup is finitely presented. Note that free groups and free abelian groups are coherent. For limit groups, coherence is an instance of a more general phenomenon, as in the next lemma. Recall that a group is *slender* if every subgroup is finitely generated. Finitely generated abelian groups are slender.

Lemma 4.8 *The fundamental group of a graph of groups with coherent vertex groups and slender edge groups is coherent.*

Proof. Let Δ be a graph of groups, with coherent vertex groups and slender edge groups. Let $G = \pi_1(\Delta)$ and $H \subset G$ a finitely generated subgroup. Then H inherits a graph-of-groups decomposition from Δ given by taking the quotient of the Bass–Serre tree T of Δ by the action of H. Since H is finitely generated it is the fundamental group of some finite core $\Delta' \subset T/H$. But, by induction on the number of edges, $H = \pi_1(\Delta')$ is finitely presented. □

Lemma 4.9 (Exercise 12 in [2]) *CLGs are coherent, in particular finitely presented.*

Proof. In the case of a free decomposition the result is immediate. In the other case, the (free abelian) edge groups of Δ are finitely generated, so slender and coherent, by lemma 4.5. Therefore all vertex groups are finitely generated; in particular, abelian vertex groups are coherent. Finitely generated surface groups are also coherent. Rigid vertex groups embed into a CLG of lower level, so by lemma 4.7 they are free products of coherent groups and hence coherent by induction. The result now follows by lemma 4.8. □

4.5 Finite $K(G,1)$

That CLGs have finite $K(G,1)$ follows from the fact that graphs of aspherical spaces are aspherical.

Theorem 4.10 (Proposition 3.6 of [9]) *Let Δ be a graph of groups; suppose that for every vertex group V there exists finite $K(V,1)$, and for every edge group E there exists a finite $K(E,1)$. Then for $G = \pi_1(\Delta)$, there exists a finite $K(G,1)$.*

Surface groups and abelian groups have finite Eilenberg–Mac Lane spaces. Rigid vertices embed into a CLG of lower level, so by lemma 4.7 and induction they also have finite Eilenberg–Mac Lane spaces.

Corollary 4.11 (Exercise 13 in [2]) *If G is a CLG then there exists a finite $K(G,1)$.*

4.6 Principal cyclic splittings

A *principal cyclic splitting* of G is a one-edge splitting of G with cyclic edge group, such that the image of the edge group is maximal abelian in one of the vertex groups; further, if it is an HNN-extension then the edge group is required to be maximal abelian in the whole group. The key observation about principal cyclic splittings is that any non-cyclic abelian subgroup is elliptic with respect to them—in other words, they are precisely those cyclic splittings that feature in the conclusion of lemma 4.2. Applying lemma 4.2, to prove that every freely indecomposable, non-abelian CLG has a principal cyclic splitting it will therefore suffice to produce any non-trivial cyclic splitting (since we now know that CLGs are CSA).

Proposition 4.12 (Exercise 14 in [2]) *Every non-abelian, freely indecomposable CLG admits a principal cyclic splitting.*

Proof. Let G be a CLG. As usual, by induction it suffices to consider the cases when G splits as an amalgamated product or HNN-extension. It suffices to exhibit any cyclic splitting of G, as observed above.

Suppose
$$G = A *_C B.$$
If C is cyclic the result is immediate, so assume C is non-cyclic abelian. If either vertex group is freely decomposable then so is G, since C has no free splittings; if both vertex groups are abelian then so is G. Therefore A, say, is freely indecomposable and non-abelian so has a principal cyclic splitting, which we shall take to be of the form

$$A = A' *_{C'} B'.$$

(It might also be an HNN-extension, but this doesn't affect the proof.) Because it is principal C is conjugate into a vertex, say B'; so G now decomposes as
$$G = A' *_{C'} (B' *_C B).$$
which is a cyclic splitting as required.

The proof when $G = A*_C$ is the same. □

4.7 A criterion in free groups

To prove that a group G is ω-residually free, it suffices to show that for any finite $X \subset G \setminus 1$ there exists a homomorphism $f : G \to \mathbb{F}$ with $1 \notin f(X)$. So a criterion to show that an element of $\mathbb{F}$ is not the identity will be useful.

Lemma 4.13 *Let $z \in \mathbb{F} \setminus 1$, and consider an element g of the form*

$$g = a_0 z^{i_1} a_1 z^{i_2} a_2 \ldots a_{n-1} z^{i_n} a_n$$

where $n \geq 1$ and, whenever $0 < k < n$, $[a_k, z] \neq 1$. Then $g \neq 1$ whenever the $|i_k|$ are sufficiently large.

Choose a generating set for $\mathbb{F}$ so the corresponding Cayley graph is a tree T. An element $u \in \mathbb{F}$ specifies a geodesic $[1, u] \subset T$. Likewise, a string of elements $u_0, u_1, \ldots, u_n \in \mathbb{F}$ defines a path

$$[1, u_0] \cdot u_0[1, u_1] \cdot \ldots \cdot (u_0 \ldots u_{n-1})[1, u_n]$$

in T, where $\cdot$ denotes concatenation of paths. The key observation we will use is as follows. The length of a word $w \in \mathbb{F}$ is denoted by $|w|$.

Remark 4.14 *Suppose z is cyclically reduced and has no proper roots. Let $a \in \mathbb{F}$ be such that a and az both lie in $L \subset T$ the axis of z. If j is minimal such that z^j lies in the geodesic $[a, az]$ then, setting $u = z^j a^{-1}$ and $v = az^{1-j}$, it follows that $uv = z = vu$; in particular, either u or v is trivial and $[a, z] = 1$.*

Proof of lemma 4.13. It can be assumed that z is cyclically reduced and has no proper roots.

Assume that, for each k, $|z^{i_k}| \geq |a_{k-1}| + |a_k| + |z|$. Let $L \subset T$ be the axis of z. Denote by g_k the partial product

$$g_k = a_0 z^{i_1} a_1 z^{i_2} a_2 \ldots a_{k-1} z^{i_k} a_k.$$

The path γ corresponding to g is of the form

$$[1, a_0] \cdot g_0[1, z^{i_1}] \cdot g_0 z^{i_1}[1, a_1] \cdot \ldots \cdot g_{n-1}[1, z^{i_n}] \cdot g_{n-1} z^{i_n}[1, a_n].$$

Suppose that $g = 1$ so this path is a loop. Each section of the form $g_k[1, z^{i_k}]$ lies in a translate of L, the axis of z. Since T is a tree, for at least one such section γ enters and leaves $g_k L$ at the same point—otherwise γ is a non-trivial loop. Since $|z^{i_k}| > |a_{k-1}| + |a_k| + |z|$ it follows that both $g_k z^{i_k} a_k$ and $g_k z^{i_k} a_k z$ lie in $g_k L$ and so $[a_k, z] = 1$ by remark 4.14. □

4.8 CLGs are limit groups

Theorem 4.15 (Exercise 15 in [2]) *CLGs are ω-residually free.*

Since the freely decomposable case is immediate, let Δ, G' and ρ be as in the definition of a CLG in [2]. By induction, G' can be assumed ω-residually free. As a warm up, and for use in the subsequent induction, we first prove the result in the case of abelian and surface vertices.

Lemma 4.16 *Let A be a free abelian group and $\rho : A \to G'$ a homomorphism to a limit group. Suppose $P \subset A$ is a nontrivial subgroup of finite corank closed under taking roots, on which ρ is injective. Then for any finite subset $X \subset A \smallsetminus 1$ there exists an automorphism α of A, fixing P, so that $1 \notin \rho \circ \alpha(X)$.*

Proof. Since $\ker \rho$ is a subgroup of A of positive codimension, for given $x \in A \smallsetminus 0$ a generic automorphism α certainly satisfies $\alpha(x) \notin \ker \rho$. Since X is finite, therefore, there exists α such that $\alpha(x) \notin \ker \rho$ for any $x \in X$. □

We now consider the surface vertex case.

Proposition 4.17 *Let S be the fundamental group of a surface Σ with non-empty boundary, with $\chi(\Sigma) \leq -1$, and $\rho : S \to G'$ a homomorphism injective on each peripheral subgroup and with non-abelian image. Then for any finite subset $X \subset S \setminus 1$ there exists an automorphism α of S, induced by an automorphism of Σ fixing the boundary components pointwise, such that $1 \notin \rho \circ \alpha(X)$.*

Let the surface Σ have $b > 0$ boundary components and Euler characteristic $\chi < 0$. When Σ is cut along a two-sided simple closed curve γ, the resulting pieces either have lower genus (defined to be $1 - \frac{1}{2}(\chi + b)$) or fundamental groups of strictly lower rank, depending on whether γ was separating or not. The simplest cases all have fundamental groups that are free of rank 2.

Example 4.18 (The simplest cases) *Suppose that S is free of rank 2. By lemma 1.7, $\rho(S)$ is free or free abelian; but $\rho(S)$ is assumed non-abelian, so $\rho(S)$ is free and ρ is injective.*

For the more complicated cases, the idea is to find a suitable simple closed curve ζ along which to cut to make the surface simpler. In order to apply the proposition inductively, ζ needs the following properties:

1. $\rho(\zeta) \neq 1$;
2. the fundamental group S' of any component of $\Sigma \setminus \zeta$ must have $\rho(S')$ non-abelian.

Let's find this curve in some examples.

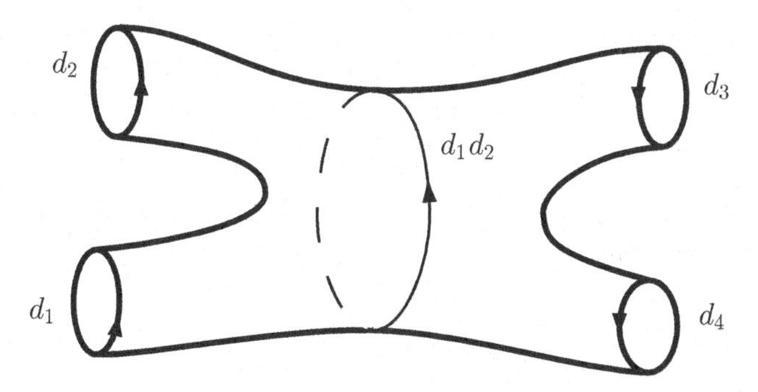

Figure 1: A four-times punctured sphere

Example 4.19 (Punctured spheres) *Suppose Σ is a punctured sphere, so*

$$S = \langle d_1, \ldots, d_n \mid \prod_j d_j \rangle.$$

Assume $n \geq 4$ and $\rho(d_i) \neq 1$ for all i. Define a relation on $\{1, \ldots, n\}$ by

$$i \sim j \Leftrightarrow \rho([d_i, d_j]) = 1.$$

Since G' is commutative transitive, $\sim$ is an equivalence relation. Because the image is non-abelian, there are at least two equivalence classes. Since

$$\prod_i d_i = 1$$

any equivalence class has at least two elements in its complement. Relabelling if necessary, it can now be assumed that

$$\rho([d_1, d_2]), \rho([d_3, d_4]) \neq 1.$$

Now if the boundary curves have been coherently oriented then $d_1 d_2$ has a representative that is a simple closed curve. Take ζ as this representative.

The case when Σ is non-orientable is closely related.

Example 4.20 (Non-orientable surfaces) *Suppose Σ is non-orientable so*

$$S = \langle c_1, \ldots, c_m, d_1, \ldots, d_n \mid \prod_i c_i^2 \prod_j d_j \rangle.$$

Exactly the same argument as in the case of a punctured sphere would work if it could be guaranteed that $\rho(c_i) \neq 1$ for all i.

Fix some c_k, therefore, and suppose $\rho(c_k) = 1$. Let γ be a simple closed curve representing c_k. Then $d_1 c_k$ has a representative δ which is a simple closed curve, and $\rho(d_1 c_k) \neq 1$. Furthermore, $\Sigma \smallsetminus \gamma$ and $\Sigma \smallsetminus \delta$ are homeomorphic surfaces, and a homeomorphism between them extends to an automorphism of Σ mapping γ to δ. This homeomorphism can be chosen not to alter any of the other c_i or the d_j.

Therefore, after an automorphism of Σ, it can be assumed that $\rho(c_i) \neq 1$ for all i, so a suitable ζ can be found as in the previous example.

Example 4.21 (Positive-genus surfaces) *Suppose Σ is an orientable surface of positive genus, so*

$$S = \langle a_1, b_1, \ldots, a_g, b_g, d_1, \ldots, d_n \mid \prod_i [a_i, b_i] \prod_j d_j \rangle.$$

Assume that $g, n \geq 1$. If, for example, $\rho(a_1) \neq 1$ then ζ can be taken to be a simple closed curve representing a_1. Otherwise, $\rho(a_1 d_1) \neq 1$ and $a_1 d_1$ has a simple closed representative. It remains to show that the single component of $\Sigma \smallsetminus \zeta$ has non-abelian image.

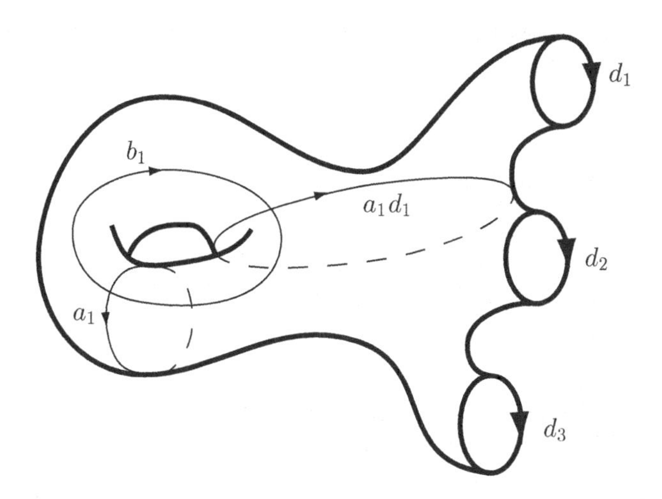

Figure 2: The positive-genus case.

Cutting along ζ expresses S as an HNN-extension:

$$S = S' *_{\mathbb{Z}} .$$

Let t be the stable letter, and suppose $\rho(S')$ is abelian. Then $\zeta \in tS't^{-1} \cap S'$ so in particular $\rho(tS't^{-1} \cap S')$ is non-trivial. But G' is a limit group and hence CSA, so $\rho(t)$ commutes with $\rho(S')$, contradicting the assumption that $\rho(S)$ is non-abelian.

Note that examples 4.19, 4.20 and 4.21 cover all the more complicated surfaces with boundary.

Proof of proposition 4.17. Example 4.18 covers all the simplest cases. Suppose therefore that Σ is more complicated. To apply the inductive hypothesis, an essential simple closed curve $\zeta \in \Sigma$ is needed such that $\rho(\zeta) \neq 1$ and, for any component S' of $\Sigma \smallsetminus \zeta$, $\rho(S')$ is non-abelian. This is provided by examples 4.19, 4.20 and 4.21.

For simplicity, assume ζ is separating. The non-separating case is similar. Then $\Sigma \smallsetminus \zeta$ has two components, Σ_1 and Σ_2. Let $S_i = \pi_1(\Sigma_i)$, and denote by X_i the *syllables* of X in S_i—the elements of S_i that occur in the normal form of some $x \in X$ with respect to the splitting over $\langle\zeta\rangle$. Because the pieces Σ_i are simpler than Σ there exists $\alpha \in \mathrm{Aut}_0(\Sigma)$ and $f : G' \to \mathbb{F}$ such that

$$1 \notin f \circ \rho \circ \alpha([\zeta, X_1 \cup X_2]).$$

Consider $\xi \in X$. The proposition follows from the claim that, for all sufficiently large k,

$$f \circ \rho \circ \delta_\zeta^k \circ \alpha(\xi) \neq 1$$

where δ_ζ is a Dehn twist in ζ. If ξ is a power of ζ then the result is immediate. Otherwise, with respect to the one-edge splitting of G over $\langle\zeta\rangle$, ξ has reduced form

$$\sigma_0\tau_0\sigma_1\tau_1\ldots\sigma_n\tau_n$$

where the $\sigma_i \in X_1$ and the $\tau_j \in X_2$. The image $x^{(k)} = f\circ\rho\circ\delta_\zeta^k\circ\alpha(\xi)$ is of the form

$$z^k s_0 z^{-k} t_0 z^k s_1 z^{-k} t_1 \ldots z^k s_n z^{-k} t_n$$

where $z = f\circ\rho(\zeta)$, $s_i = f\circ\rho\circ\alpha(\sigma_i)$ and $t_i = f\circ\rho\circ\alpha(\tau_i)$. This expression for $x^{(k)}$ satisfies the hypotheses of lemma 4.13, so $x^{(k)} \neq 1$ for all sufficiently large k. □

The proof of theorem 4.15 is very similar to the proof of proposition 4.17. The theorem follows from the following proposition, by induction on level.

Proposition 4.22 *Let G be a freely indecomposable CLG, let Δ and ρ be as usual, and let G' be ω-residually free. For any finite subset $X \subset G\smallsetminus 1$ there exists a modular automorphism α of G such that $1 \notin \rho\circ\alpha(X)$.*

Proof. As usual, the proposition is proved by induction on the number of edges of Δ. The case of Δ having no edges follows from lemmas 4.16 and 4.17, and the fact that ρ is injective on rigid vertices. By induction on level, G' is a limit group.

Now suppose Δ has an edge group E. For simplicity, assume E is separating. The non-separating case is similar. Then removing the edge corresponding to E divides Δ into two subgraphs Δ_1 and Δ_2. Let $G_i = \pi_1(\Delta_i)$, and denote by X_i the syllables of X in G_i. Without loss assume E is maximal abelian in G_1. Fix non-trivial $\zeta \in E$. By induction there exists $\alpha \in \mathrm{Mod}(\Delta)$ and $f : G' \to \mathbb{F}$ such that

$$1 \notin f\circ\rho\circ\alpha([\zeta, X_1]\cup X_2).$$

Consider $\xi \in X$. The proposition follows from the claim that, for all sufficiently large k,

$$f\circ\rho\circ\delta_\zeta^k\circ\alpha(\xi) \neq 1.$$

If $\xi \in E$ then the result is immediate. Otherwise, with respect to the one-edge splitting of G over E, ξ has reduced form

$$\sigma_0\tau_0\sigma_1\tau_1\ldots\sigma_n\tau_n$$

where the $\sigma_i \in X_1$ and the $\tau_j \in X_2$. The image $x^{(k)} = f\circ\rho\circ\delta_\zeta^k\circ\alpha(\xi)$ is of the form

$$z^k s_0 z^{-k} t_0 z^k s_1 z^{-k} t_1 \ldots z^k s_n z^{-k} t_n$$

where $z = f \circ \rho(\zeta)$, $s_i = f \circ \rho \circ \alpha(\sigma_i)$ and $t_i = f \circ \rho \circ \alpha(\tau_i)$. In particular, canceling across those t_i that commute with z, we have

$$x^{(k)} = u_0 z^{k\epsilon_1} u_1 \ldots u_{n-1} z^{k\epsilon_n} u_n$$

where $\epsilon_i = \pm 1$ and u_i don't commute with z for $0 < i < n$. This second expression for $x^{(k)}$ satisfies the hypotheses of lemma 4.13, so $x^{(k)} \neq 1$ for all sufficiently large k. □

5 The Shortening Argument

We consider a sequence of G-trees T_i, arising from homomorphisms $f_i : G \to \mathbb{F}$, that converge in the Gromov topology to a G-tree T. By the results of section 3 of [2], if the action of G on the limit tree T is faithful then it gives rise to a generalized abelian decomposition for G. This section is entirely devoted to the solution of exercise 16, which is essentially Rips and Sela's shortening argument—an ingenious means of using this generalized abelian decomposition to force the action on the limit tree to be unfaithful.

5.1 Preliminary ideas

Once again, fix a generating set for $\mathbb{F}$ so that the corresponding Cayley graph is a tree, and let $|w|$ denote the length of a word $w \in \mathbb{F}$. Fix a generating set S for G. For $f : G \to \mathbb{F}$, let

$$|f| = \max_{g \in S} |f(g)|.$$

A homomorphism is *short* if

$$|f| \leq |\iota \circ f \circ \alpha|$$

whenever α is a modular automorphism of G and ι is an inner automorphism of $\mathbb{F}$.

Theorem 5.1 (Exercise 16 of [2]) *Suppose every f_i is short. Then the action on T is not faithful.*

The proof is by contradiction. We assume therefore, for the rest of section 5, that the action is faithful. By the results summarized in section 3 of [2], the action of G on T gives a GAD Δ for G. The idea is, if T_i are the limiting trees with basepoints x_i, to construct modular automorphisms ϕ_n so that

$$d_i(x_i, f_i \circ \phi_i(g) x_i) < d_i(x_i, f_i(g) x_i)$$

for all sufficiently large i. Then apply these automorphisms to carefully chosen basepoints.

All constructions of the limit tree T, such as the asymptotic cone [11], use some form of based convergence: basepoints $x_i \in T_i$ are fixed, and converge to a basepoint $[x_i] \in T$. Because the f_i are short,

$$\max_{g \in S} d_{\mathbb{F}}(1, f_i(g)) \leq \max_{g \in S} d_{\mathbb{F}}(t, f_i(g)t)$$

for all $t \in T_{\mathbb{F}}$; otherwise, conjugation by the element of $\mathbb{F}$ nearest to t leads to a shorter equivalent homomorphism. It follows that $1 \in T_i$ is always a valid basepoint; we set $x = [1] \in T$ to be the basepoint for T.

The proof of theorem 5.1 goes on a case-by-case basis, depending on whether $[x, gx]$ intersects a simplicial part or a minimal part of T.

5.2 The abelian part

The next proposition is a prototypical shortening result for a minimal vertex.

Proposition 5.2 *Let V be an abelian vertex group of Δ. For $g \in G$, let $l(g)$ be the translation length of g on T. Fix $\epsilon > 0$. Then for any finite subset $S \subset V$ there exists a modular automorphism ϕ of G such that*

$$\max_{g \in S} l(\phi(g)) < \epsilon.$$

Proof. The minimal V-invariant subtree T_V is a line in T, on which V acts indiscretely. Since S is finite, V can be assumed finitely generated. It suffices to prove the theorem in the case where S is a basis for V. Assume furthermore that each element of S translates T_V in the same direction.

Suppose the action of V on T_V is free. Let $S = \{g_1, \ldots, g_n\}$, ordered so that

$$l(g_1) > l(g_2) > \ldots > l(g_n) > 0.$$

Since the action is indiscrete, there exists an integer λ such that

$$l(g_1) - \lambda l(g_2) < \frac{1}{2} l(g_2).$$

Applying the automorphism that maps $g_1 \mapsto g_1 - \lambda g_2$ and proceeding inductively, we can make $l(g_1)$ as short as we like.

If the action of V is not free then $V = V' \oplus V_0$ where V' acts freely on T_V and V_0 fixes T_V pointwise. Applying the free case to V' gives the result. □

The aim is to prove the following theorem.

Theorem 5.3 *Let V be an abelian vertex. Then for any finite subset $S \subset G$ there exists a modular automorphism ϕ such that for any $g \in S$:*

1. *if $[x, gx]$ intersects a translate of T_V in a segment of positive length then*

$$d(x, \phi(g)x) < d(x, gx);$$

2. *otherwise,* $\phi(g) = g$.

Proof. By a result of J. Morgan (claim 3.3 of [6]—the article is phrased in terms of laminations), the path $[x, gx]$ intersects finitely many translates of T_V in non-trivial segments. Let ϵ be the minimal length of all such segments across all $g \in S$. Assume that $g \in S$ is such that $[x, gx]$ intersects a translate of T_V non-trivially.

Suppose first that x lies in a translate of T_V, so without loss of generality $x \in T_V$. Then g has a non-trivial decomposition in the GAD provided by corollary 3.16 of [2] of the form

$$g = a_0 b_1 a_1 \dots a_n$$

where the a_i lie in V and the b_i are products of elements of other vertices and loop elements. Write $g_i = a_0 b_1 \dots b_{i-1} a_{i-1}$. The decomposition can be chosen so that each component of the geodesic $[x, gx]$ that lies in $g_i T_V$ is non-trivial. For each i, decompose $[x, b_i x]$ as

$$[x, s_i] \cdot [s_i, t_i] \cdot [t_i, b_i x]$$

where $[x, s_i]$ and $[t_i, b_i x]$ are maximal segments in T_V and $b_i T_V$ respectively. Then

$$[x, gx] = [x, a_0 s_1] \cdot g_1 [s_1, t_1] \cdot g_1 [t_1, b_1 a_1 s_2] \cdot \ldots \cdot g_n [s_n, t_n] \cdot g_n [t_n, b_n a_n x]$$

where each $[s_i, t_i]$ and $[t_i, b_i a_i s_{i+1}]$ is a non-trivial segment. Therefore

$$\begin{aligned} d(x, gx) &= d(x, a_0 s_1) + \sum_{i=1}^{n} d(s_i, t_i) + \sum_{i=1}^{n-1} d(t_i, b_i a_i s_{i+1}) + d(t_n, b_n a_n x) \\ &\geq \sum_{i=1}^{n} d(s_i, t_i) + (n+1)\epsilon. \end{aligned}$$

Since V acts indiscretely on the line T_V, by modifying the b_i by elements of V it can be assumed that

$$d(x, s_i), d(t_i, b_i x) < \frac{1}{4}\epsilon.$$

By proposition 5.2 there exists $\phi \in \mathrm{Mod}(G)$ such that $\phi(b_i) = b_i$ for all b_i and

$$d(x, \phi(a_i)x) < \frac{1}{2}\epsilon$$

for all a_i. Now as before $[x, \phi(g)x]$ decomposes as

$$[x, \phi(a_0)s_1] \cdot \phi(g_1)[s_1, t_1] \cdot \ldots \cdot \phi(g_n)[t_n, b_n \phi(a_n)x]$$

so

$$\begin{aligned} d(x,\phi(g)x) &= d(x,\phi(a_0)s_1) + \sum_{i=1}^{n} d(s_i,t_i) + \sum_{i=1}^{n-1} d(t_i, b_i\phi(a_i)s_{i+1}) \\ &\quad + d(t_n, b_n\phi(a_n)x) \\ &< d(x,\phi(a_0)x) + \sum_{i=1}^{n} d(s_i,t_i) + \sum_{i=1}^{n-1} d(b_i x, b_i\phi(a_i)x) \\ &\quad + d(b_n x, b_n\phi(a_n)x) + \frac{n}{2}\epsilon \\ &< \sum_{i=1}^{n} d(s_i,t_i) + (n+\frac{1}{2})\epsilon. \end{aligned}$$

Therefore, $d(x,\phi(g)x) < d(x,gx) - \frac{1}{2}\epsilon$ and in particular the result follows.

Now suppose x does not lie in a translate of T_V. Then g has a non-trivial decomposition in the corresponding GAD of the form

$$g = b_0 a_1 b_1 \ldots b_n$$

where the a_i lie in V and the b_i are products of elements of other vertices and loop elements. Let $g' = a_1 b_1 \ldots b_{n-1} a_n$. Let x' be the first point on $[x, gx]$ in a translate of T_V, so $x' = b_0 y \in b_0 T_V$ for some $y \in T_V$. Likewise let x'' be the last point on $[x, gx]$, so $x'' = b_0 g' z \in b_0 g' T_V$ for some $z \in T_V$. Since the action of V on T_V is indiscrete we can modify b_n by an element of V and assume that $d(y,z) < \frac{1}{4}\epsilon$.

Then the geodesic $[x, gx]$ decomposes as

$$[x,gx] = [x, b_0 y] \cdot [b_0 y, b_0 g' z] \cdot [b_0 g' z, gx]$$

so

$$d(x,gx) > d(x,b_0 y) + d(y,g'y) + d(z,b_n x) - \frac{1}{4}\epsilon.$$

Applying the first case to g' and y we obtain $\phi \in \mathrm{Mod}(G)$ such that

$$d(y,\phi(g')y) < d(y,g'y) - \frac{1}{2}\epsilon$$

so

$$\begin{aligned} d(x,\phi(g)x) &< d(x,b_0 y) + d(y,\phi(g')y) + d(z,b_n x) + \frac{1}{4}\epsilon \\ &< d(x,b_0 y) + d(y,g'y) - \frac{1}{2}\epsilon + d(z,b_n x) + \frac{1}{4}\epsilon \\ &< d(x,gx) \end{aligned}$$

as required. □

5.3 The surface part

The surface part is dealt with by Rips and Sela, in [8], in the following theorem.

Theorem 5.4 (Theorem 5.1 of [8]) *Let V be a surface vertex. Then for any finite subset $S \subset G$ there exists a modular automorphism ϕ such that for any $g \in S$:*

1. *if $[x, gx]$ intersects a translate T_V in a segment of positive length then*

$$d(x, \phi(g)x) < d(x, gx);$$

2. *otherwise, $\phi(g) = g$.*

Rips and Sela use the notion of groups of interval exchange transformations, which are equivalent to surface groups, and prove an analogous result to proposition 5.2. The rest of the proof is the same as that of theorem 5.3.

5.4 The simplicial part

It remains to consider the case where $[x, gx]$ is contained in the simplicial part of T.

Theorem 5.5 *Let $S \subset G$ be finite and let $x \in T$. Then there exist $\phi_n \in$ Mod(Δ) such that, for all $g \in S$,*

$$d(x, \phi_n(g)x) = d(x, gx);$$

furthermore, for all $g \in S$ that do not fix x, and for all sufficiently large n,

$$d_n(x_n, f_n \circ \phi_n(g)x_n) < d_n(x_n, f_n(g)x_n).$$

Let e be a closed simplicial edge containing x. The proof of the theorem is divided into cases, depending on whether the image of e is separating in T/G. In both cases, the following lemma will prove useful.

Lemma 5.6 *Let A be a vertex group of the splitting over e. Let T_A be the minimal A-invariant subtree of T; conjugating A, we can assume that $T_A \cap e$ is precisely one point, y. Fix any non-trivial $c \in C =$ Stab(e). Then there exists a sequence of integers m_n such that, for any $a \in A$,*

$$d_n(x_n, f_n(c^{-m_n} a c^{m_n})x_n) \to d(y, ay)$$

as $n \longrightarrow \infty$.

Proof. The key observation is that

$$2d_n(x_n, \text{Axis}(f_n(c))) < d_n(x_n, f_n(c)x_n) \to 0,$$

and the same holds for the y_n. Let x'_n be the nearest point on $\text{Axis}(f_n(c))$ to x_n; likewise, let y'_n be the nearest point on $\text{Axis}(f_n(c))$ to y_n. Then, for each n, there exists m_n such that

$$d_n(f_n(c^{m_n})x'_n, y'_n) < l(f_n(c)) \to 0$$

as $n \to \infty$. Therefore

$$d_n(f_n(c^{m_n})x_n, y_n) \to 0$$

as $n \to \infty$, and the result follows. □

The next lemma helps with the case where the image of e is separating.

Lemma 5.7 *Assume the image of* e *is separating, so the induced splitting is*

$$G = A *_C B.$$

Assume furthermore that, with the notation of the previous lemma, $x \neq y$. Then there exists $\alpha_n \in \text{Mod}(\Delta)$ such that, for all $g \in S$:

1. *if $g \in A$ then $\alpha_n(g) = g$;*
2. *if $g \notin A$ then*

$$d_n(x_n, f_n \circ \alpha_n(g)x_n) < d_n(x_n, f_n(g)x_n).$$

Proof. Fix a non-trivial $c \in C$, and let δ_c be the Dehn twist in c that is the identity when restricted to A. Let $\alpha_n = \delta_c^{m_n}$ where m_n are the integers given by lemma 5.6. Any $g \notin A$ has normal form

$$g = a_0 b_1 a_1 \ldots b_l a_l$$

with the $a_i \in A \smallsetminus C$ and the $b_i \in B \smallsetminus C$, except for a_0 and a_l which may be trivial. Therefore $d(x, gx) = \sum_i d(x, b_i x) + \sum_i d(x, a_i x)$. Fix $\epsilon > 0$. If a_i is non-trivial then $a_i \notin C$ and so

$$d(x, a_i x) = 2d(x, y) + d(y, a_i y).$$

Let k be the number of a_i that are non-trivial (so $l-1 \leq k \leq l+1$). Therefore, for all sufficiently large n,

$$d_n(x_n, f_n(g)x_n) > \sum_i d(x, b_i x) + \sum_i d(y, a_i y) + 2kd(x, y) - \epsilon.$$

By contrast, for all sufficiently large n,

$$d_n(x_n, f_n \circ \alpha_n(g)x_n) < \sum_i d(x, b_i x) + \sum_i d(y, a_i y) + \epsilon$$

by lemma 5.6. By assumption $x \neq y$, so $d(x,y) > 0$. Therefore taking $\epsilon < kd(x,y)$ gives the result. □

We now turn to the non-separating case.

Lemma 5.8 *Assume the image of e is non-separating, so the splitting induced by e is*

$$G = A *_C .$$

Let t be a stable letter. As before, conjugate A so that $T_A \cap e$ is precisely one point y. Fix any non-trivial $c \in C = \mathrm{Stab}(e)$. Then there exists a sequence of integers p_n such that

$$d_n(y_n, f_n(tc^{p_n})y_n) \to 0$$

as $n \to \infty$. Therefore, for any fixed integer j,

$$d_n(y_n, f_n(tc^{p_n})^j y_n) \to 0$$

as $n \to \infty$.

Proof. As in the proof of lemma 5.6, by the definition of Gromov convergence,

$$2d_n(y_n, \mathrm{Axis}(f_n(c))) < d_n(y_n, f_n(c)y_n) \to 0$$

as $n \to \infty$, and similarly,

$$2d_n(f_n(t^{-1})y_n, \mathrm{Axis}(f_n(c))) \to 0$$

as $n \to \infty$. Let y_n' be the nearest point on $\mathrm{Axis}(f_n(c))$ to y_n, and let y_n'' be the nearest point on $\mathrm{Axis}(f_n(c))$ to $f_n(t^{-1})y_n$. Then there exist integers p_n such that

$$d_n(f_n(c^{p_n})y_n', y_n'') \to 0$$

as $n \to \infty$. The result now follows. □

Lemma 5.9 *Assume the situation is in lemma 5.8. Then there exists $\alpha_n \in \mathrm{Mod}(\Delta)$ such that, for all $g \in S$:*

1. *if $g \in C$ then $\alpha_n(g) = g$;*

2. *if $g \notin C$ then*

$$d_n(x_n, f_n \circ \alpha_n(g)x_n) < d_n(x_n, f_n(g)x_n).$$

Proof. Fix a stable letter t that translates x away from y. Fix a non-trivial $c \in C$ and let $i_c \in \mathrm{Mod}(G)$ be conjugation by c. Set $\alpha_n = i_c^{m_n} \circ \delta_c^{p_n}$, where m_n are integers given by lemma 5.6 and p_n are given by lemma 5.8. Any g is of the form

$$g = a_0 t^{j_1} a_1 \ldots t^{j_l} a_l$$

with $j_i \neq 0$ and the $a_i \in A \setminus C$ except for a_0 and a_l which may be trivial. Unlike in the case of a separating edge, we have to be a little more careful in estimating $d(x, gx)$ because the natural path from x to gx given by the decomposition of g may backtrack. To be precise, backtracking occurs when $a_i \neq 1$ and $j_{i+1} < 0$ and also when $j_i > 0$ and $a_{i+1} \neq 1$. Let k be the number of i for which backtracking does not occur, so $0 \leq k \leq 2$. Then

$$d(x, gx) = \sum_i d(y, a_i y) + \sum_i d(x, t^{j_i} x) + 2kd(x, y).$$

Fix $\epsilon > 0$. Then for all sufficiently large n,

$$d_n(x_n, f_n(g)x_n) > \sum_i d(y, a_i y) + 2kd(x, y) + \sum_i d(x, t^{j_i} x) - \epsilon.$$

Now for each i,

$$d_n(f_n(c^{m_n})x_n, f_n((tc^{p_n})^{j_i}) f_n(c^{m_n})x_n) \to 0$$

and

$$d_n(f_n(c^{m_n})x_n, f_n(a_i c^{m_n})x_n) \to d(y, a_i y).$$

So for all sufficiently large n,

$$d_n(x_n, f_n \circ \alpha_n(g)x_n) < \sum_i d(y, a_i y) + \epsilon.$$

Taking $2\epsilon < 2kd(x, y) + \sum_i d(x, t^{j_i} x)$ gives the result. □

We are now ready to prove the theorem.

Proof of theorem 5.5. Suppose first that x lies in the interior of an edge e. If e has separating image in the quotient then lemma 5.7 can be applied both ways round, giving rise to modular automorphisms α_n and β_n. The theorem is then proved by taking $\phi_n = \alpha_n \circ \beta_n$. If e is non-separating then applying lemma 5.9 and taking $\phi_n = \alpha_n$ gives the result.

Suppose now that x is a vertex. For each orbit of edges $[e]$ adjoining x, let α_n^e be the result of applying lemma 5.7 or lemma 5.9 as appropriate to e. Now taking

$$\phi_n = \alpha_n^{e_1} \circ \ldots \circ \alpha_n^{e_p}$$

where $[e_1], \ldots, [e_p]$ are the orbits adjoining x gives the required automorphism. □

This is the final piece of the shortening argument.

Proof of theorem 5.1. Fix a generating set S for G. Let $f_i : G \to \mathbb{F}$ be a sequence of short homomorphisms corresponding to the convergent sequence of G-trees T_i. Let T be the limiting G-tree and suppose that the action of G on T is faithful. By corollary 3.16 of [2] this induces a GAD Δ for G. Let $x \in T$ be the basepoint fixed in subsection 5.1.

Composing the automorphisms given by theorems 5.3 and 5.4 there exists $\alpha \in \mathrm{Mod}(\Delta)$ such that, for any $g \in G$,

$$d(x, \phi(g)x) < d(x, gx)$$

if $[x, gx]$ intersects an abelian or surface component of T and $\phi(g) = g$ otherwise. By theorem 5.5, for all sufficiently large i there exist $\beta_i \in \mathrm{Mod}(\Delta)$ such that $d(x, \beta_i(g)x) = d(x, gx)$ and, furthermore,

$$d_i(1, f_i \circ \beta_i(g)) < d_i(1, f_i(g))$$

whenever $[x, gx]$ is a non-trivial arc in the simplicial part of the tree. It follows that for $\phi_i = \beta_i \circ \alpha$,

$$d_i(1, f_i \circ \phi_i(g)) < d_i(1, f_i(g))$$

for all $g \in S$ and all sufficiently large i. This contradicts the assumption that the f_i were short. □

6 Bestvina and Feighn's geometric approach

In section 7 of [2], Bestvina and Feighn provide a more geometric proof of their Main Proposition. In this section we provide proofs of the exercises needed in this argument.

6.1 The space of laminations

Recall that $\mathcal{ML}(K)$ is the space of measured laminations on K, and $\mathbb{P}\mathcal{ML}(K)$ is its quotient by the action of $\mathbb{R}_+$. Let E be the set of edges of K.

Proposition 6.1 (Exercise 18 of [2]) *The space of measured laminations on K can be identified with a closed cone in $\mathbb{R}^E_+ - \{0\}$, given by the triangle inequality for each 2-cell of K. Hence, when $\mathcal{ML}(K)$ is endowed with the corresponding topology, $\mathbb{P}\mathcal{ML}(K)$ is compact.*

Proof. Recall that two laminations are considered equivalent if they assign the same measure to each edge. Therefore it suffices to show existence of a lamination with the prescribed values on the edges. First, for each edge e

with $\int_e \mu > 0$, fix a closed proper subinterval I_e contained in the interior of e. Now fix a Cantor function $c_e : I_e \to [0, \int_e \mu]$. This gives a measure μ on e, given by

$$\int_J \mu = \int_{I_e \cap J} c_e d\lambda$$

where $d\lambda$ is Lebesgue measure on $\mathbb{R}$. Now suppose e_1, e_2, e_3 are the edges of a simplex in K. Divide e_1 into intervals e_1^2 and e_1^3 so that

$$2\int_{e_1^2} d\mu = \int_{e_1} d\mu + \int_{e_2} d\mu - \int_{e_3} d\mu$$

and e_1^2 shares a vertex with e_2, and similarly for e_1^3. Divide e_2 and e_3 likewise. Fix a Cantor set in each e_i^j. Now for each distinct i, j inscribe a lamination between e_i^j and e_j^i. Since any path transverse to this lamination can be homotoped to an edge path respecting the lamination, the measure on the edges determines a transverse measure to the lamination. □

6.2 Matching resolutions in the limit

A measured lamination on K defines a G-tree. The next exercise shows the close relation between the topology on the space of laminations and the topology on the space of trees. For the definition of a resolution, see [2]. The solution is most easily phrased in terms of some explicit construction of the limiting tree. I shall use the asymptotic cone, T_ω; T can be realized as the minimal G-invariant subtree of T_ω. For the definition of the asymptotic cone see, for example, [11]. To see how to choose basepoints and scaling to ensure that the action is non-trivial see, for example, [7].

Proposition 6.2 (Exercise 19 in [2]) *Consider f_i-equivariant resolutions*

$$\phi_i : \tilde{K} \to T_{\mathbb{F}}.$$

Suppose $\lim T_{f_i} = T$, $\lim \Lambda_{\phi_i} = \Lambda$ *and the sequences* $(|f_i|)$ *and* $(||\phi_i||_K)$ *are comparable. Then there is a resolution that sends lifts of leaves of* Λ *to points of* T *and is a Cantor function on edges of* $\tilde{K}$.

Proof. A resolution $\phi : \tilde{K} \to T$ is determined by a choice of $\phi(\tilde{v})$ for a lift $\tilde{v}$ of each vertex v of K.

First, define a resolution $\phi' : \tilde{K} \to T_\omega$ by setting $\phi'(\tilde{v}) = [\phi_i(\tilde{v})]$. Since $(|f_i|)$ and $(||\phi_i||_K)$ are comparable, $\phi'(\tilde{v})$ is a valid point of T_ω. The resolution ϕ' maps leaves of Λ to points, and is a Cantor function on edges. However, T_ω is far from minimal. Let $\pi : T_\omega \to T$ be closest-point projection to the minimal invariant subtree, which is equivariantly isomorphic to T. Now let $\phi = \pi \circ \phi''$; this is a resolution that still maps leaves of Λ to points, and is a Cantor function on edges, as required. □

6.3 Finding kernel elements carried by leaves

Exercise 20 of [2] relies heavily on the results of [1]. The most important result is a structure theorem for resolutions of stable actions on real trees, summarized in the following theorem.

Theorem 6.3 (Theorems 9.4 and 9.5 of [1]) *Let Λ be a lamination on a 2-complex K, resolving a stable action of $G = \pi_1(K)$ on a real tree T. Then*

$$\Lambda = \Lambda_1 \sqcup \ldots \sqcup \Lambda_k.$$

Each component has a standard neighbourhood N_i carrying a subgroup $H_i \subset G$. Let T_i be the minimal H_i-invariant subtree of T. Each component is of one of the following types.

1. ***Surface type.*** *N_i is a cone-type 2-orbifold, with some annuli attached. H_i fits into a short exact sequence*

$$1 \to \ker T_i \to H_i \to \pi_1(O) \to 1$$

where O is a cone-type 2-orbifold.

2. ***Toral type.*** *T_i is a line, and H_i fits into a short exact sequence*

$$1 \to \ker T_i \to H_i \to A \to 1$$

where $A \subset \mathrm{Isom}(\mathbb{R})$.

3. ***Thin type.*** *H_i splits over an arc stabilizer, carried by a leaf of Λ_i.*

4. ***Simplicial type.*** *All the leaves of Λ_i are compact, and N_i is an interval bundle over a leaf. H_i fits into a short exact sequence satisfying*

$$1 \to \ker T_i \to H_i \to C \to 1$$

where C is finite.

Furthermore, if E is a subgroup carried by a leaf, E fits into a short exact sequence of the form

$$1 \to \kappa \to E \to C \to 1$$

where κ fixes an arc of T and C is finite or cyclic.

In particular, the standard neighbourhoods induce a graph-of-spaces decomposition for K, and a corresponding graph-of-groups decomposition for G. The vertex spaces are the N_i and the closures of the components of $K - \cup_i N_i$. The edge spaces are boundary components of the N_i, and are all contained in a leaf. See theorem 5.13 of [1].

The proof of this exercise will also make use of the following result.

Proposition 6.4 (Corollary 5.9 of [1]) *If $h \in H_i$ fixes an arc of T_i then $h \in \ker T_i$.*

We are now ready to prove the exercise.

Theorem 6.5 (Exercise 20 of [2]) *In the situation describe in the exercise, the lamination Λ has a leaf carrying non-trivial elements of the kernel.*

Proof. Note that $G/\ker T$ is a limit group. Suppose no elements of $\ker T$ are carried by a leaf of Λ.

Consider Γ the graph of groups for G induced by Λ. The aim is to show that Γ really is a GAD. Since a GAD decomposition can be used to shorten, this contradicts the assumption that the f_i are short. We deal with each sort of vertex in turn.

1. Suppose Λ_i is of surface type. Then N_i is a cone-type 2-orbifold, with some annuli attached. Suppose $g \in H_i$ is carried by an annulus. Then g fixes an arc of T_i, so by proposition 6.4, $g \in \ker T_i$. But T_i contains a tripod, and tripod stabilizers are trivial, so $g \in \ker T$ contradicting the assumption. Therefore N_i can be assumed to have no attached annuli. Consider an element $g \in H_i$ carried by the leaf corresponding to a cone-point. Then g has finite order, so $g \in \ker T$, since $G/\ker T$ is a limit group. This contradicts the assumption, so N_i has no cone-points. Therefore N_i is genuinely a surface. Moreover, N_i carries a pseudo-Anosov homeomorphism, since it carries a minimal lamination.

2. If Λ_i is toral, then H_i is an extension

$$1 \to \ker T_i \to H_i \to A \to 1$$

for $A \subset \mathrm{Isom}\mathbb{R}$. The elements of $\ker T_i$ are carried by annuli in N_i. But $\ker T_i$ itself fits into an exact sequence

$$1 \to \kappa \to \ker T_i \to A' \to 1$$

where $\kappa \subset \ker T$ and A' is abelian. In order not to contradict the assumption that no elements of the kernel are carried by a leaf, therefore, κ must be trivial; so we have

$$1 \to A' \to H_i \to A \to 1.$$

and H_i acts faithfully on T. In particular, H_i embeds in the limit group $G/\ker T$, and so is a limit group. But A' is normal; since limit groups are torsion-free and CSA, it follows that H_i is free abelian.

3. If Λ_i is thin, then G splits over a subgroup H fixing an arc of T_i. By proposition 6.4, $H \subset \ker T_i$. But T_i contains a tripod, and tripod stabilizers are trivial, so $H \subset \ker T$; since H is carried by a leaf, H must be trivial by assumption. But this contradicts the assumption that G is freely indecomposable.

4. If Λ_i is simplicial, then H_i fits into the short exact sequence

$$1 \to \ker T_i \to H_i \to C \to 1.$$

for finite C. As in the toral case, the assumption implies that $\ker T_i$ is abelian, and H_i embeds in $G/\ker T$, and so is a limit group. But, again, H_i is torsion-free and CSA; so C is trivial, and H_i is abelian and fixes an arc of T.

Now consider an edge-group E of Γ. Then E is carried by a leaf, and satisfies

$$1 \to \kappa \to E \to C \to 1$$

where C is cyclic and κ fixes an arc of T. Then κ fits into a short exact sequence

$$1 \to \kappa' \to \kappa \to A \to 1$$

where $\kappa' \subset \ker T$ and A is abelian. By assumption, therefore, κ' is trivial and κ is abelian; furthermore, E acts faithfully on T, so embeds in $G/\ker T$ and is a limit group. Therefore E is free abelian.

In conclusion, Γ is a GAD. Just as in the proof of theorem 5.1, this contradicts the assumption that the f_i are all short. □

6.4 Examples of limit groups

To complete their argument, Bestvina and Feighn need some elementary examples of limit groups. This theorem is required.

Theorem 6.6 (Exercise 21 of [2]) *Let Δ be a 1-edged GAD of a group G with a homomorphism q to a limit group Γ. Suppose:*

1. *the vertex groups of Δ are non-abelian,*

2. *the edge group of Δ is maximal abelian in each vertex group, and*

3. *q is injective on vertex groups of Δ. Then G is a limit group.*

This theorem is just a special case of proposition 4.22.

References

[1] M. Bestvina and M. Feighn. Stable actions of groups on real trees. *Invent. Math.*, 121, 1995.

[2] M. Bestvina and M. Feighn. Notes on Sela's work: Limit groups and Makanin-Razborov diagrams, 2003. Preprint.

[3] J. Bochnak, M. Coste, and M-F. Roy. *Géométrie algébrique réelle*, volume 3 of *Ergebnisse der Mathematik und ihrer Grenzgebiete*. Springer-Verlag, Berlin, 1987.

[4] Andrew J. Casson and Steven A. Bleiler. *Automorphisms of surfaces after Nielsen and Thurston.* Number 9 in London Mathematical Society Student Texts. Cambridge University Press, 1988.

[5] R. S. Lyndon. The equation $a^2b^2 = c^2$ in free groups. *Michigan Math. J*, 6, 1959.

[6] J. Morgan. Ergodic theory and free actions of groups on R-trees. *Invent. Math.*, 94, 1988.

[7] F. Paulin. Topologie de Gromov équivariante, structures hyperboliques et arbes réels. *Invent. Math.*, 94(1), 1988.

[8] E. Rips and Z. Sela. Structure and rigidity in hyperbolic groups I. *GAFA*, 4(3), 1994.

[9] Peter Scott and Terry Wall. Topological methods in group theory. In *Proc. Sympos., Durham, 1977*, volume 36 of *London Math. Soc. Lecture Note Ser.*, pages 137–203. Cambridge University Press, 1979.

[10] Z. Sela. Diophantine geometry over groups I: Makanin-Razborov diagrams. *Publ. Inst. Hautes Études Sci.*, 93, 2001.

[11] L. van den Dries and A. Wilkie. On Gromov's theorem concerning groups of polynomial growth and elementary logic. *J. Algebra*, 89, 1984.

L^2-Invariants from the algebraic point of view

Wolfgang Lück*
Fachbereich Mathematik
Universität Münster
Einsteinstr. 62
48149 Münster
Germany

Abstract

We give a survey on L^2-invariants such as L^2-Betti numbers and L^2-torsion taking an algebraic point of view. We discuss their basic definitions, properties and applications to problems arising in topology, geometry, group theory and K-theory.

Key words: dimensions theory over finite von Neumann algebras, L^2-Betti numbers, Novikov Shubin invariants, L^2-torsion, Atiyah Conjecture, Singer Conjecture, algebraic K-theory, geometric group theory, measure theory.

MSC 2000: 57S99, 46L99, 18G15, 19A99, 19B99, 20C07, 20F25.

0 Introduction

The purpose of this survey article is to present an algebraic approach to L^2-invariants such as L^2-Betti numbers and L^2-torsion. Originally these were defined analytically in terms of heat kernels. Since it was discovered that they have simplicial and homological algebraic counterparts, there have been many applications to various problems in topology, geometry, group theory and algebraic K-theory, which on the first glance do not involve any L^2-notions. Therefore it seems to be useful to give a quick and friendly introduction to these notions in particular for mathematicians who have a more algebraic than analytic background. This does not mean at all that the analytic aspects are less important, but for certain applications it is not necessary to know the analytic approach and it is possible and easier to focus on the algebraic aspects. Moreover, questions about L^2-invariants of heat kernels such

*email: lueck@math.uni-muenster.de
www: http://www.math.uni-muenster.de/u/lueck/
FAX: 49 251 8338370

as the Atiyah Conjecture or the zero-in-the-spectrum-Conjecture turn out to be strongly related to algebraic questions about modules over group rings.

The hope of the author is that more people take notice of L^2-invariants and L^2-methods, and may be able to apply them to their favourite problems, which a priori do not necessarily come from an L^2-setting. Typical examples of such instances will be discussed in this survey article. There are many open questions and conjectures which have the potential to stimulate further activities.

The author has tried to write this article in a way which makes it possible to quickly pick out specific topics of interest and read them locally without having to study too much of the previous text.

These notes are based on a series of lectures which were presented by the author at the LMS Durham Symposium on Geometry and Cohomology in Group Theory in July 2003. The author wants to thank the organizers Martin Bridson, Peter Kropholler and Ian Leary and the London Mathematical Society for this wonderful symposium and Michael Weiermann for proof reading the manuscript.

In the sequel ring will always mean associative ring with unit and R-module will mean left R-module unless explicitly stated otherwise. The letter G denotes a discrete group. Actions of G on spaces are always from the left.

Contents

1 Group von Neumann Algebras

The integral group ring $\mathbb{Z}G$ plays an important role in topology and geometry, since for a G-space its singular chain complex or for a G-CW-complex its cellular chain complex are $\mathbb{Z}G$-chain complexes. However, this ring is rather complicated and does not have some of the useful properties which other rings such as fields or semisimple rings have. Therefore it is very hard to analyse modules over $\mathbb{Z}G$. Often in algebra one studies a complicated ring by investigating certain localizations or completions of it which do have nice properties. They still contain and focus on useful information about the original ring, which now becomes accessible. Examples are the quotient field of an integral domain, the p-adic completion of the integers or the algebraic closure of a field. In this section we present a kind of completion of the complex group ring $\mathbb{C}G$ given by the group von Neumann algebra and discuss its ring theoretic properties.

1.1 The Definition of the Group von Neumann Algebra

Denote by $l^2(G)$ the Hilbert space $l^2(G)$ consisting of formal sums $\sum_{g\in G} \lambda_g \cdot g$ for complex numbers λ_g such that $\sum_{g\in G} |\lambda_g|^2 < \infty$. The scalar product is defined by

$$\left\langle \sum_{g\in G} \lambda_g \cdot g, \sum_{g\in G} \mu_g \cdot g \right\rangle \quad := \quad \sum_{g\in G} \lambda_g \cdot \overline{\mu_g}.$$

This is the same as the Hilbert space completion of the complex group ring $\mathbb{C}G$ with respect to the pre-Hilbert space structure for which G is an orthonormal basis. Notice that left multiplication with elements in G induces an isometric G-action on $l^2(G)$. Given a Hilbert space H, denote by $\mathcal{B}(H)$ the C^*-algebra of bounded (linear) operators from H to itself, where the norm is the operator norm and the involution is given by taking adjoints.

Definition 1.1 (Group von Neumann algebra). *The* group von Neumann algebra $\mathcal{N}(G)$ *of the group G is defined as the algebra of G-equivariant bounded operators from* $l^2(G)$ *to* $l^2(G)$

$$\mathcal{N}(G) \quad := \quad \mathcal{B}(l^2(G))^G.$$

In the sequel we will view the complex group ring $\mathbb{C}G$ as a subring of $\mathcal{N}(G)$ by the embedding of $\mathbb{C}$-algebras $\rho_r : \mathbb{C}G \to \mathcal{N}(G)$ which sends $g \in G$ to the G-equivariant operator $r_{g^{-1}} : l^2(G) \to l^2(G)$ given by right multiplication with g^{-1}.

Remark 1.2 (The general definition of von Neumann algebras). In general a *von Neumann algebra* $\mathcal{A}$ is a sub-$*$-algebra of $\mathcal{B}(H)$ for some Hilbert space H, which is closed in the weak topology and contains $\mathrm{id}\colon H \to H$. Often in the literature the group von Neumann algebra $\mathcal{N}(G)$ is defined as the closure in the weak topology of the complex group ring $\mathbb{C}G$ considered as $*$-subalgebra of $\mathcal{B}(l^2(G))$. This definition and Definition 1.1 agree (see [60, Theorem 6.7.2 on page 434]).

Example 1.3 (The von Neumann algebra of a finite group). If G is finite, then nothing happens, namely $\mathbb{C}G = l^2(G) = \mathcal{N}(G)$.

Example 1.4 (The von Neumann algebra of $\mathbb{Z}^n$). In general there is no concrete model for $\mathcal{N}(G)$. However, for $G = \mathbb{Z}^n$, there is the following illuminating model for the group von Neumann algebra $\mathcal{N}(\mathbb{Z}^n)$. Let $L^2(T^n)$ be the Hilbert space of equivalence classes of L^2-integrable complex-valued functions on the n-dimensional torus T^n, where two such functions are called equivalent if they differ only on a subset of measure zero. Define the ring $L^\infty(T^n)$ by equivalence classes of essentially bounded measurable functions $f\colon T^n \to \mathbb{C}$, where essentially bounded means that there is a constant $C > 0$ such that the set $\{x \in T^n \mid |f(x)| \geq C\}$ has measure zero. An element $(k_1, \ldots, k_n)$ in $\mathbb{Z}^n$ acts isometrically on $L^2(T^n)$ by pointwise multiplication with the function $T^n \to \mathbb{C}$, which maps $(z_1, z_2, \ldots, z_n)$ to $z_1^{k_1} \cdot \ldots \cdot z_n^{k_n}$. Fourier transform yields an isometric $\mathbb{Z}^n$-equivariant isomorphism $l^2(\mathbb{Z}^n) \xrightarrow{\cong} L^2(T^n)$. Hence $\mathcal{N}(\mathbb{Z}^n) = \mathcal{B}(L^2(T^n))^{\mathbb{Z}^n}$. We obtain an isomorphism (of C^*-algebras)

$$L^\infty(T^n) \xrightarrow{\cong} \mathcal{N}(\mathbb{Z}^n)$$

by sending $f \in L^\infty(T^n)$ to the $\mathbb{Z}^n$-equivariant operator

$$M_f\colon L^2(T^n) \to L^2(T^n), \quad g \mapsto g \cdot f,$$

where $g \cdot f(x)$ is defined by $g(x) \cdot f(x)$.

Let $i\colon H \to G$ be an injective group homomorphism. It induces a ring homomorphism $\mathbb{C}i\colon \mathbb{C}H \to \mathbb{C}G$, which extends to a ring homomorphism

$$\mathcal{N}(i)\colon \mathcal{N}(H) \to \mathcal{N}(G) \tag{1.5}$$

as follows. Let $g\colon l^2(H) \to l^2(H)$ be a H-equivariant bounded operator. Then $\mathbb{C}G \otimes_{\mathbb{C}H} l^2(H) \subseteq l^2(G)$ is a dense G-invariant subspace and

$$\mathrm{id}_{\mathbb{C}G} \otimes_{\mathbb{C}H} g\colon \mathbb{C}G \otimes_{\mathbb{C}H} l^2(H) \to \mathbb{C}G \otimes_{\mathbb{C}H} l^2(H)$$

is a G-equivariant linear map, which is bounded with respect to the norm coming from $l^2(G)$. Hence it induces a G-equivariant bounded operator $l^2(G) \to l^2(G)$, which is by definition the image of $g \in \mathcal{N}(H)$ under $\mathcal{N}(i)$.

In the sequel we will ignore the functional analytic aspects of $\mathcal{N}(G)$ and will only consider its algebraic properties as a ring.

1.2 Ring Theoretic Properties of the Group von Neumann Algebra

On the first glance the von Neumann algebra $\mathcal{N}(G)$ looks not very nice as a ring. It is an *integral domain,* i.e. has no non-trivial zero-divisors if and only if G is trivial. It is Noetherian if and only if G is finite (see [80, Exercise 9.11]). It is for instance easy to see that $\mathcal{N}(\mathbb{Z}^n) \cong L^\infty(T^n)$ does contain non-trivial zero-divisors and is not Noetherian. The main advantage of $\mathcal{N}(G)$ is that it contains many more idempotents than $\mathbb{C}G$. This has the effect that $\mathcal{N}(G)$ has the following ring theoretic property. A ring R is called *semihereditary* if every finitely generated submodule of a projective module is again projective. This implies that the category of finitely presented R-modules is an abelian category.

Theorem 1.6 (Von Neumann algebras are semihereditary). *Any von Neumann algebra $\mathcal{A}$ is semihereditary.*

Proof. This follows from the facts that any von Neumann algebra is a Baer $*$-ring and hence in particular a Rickart C^*-algebra [5, Definition 1, Definition 2 and Proposition 9 in Chapter 1.4] and that a C^*-algebra is semihereditary if and only if it is Rickart [1, Corollary 3.7 on page 270]. □

Remark 1.7 (Group von Neumann algebras are semihereditary). It is quite useful to study the following elementary proof of Theorem 1.6 in the special case of a group von Neumann algebra $\mathcal{N}(G)$. One easily checks that it suffices to show for a finitely generated submodule $M \subseteq \mathcal{N}(G)^n$ that M is projective. Let $f\colon \mathcal{N}(G)^m \to \mathcal{N}(G)^n$ be an $\mathcal{N}(G)$-linear map. Choose a matrix $A \in$

$M(m,n;\mathcal{N}(G))$ such that f is given by right multiplication with A. Because of $\mathcal{N}(G) = \mathcal{B}(l^2(G))^G$ we can define a G-equivariant bounded operator

$$\nu(f)\colon l^2(G)^m \to l^2(G)^n, \quad (u_1,\ldots,u_m) \mapsto \left(\sum_{i=1}^m \overline{a_{i,1}^*(\overline{u_i})},\ldots,\sum_{i=1}^m \overline{a_{i,n}^*(\overline{u_i})}\right),$$

where by definition $\overline{\sum_{g\in G}\lambda_g \cdot g} := \sum_{g\in G}\overline{\lambda_g}\cdot g$ and $a_{i,j}^*$ denotes the adjoint of $a_{i,j}$. With these conventions $\nu(\mathrm{id}) = \mathrm{id}$, $\nu(r\cdot f + s\cdot g) = r\cdot\nu(f) + s\cdot\nu(g)$ and $\nu(g\circ f) = \nu(g)\circ\nu(f)$ for $r,s\in\mathbb{C}$ and $\mathcal{N}(G)$-linear maps f and g. Moreover we have $\nu(f)^* = \nu(f^*)$ for an $\mathcal{N}(G)$-map $f\colon \mathcal{N}(G)^m \to \mathcal{N}(G)^n$, where $f^*\colon \mathcal{N}(G)^n \to \mathcal{N}(G)^m$ is given by right multiplication with the matrix $(a_{j,i}^*)$, if f is given by right multiplication with the matrix $(a_{i,j})$, and $\nu(f)^*$ is the adjoint of the operator $\nu(f)$.

Every equivariant bounded operator $l^2(G)^m \to l^2(G)^n$ can be written as $\nu(f)$ for a unique f. Moreover, the sequence $\mathcal{N}(G)^m \xrightarrow{f} \mathcal{N}(G)^n \xrightarrow{g} \mathcal{N}(G)^p$ of $\mathcal{N}(G)$-modules is exact if and only if the sequence of bounded G-equivariant operators $l^2(G)^m \xrightarrow{\nu(f)} l^2(G)^n \xrightarrow{\nu(g)} l^2(G)^p$ is exact. More details and explanations for the last two statements can be found in [80, Section 6.2].

Consider the finitely generated $\mathcal{N}(G)$-submodule $M \subseteq \mathcal{N}(G)^n$. Choose an $\mathcal{N}(G)$-linear map $f\colon \mathcal{N}(G)^m \to \mathcal{N}(G)^n$ with image M. The kernel of $\nu(f)$ is a closed G-invariant linear subspace of $l^2(G)^m$. Hence there is an $\mathcal{N}(G)$-map $p\colon \mathcal{N}(G)^m \to \mathcal{N}(G)^m$ such that $\nu(p)$ is a G-equivariant projection, whose image is $\ker(\nu(f))$. Now $\nu(p)\circ\nu(p) = \nu(p)$ implies $p\circ p = p$ and $\mathrm{im}(\nu(p)) = \ker(\nu(f))$ implies $\mathrm{im}(p) = \ker(f)$. Hence $\ker(f)$ is a direct summand in $\mathcal{N}(G)^m$ and $\mathrm{im}(f) = M$ is projective.

The point is that in order to get the desired projection p one passes to the interpretation by Hilbert spaces and uses orthogonal projections there. We have enlarged the group ring $\mathbb{C}G$ to the group von Neumann algebra $\mathcal{N}(G)$, which does contain these orthogonal projections in contrast to $\mathbb{C}G$.

1.3 Dimension Theory over the Group von Neumann Algebra

An important feature of the group von Neumann algebra is its trace.

Definition 1.8 (Von Neumann trace). *The* von Neumann trace *on $\mathcal{N}(G)$ is defined by*

$$\mathrm{tr}_{\mathcal{N}(G)}\colon \mathcal{N}(G) \to \mathbb{C}, \quad f \mapsto \langle f(e), e\rangle_{l^2(G)},$$

where $e \in G \subseteq l^2(G)$ is the unit element.

It enables us to define a dimension for finitely generated projective $\mathcal{N}(G)$-modules.

Definition 1.9 (Von Neumann dimension for finitely generated projective $\mathcal{N}(G)$-modules). *Let P be a finitely generated projective $\mathcal{N}(G)$-module, and choose a matrix $A = (a_{i,j}) \in M(n,n;\mathcal{N}(G))$ with $A^2 = A$ such that the image of the $\mathcal{N}(G)$-linear map $r_A \colon \mathcal{N}(G)^n \to \mathcal{N}(G)^n$ given by right multiplication with A is $\mathcal{N}(G)$-isomorphic to P. Define the* von Neumann dimension *of P by*

$$\dim_{\mathcal{N}(G)}(P) \quad := \quad \sum_{i=1}^{n} \operatorname{tr}_{\mathcal{N}(G)}(a_{i,i}) \quad \in [0,\infty).$$

We omit the standard proof that $\dim_{\mathcal{N}(G)}(P)$ depends only on the isomorphism class of P but not on the choice of the matrix A. Obviously

$$\dim_{\mathcal{N}(G)}(P \oplus Q) = \dim_{\mathcal{N}(G)}(P) + \dim_{\mathcal{N}(G)}(Q).$$

It is not hard to show that $\dim_{\mathcal{N}(G)}$ is *faithful*, i.e. $\dim_{\mathcal{N}(G)}(P) = 0 \Leftrightarrow P = 0$ holds for any finitely generated projective $\mathcal{N}(G)$-module P.

Recall that the dual M^* of a left or right R-module M is the right or left R-module $\hom_R(M,R)$ respectively, where the R-multiplication is given by $(fr)(x) = f(x)r$ or $(rf)(x) = rf(x)$ respectively for $f \in M^*$, $x \in M$ and $r \in R$.

Definition 1.10 (Closure of a submodule). *Let M be an R-submodule of N. Define the* closure *of M in N to be the R-submodule of N*

$$\overline{M} \;:=\; \{x \in N \mid f(x) = 0 \text{ for all } f \in N^* \text{ with } M \subseteq \ker(f)\}.$$

For an R-module M define the R-submodule $\mathbf{T}M$ and the quotient R-module $\mathbf{P}M$ by

$$\begin{aligned} \mathbf{T}M \;&:=\; \{x \in M \mid f(x) = 0 \text{ for all } f \in M^*\}; \\ \mathbf{P}M \;&:=\; M/\mathbf{T}M. \end{aligned}$$

Notice that $\mathbf{T}M$ is the closure of the trivial submodule in M. It can also be described as the kernel of the canonical map $i(M) \colon M \to (M^*)^*$, which sends $x \in M$ to the map $M^* \to R,\; f \mapsto f(x)$. Notice that $\mathbf{TP}M = 0$, $\mathbf{PP}M = \mathbf{P}M$, $M^* = (\mathbf{P}M)^*$ and that $\mathbf{P}M = 0$ is equivalent to $M^* = 0$.

The next result is the key ingredient in the definition of L^2-Betti numbers for G-spaces. Its proof can be found in [76, Theorem 0.6], [80, Theorem 6.7].

Theorem 1.11. (Dimension function for arbitrary $\mathcal{N}(G)$-modules).

(i) *If $K \subseteq M$ is a submodule of the finitely generated $\mathcal{N}(G)$-module M, then $M/\overline{K}$ is finitely generated projective and $\overline{K}$ is a direct summand in M;*

(ii) *If M is a finitely generated $\mathcal{N}(G)$-module, then $\mathbf{P}M$ is finitely generated projective, there is an exact sequence $0 \to \mathcal{N}(G)^n \to \mathcal{N}(G)^n \to \mathbf{T}M \to 0$ and*

$$M \cong \mathbf{P}M \oplus \mathbf{T}M;$$

(iii) *There exists precisely one dimension function*

$$\dim_{\mathcal{N}(G)} : \{\mathcal{N}(G)\text{-modules}\} \rightarrow [0,\infty] := \{r \in \mathbb{R} \mid r \geq 0\} \amalg \{\infty\}$$

which satisfies:

(a) *Extension Property*
If M is a finitely generated projective $\mathcal{N}(G)$-module, then equality holds between $\dim_{\mathcal{N}(G)}(M)$ and the expression introduced Definition 1.9;

(b) *Additivity*
If $0 \to M_0 \to M_1 \to M_2 \to 0$ is an exact sequence of $\mathcal{N}(G)$-modules, then

$$\dim_{\mathcal{N}(G)}(M_1) = \dim_{\mathcal{N}(G)}(M_0) + \dim_{\mathcal{N}(G)}(M_2),$$

where for $r, s \in [0,\infty]$ we define $r + s$ by the ordinary sum of two real numbers if both r and s are not ∞, and by ∞ otherwise;

(c) *Cofinality*
Let $\{M_i \mid i \in I\}$ be a cofinal system of submodules of M, i.e. $M = \bigcup_{i\in I} M_i$ and for two indices i and j there is an index k in I satisfying $M_i, M_j \subseteq M_k$. Then

$$\dim_{\mathcal{N}(G)}(M) = \sup\{\dim_{\mathcal{N}(G)}(M_i) \mid i \in I\};$$

(d) *Continuity*
If $K \subseteq M$ is a submodule of the finitely generated $\mathcal{N}(G)$-module M, then

$$\dim_{\mathcal{N}(G)}(K) = \dim_{\mathcal{N}(G)}(\overline{K}).$$

Definition 1.12 (Von Neumann dimension for arbitrary $\mathcal{N}(G)$-modules). *In the sequel we mean for an (arbitrary) $\mathcal{N}(G)$-module M by $\dim_{\mathcal{N}(G)}(M)$ the value of the dimension function appearing in Theorem 1.11 and call it the* von Neumann dimension *of M.*

Remark 1.13 (Uniqueness of the dimension function). There is only one possible definition for the dimension function appearing in Theorem 1.11, namely one must have

$$\dim_{\mathcal{N}(G)}(M) := \sup\{\dim_{\mathcal{N}(G)}(P) \mid P \subseteq M \text{ finitely generated projective submodule}\} \in [0,\infty].$$

Namely, consider the directed system of finitely generated $\mathcal{N}(G)$-submodules $\{M_i \mid i \in I\}$ of M which is directed by inclusion. By Cofinality

$$\dim_{\mathcal{N}(G)}(M) = \sup\{\dim_{\mathcal{N}(G)}(M_i) \mid i \in I\}.$$

From Additivity and Theorem 1.11 (ii) we conclude

$$\dim_{\mathcal{N}(G)}(M_i) = \dim_{\mathcal{N}(G)}(\mathbf{P}M_i)$$

and that $\mathbf{P}M_i$ is finitely generated projective. This shows uniqueness of $\dim_{\mathcal{N}(G)}$. The hard part in the proof of Theorem 1.11 (iii) is to show that the definition above does have all the desired properties.

We also see what $\dim_{\mathcal{N}(G)}(M) = 0$ means. It is equivalent to the condition that M contains no non-trivial projective $\mathcal{N}(G)$-submodule, or, equivalently, no non-trivial finitely generated projective $\mathcal{N}(G)$-submodule.

Example 1.14 (The von Neumann dimension for finite groups)**.** If G is finite, then $\mathcal{N}(G) = \mathbb{C}G$ and $\operatorname{tr}_{\mathcal{N}(G)}\left(\sum_{g \in G} \lambda_g \cdot g\right)$ is the coefficient λ_e of the unit element $e \in G$. For an $\mathcal{N}(G)$-module M its von Neumann dimension $\dim_{\mathcal{N}(G)}(V)$ is $\frac{1}{|G|}$-times the complex dimension of the underlying complex vector space M.

The next example implies that $\dim_{\mathcal{N}(G)}(P)$ for a finitely generated projective $\mathcal{N}(G)$-module can be any non-negative real number.

Example 1.15 (The von Neumann dimension for $\mathbb{Z}^n$)**.** Consider $G = \mathbb{Z}^n$. Recall that $\mathcal{N}(\mathbb{Z}^n) = L^\infty(T^n)$. Under this identification we get for the von Neumann trace

$$\operatorname{tr}_{\mathcal{N}(\mathbb{Z}^n)} : \mathcal{N}(\mathbb{Z}^n) \to \mathbb{C}, \quad f \mapsto \int_{T^n} f d\mu,$$

where μ is the standard Lebesgue measure on T^n.

Let $X \subseteq T^n$ be any measurable set and $\chi_X \in L^\infty(T^n)$ be its characteristic function. Denote by $M_{\chi_X} : L^2(T^n) \to L^2(T^n)$ the $\mathbb{Z}^n$-equivariant unitary projection given by multiplication with χ_X. Its image P is a finitely generated projective $\mathcal{N}(\mathbb{Z}^n)$-module, whose von Neumann dimension $\dim_{\mathcal{N}(\mathbb{Z}^n)}(P)$ is the volume $\mu(X)$ of X.

In view of the results above the following slogan makes sense.

Slogan 1.16. The group von Neumann algebra $\mathcal{N}(G)$ behaves like the ring of integers $\mathbb{Z}$ provided one ignores the properties integral domain and Noetherian.

Namely, Theorem 1.11 (ii) corresponds to the statement that a finitely generated $\mathbb{Z}$-module M decomposes into $M = M/\operatorname{tors}(M) \oplus \operatorname{tors}(M)$ and that there exists an exact sequence of $\mathbb{Z}$-modules $0 \to \mathbb{Z}^n \to \mathbb{Z}^n \to \operatorname{tors}(M) \to 0$, where $\operatorname{tors}(M)$ is the $\mathbb{Z}$-module consisting of torsion elements. One obtains the obvious analog of Theorem 1.11 (iii) if one considers

$$\{\mathbb{Z}\text{-modules}\} \to [0,\infty], \quad M \mapsto \dim_{\mathbb{Q}}(\mathbb{Q} \otimes_{\mathbb{Z}} M).$$

One basic difference between the case $\mathbb{Z}$ and $\mathcal{N}(G)$ is that there exist projective $\mathcal{N}(G)$-modules with finite dimension which are not finitely generated, which is not true over $\mathbb{Z}$. For instance take the direct sum $P = \bigoplus_{i=1}^{\infty} P_i$ of $\mathcal{N}(\mathbb{Z}^n)$-modules P_i appearing in Example 1.15 with $\dim_{\mathcal{N}(\mathbb{Z}^n)}(P_i) = 2^{-i}$. Then P is projective but not finitely generated and satisfies $\dim_{\mathcal{N}(\mathbb{Z}^n)}(P) = 1$.

The proof of the following two results is given in [80, Theorem 6.13 and Theorem 6.39].

Theorem 1.17 (Dimension and colimits). *Let $\{M_i \mid i \in I\}$ be a directed system of $\mathcal{N}(G)$-modules over the directed set I. For $i \leq j$ let $\phi_{i,j} \colon M_i \to M_j$ be the associated morphism of $\mathcal{N}(G)$-modules. For $i \in I$ let $\psi_i \colon M_i \to \operatorname{colim}_{i\in I} M_i$ be the canonical morphism of $\mathcal{N}(G)$-modules. Then:*

(i) We get for the dimension of the $\mathcal{N}(G)$-module given by the colimit

$$\dim_{\mathcal{N}(G)}\left(\operatorname{colim}_{i\in I} M_i\right) = \sup\left\{\dim_{\mathcal{N}(G)}(\operatorname{im}(\psi_i)) \mid i \in I\right\};$$

(ii) Suppose for each $i \in I$ that there exists $i_0 \in I$ with $i \leq i_0$ such that $\dim_{\mathcal{N}(G)}(\operatorname{im}(\phi_{i,i_0})) < \infty$ holds. Then

$$\begin{aligned}&\dim_{\mathcal{N}(G)}\left(\operatorname{colim}_{i\in I} M_i\right)\\ &= \sup\left\{\inf\left\{\dim_{\mathcal{N}(G)}(\operatorname{im}(\phi_{i,j} \colon M_i \to M_j)) \mid j \in I, i \leq j\right\} \mid i \in I\right\}.\end{aligned}$$

Theorem 1.18 (Induction and dimension). *Let $i \colon H \to G$ be an injective group homomorphism. Then*

(i) Induction with $\mathcal{N}(i)\colon \mathcal{N}(H) \to \mathcal{N}(G)$ is a faithfully flat functor $M \mapsto i_*M := \mathcal{N}(G) \otimes_{\mathcal{N}(i)} M$ from the category of $\mathcal{N}(H)$-modules to the category of $\mathcal{N}(G)$-modules, i.e. a sequence of $\mathcal{N}(H)$-modules $M_0 \to M_1 \to M_2$ is exact at M_1 if and only if the induced sequence of $\mathcal{N}(G)$-modules $i_*M_0 \to i_*M_1 \to i_*M_2$ is exact at i_*M_1;

(ii) For any $\mathcal{N}(H)$-module M we have:

$$\dim_{\mathcal{N}(H)}(M) = \dim_{\mathcal{N}(G)}(i_*M).$$

Example 1.19 (The von Neumann dimension and $\mathbb{C}[\mathbb{Z}^n]$-modules). Consider the case $G = \mathbb{Z}^n$. Then $\mathbb{C}[\mathbb{Z}^n]$ is a commutative integral domain and hence has a quotient field $\mathbb{C}[\mathbb{Z}^n]_{(0)}$. Let $\dim_{\mathbb{C}[\mathbb{Z}^n]_{(0)}}$ denote the usual dimension for vector spaces over $\mathbb{C}[\mathbb{Z}^n]_{(0)}$. Let M be a $\mathbb{C}[\mathbb{Z}^n]$-module. Then

$$\dim_{\mathcal{N}(\mathbb{Z}^n)}\left(\mathcal{N}(\mathbb{Z}^n) \otimes_{\mathbb{C}[\mathbb{Z}^n]} M\right) = \dim_{\mathbb{C}[\mathbb{Z}^n]_{(0)}}\left(\mathbb{C}[\mathbb{Z}^n]_{(0)} \otimes_{\mathbb{C}[\mathbb{Z}^n]} M\right). \quad (1.20)$$

This follows from the following considerations. Let $\{M_i \mid i \in I\}$ be the directed system of finitely generated submodules of M. Then $M = \operatorname{colim}_{i \in I} M_i$. Since the tensor product has a right adjoint, it is compatible with colimits. This implies together with Theorem 1.17

$$\begin{aligned} \dim_{\mathcal{N}(\mathbb{Z}^n)}\left(\mathcal{N}(\mathbb{Z}^n) \otimes_{\mathbb{C}[\mathbb{Z}^n]} M\right) &= \sup\left\{\dim_{\mathcal{N}(\mathbb{Z}^n)}\left(\mathcal{N}(\mathbb{Z}^n) \otimes_{\mathbb{C}[\mathbb{Z}^n]} M_i\right)\right\}; \\ \dim_{\mathbb{C}[\mathbb{Z}^n]_{(0)}}\left(\mathbb{C}[\mathbb{Z}^n]_{(0)} \otimes_{\mathbb{C}[\mathbb{Z}^n]} M\right) &= \sup\left\{\dim_{\mathbb{C}[\mathbb{Z}^n]_{(0)}}\left(\mathbb{C}[\mathbb{Z}^n]_{(0)} \otimes_{\mathbb{C}[\mathbb{Z}^n]} M_i\right)\right\}. \end{aligned}$$

Hence it suffices to prove the claim for a finitely generated $\mathbb{C}[\mathbb{Z}^n]$-module N. The case $n = 1$ is easy. Then $\mathbb{C}[\mathbb{Z}]$ is a principal integral domain and we can write

$$N = \mathbb{C}[\mathbb{Z}]^r \oplus \bigoplus_{i=1}^{k} \mathbb{C}[\mathbb{Z}]/(u_i)$$

for non-trivial elements $u_i \in \mathbb{C}[\mathbb{Z}]$ and some non-negative integers k and r. One easily checks that there is an exact $\mathcal{N}(\mathbb{Z})$-sequence

$$0 \to \mathcal{N}(\mathbb{Z}) \xrightarrow{r_{u_i}} \mathcal{N}(\mathbb{Z}) \to \mathcal{N}(\mathbb{Z}) \otimes_{\mathbb{C}[\mathbb{Z}]} \mathbb{C}[\mathbb{Z}]/(u_i) \to 0$$

using the identification $\mathcal{N}(\mathbb{Z}) = L^\infty(S^1)$ from Example 1.4 to show injectivity of the map r_{u_i} given by multiplication with u_i. This implies

$$\dim_{\mathcal{N}(\mathbb{Z})}\left(\mathcal{N}(\mathbb{Z}) \otimes_{\mathbb{C}[\mathbb{Z}]} N\right) = r = \dim_{\mathbb{C}[\mathbb{Z}]_{(0)}}\left(\mathbb{C}[\mathbb{Z}]_{(0)} \otimes_{\mathbb{C}[\mathbb{Z}]} N\right).$$

In the general case $n \geq 1$ one knows that there exists a finite free $\mathbb{C}[\mathbb{Z}^n]$-resolution of N. Now the claim follows from [80, Lemma 1.34].

This example is the commutative version of a general setup for arbitrary groups, which will be discussed in Subsection 4.2.

A center-valued dimension function for finitely generated projective modules will be introduced in Definition 7.3. It can be used to classify finitely generated projective $\mathcal{N}(G)$-modules (see Theorem 7.5) and shows that the representation theory of finite dimensional representations over a finite group extends to infinite groups if one works with $\mathcal{N}(G)$ (see Remark 7.6).

2 Definition and Basic Properties of L^2-Betti Numbers

In this section we define L^2-Betti numbers for arbitrary G-spaces and study their basic properties. Our general algebraic definition is very general and is very flexible. This allows to apply standard techniques such as spectral sequences and Mayer-Vietoris arguments directly. The original analytic definition for free proper smooth G-manifolds with G-invariant Riemannian metrics is due to Atiyah and will be briefly discussed in Subsection 2.3.

2.1 The Definition of L^2-Betti Numbers

Definition 2.1 (L^2-Betti numbers of G-spaces). *Let X be a (left) G-space. Equip $\mathcal{N}(G)$ with the obvious $\mathcal{N}(G)$-$\mathbb{Z}G$-bimodule structure. The* singular homology $H_p^G(X;\mathcal{N}(G))$ of X with coefficients in $\mathcal{N}(G)$ *is the homology of the $\mathcal{N}(G)$-chain complex $\mathcal{N}(G) \otimes_{\mathbb{Z}G} C_*^{\text{sing}}(X)$, where $C_*^{\text{sing}}(X)$ is the singular chain complex of X with the induced $\mathbb{Z}G$-structure. Define the* p-th L^2-Betti number of X *by*

$$b_p^{(2)}(X;\mathcal{N}(G)) \; := \; \dim_{\mathcal{N}(G)}\left(H_p^G(X;\mathcal{N}(G))\right) \quad \in [0,\infty],$$

where $\dim_{\mathcal{N}(G)}$ *is the dimension function of Definition 1.12.*

If G and its action on X are clear from the context, we often omit $\mathcal{N}(G)$ in the notation above. For instance, for a connected CW-complex X we denote by $b_p^{(2)}(\widetilde{X})$ the L^2-Betti number $b_p^{(2)}(\widetilde{X};\mathcal{N}(\pi_1(X)))$ of its universal covering $\widetilde{X}$ with respect to the obvious $\pi_1(X)$-action.

Notice that we have *no* assumptions on the G-action or on the topology on X, we do *not* need to require that the operation is free, proper, simplicial or cocompact. Thus we can apply this definition to the *classifying space for free proper G-actions EG*, which is a free G-CW-complex which is contractible (after forgetting the group action). Recall that EG is unique up to G-homotopy. Its quotient $BG = G\backslash EG$ is a connected CW-complex, which is up to homotopy uniquely determined by the property that $\pi_n(BG) = \{1\}$ for $n \geq 2$ and $\pi_1(BG) \cong G$ holds, and called *classifying space of G*. Moreover, $G \to EG \to BG$ is the universal G-principal bundle.

Definition 2.2 (L^2-Betti numbers of groups). *Define for any (discrete) group G its* p-th L^2-Betti number *by*

$$b_p^{(2)}(G) \; := \; b_p^{(2)}(EG,\mathcal{N}(G)).$$

Remark 2.3 (Comparison with the approach by Cheeger and Gromov). A detailed comparison of our approach with the one by Cheeger and Gromov [15, section 2] can be found in [80, Remark 6.76]. Cheeger and Gromov [15, Section 2] define L^2-cohomology and L^2-Betti numbers of a G-space X by considering the category whose objects are G-maps $f\colon Y \to X$ for a simplicial complex Y with cocompact free simplicial G-action and then using inverse limits to extend the classical notions for finite free G-CW-complexes such as Y to X. Their approach is technically more complicated because for instance they work with cohomology instead of homology and therefore have to deal with inverse limits instead of directed limits. Our approach is closer to standard notions, the only non-standard part is the verification of the properties of the extended dimension function (Theorem 1.11).

Remark 2.4 (L^2-Betti numbers for von Neumann algebras). The algebraic approach to L^2-Betti numbers of groups as

$$b_p^{(2)}(G) = \dim_{\mathcal{N}(G)}\left(\operatorname{Tor}_p^{\mathbb{C}G}(\mathbb{C}, \mathcal{N}(G))\right)$$

based on the dimension function for arbitrary modules and homological algebra plays a role in the definition of L^2-Betti numbers for certain von Neumann algebras by Connes-Shlyakhtenko [18]. The point of their construction is to introduce invariants which depend on the group von Neumann algebra $\mathcal{N}(G)$ only. If one could show that their invariants applied to $\mathcal{N}(G)$ agree with the L^2-Betti numbers of G, one would get a positive answer to the open problem, whether the von Neumann algebras of two finitely generated free groups F_1 and F_2 are isomorphic as von Neumann algebras if and only if the groups F_1 and F_2 are isomorphic.

Definition 2.5 (G-CW-complex). *A G-CW-complex X is a G-space together with a G-invariant filtration*

$$\emptyset = X_{-1} \subseteq X_0 \subseteq X_1 \subseteq \ldots \subseteq X_n \subseteq \ldots \subseteq \bigcup_{n \geq 0} X_n = X$$

such that X carries the colimit topology with respect to this filtration (i.e. a set $C \subseteq X$ is closed if and only if $C \cap X_n$ is closed in X_n for all $n \geq 0$) and X_n is obtained from X_{n-1} for each $n \geq 0$ by attaching equivariant n-dimensional cells, i.e. there exists a G-pushout

$$\begin{array}{ccc} \coprod_{i \in I_n} G/H_i \times S^{n-1} & \xrightarrow{\coprod_{i \in I_n} q_i} & X_{n-1} \\ \downarrow & & \downarrow \\ \coprod_{i \in I_n} G/H_i \times D^n & \xrightarrow[\coprod_{i \in I_n} Q_i]{} & X_n \end{array}$$

The space X_n is called the *n-skeleton* of X. A G-CW-complex X is *proper* if and only if all its isotropy groups are finite. A G-space is called *cocompact* if $G\backslash X$ is compact. A G-CW-complex X is *finite* if X has only finitely many equivariant cells. A G-CW-complex is finite if and only if it is cocompact. A G-CW-complex X is *of finite type* if each n-skeleton is finite. It is called *of dimension* $\leq n$ if $X = X_n$ and *finite dimensional* if it is of dimension $\leq n$ for some integer n. A free G-CW-complex X is the same as a regular covering $X \to Y$ of a CW-complex Y with G as group of deck transformations.

Notice that Definition 2.5 also makes sense in the case where G is a topological group. Every proper smooth cocompact G-manifold is a proper G-CW-complex by means of an equivariant triangulation.

For a G-CW-complex one can use the cellular $\mathbb{Z}G$-chain complex instead of the singular chain complex in the definition of L^2-Betti numbers by the next result. Its proof can be found in [76, Lemma 4.2]. For more information about G-CW-complexes we refer for instance to [104, Sections II.1 and II.2], [71, Sections 1 and 2], [80, Subsection 1.2.1].

Lemma 2.6. *Let X be a G-CW-complex. Let $C_*^c(X)$ be its cellular $\mathbb{Z}G$-chain complex. Then there is a $\mathbb{Z}G$-chain homotopy equivalence $C_*^{\mathrm{sing}}(X) \to C_*^c(X)$ and we get*

$$b_p^{(2)}(X;\mathcal{N}(G)) \quad = \quad \dim_{\mathcal{N}(G)}\left(H_p\left(\mathcal{N}(G) \otimes_{\mathbb{Z}G} C_*^c(X)\right)\right).$$

The definition of $b_p^{(2)}(X;\mathcal{N}(G))$ and the above lemma extend in the obvious way to pairs (X, A).

2.2 Basic Properties of L^2-Betti Numbers

The basic properties of L^2-Betti numbers are summarized in the following theorem. Its proof can be found in [80, Theorem 1.35 and Theorem 6.54] except for assertion (viii) which follows from [80, Lemma 13.45].

Theorem 2.7 (L^2-Betti numbers for arbitrary spaces).

(i) *Homology invariance*
We have for a G-map $f\colon X \to Y$:

(a) *Suppose for $n \geq 1$ that for each subgroup $H \subseteq G$ the induced map $f^H : X^H \to Y^H$ is $\mathbb{C}$-homologically n-connected, i.e. the map*

$$H_p^{\mathrm{sing}}(f^H;\mathbb{C})\colon H_p^{\mathrm{sing}}(X^H;\mathbb{C}) \to H_p^{\mathrm{sing}}(Y^H;\mathbb{C})$$

induced by f^H on singular homology with complex coefficients is bijective for $p < n$ and surjective for $p = n$. Then

$$\begin{array}{rcll} b_p^{(2)}(X) & = & b_p^{(2)}(Y) & \quad \text{for } p < n; \\ b_p^{(2)}(X) & \geq & b_p^{(2)}(Y) & \quad \text{for } p = n; \end{array}$$

(b) *Suppose that for each subgroup $H \subseteq G$ the induced map $f^H : X^H \to Y^H$ is a $\mathbb{C}$-homology equivalence, i.e. $H_p^{\mathrm{sing}}(f^H;\mathbb{C})$ is bijective for $p \geq 0$. Then*

$$b_p^{(2)}(X) \;=\; b_p^{(2)}(Y) \qquad \text{for } p \geq 0;$$

(ii) *Comparison with the Borel construction*
Let X be a G-CW-complex. Suppose that for all $x \in X$ the isotropy group G_x is finite or satisfies $b_p^{(2)}(G_x) = 0$ for all $p \geq 0$. Then

$$b_p^{(2)}(X;\mathcal{N}(G)) \quad = \quad b_p^{(2)}(EG \times X;\mathcal{N}(G)) \qquad \text{for } p \geq 0,$$

where G acts diagonally on $EG \times X$;

(iii) *Invariance under non-equivariant $\mathbb{C}$-homology equivalences*
Suppose that $f\colon X \to Y$ is a G-equivariant map of G-CW-complexes such that the induced map $H_p^{\mathrm{sing}}(f;\mathbb{C})$ on singular homology with complex coefficients is bijective for all p. Suppose that for all $x \in X$ the isotropy group G_x is finite or satisfies $b_p^{(2)}(G_x) = 0$ for all $p \geq 0$, and analogously for all $y \in Y$. Then we have for all $p \geq 0$

$$b_p^{(2)}(X;\mathcal{N}(G)) \;=\; b_p^{(2)}(Y;\mathcal{N}(G));$$

(iv) *Independence of equivariant cells with infinite isotropy*
Let X be a G-CW-complex. Let $X[\infty]$ be the G-CW-subcomplex consisting of those points whose isotropy subgroups are infinite. Then we get for all $p \geq 0$

$$b_p^{(2)}(X;\mathcal{N}(G)) \;=\; b_p^{(2)}(X,X[\infty];\mathcal{N}(G));$$

(v) *Künneth formula*
Let X be a G-space and Y be an H-space. Then $X \times Y$ is a $G\times H$-space and we get for all $n \geq 0$

$$b_n^{(2)}(X \times Y) \;=\; \sum_{p+q=n} b_p^{(2)}(X) \cdot b_q^{(2)}(Y),$$

where we use the convention that $0 \cdot \infty = 0$, $r \cdot \infty = \infty$ for $r \in (0,\infty]$ and $r + \infty = \infty$ for $r \in [0,\infty]$;

(vi) *Induction*
Let $i\colon H \to G$ be an inclusion of groups and let X be an H-space. Let $\mathcal{N}(i)\colon \mathcal{N}(H) \to \mathcal{N}(G)$ be the induced ring homomorphism (see (1.5)*). Then:*

$$H_p^G(G \times_H X;\mathcal{N}(G)) \;=\; \mathcal{N}(G) \otimes_{\mathcal{N}(i)} H_p^H(X;\mathcal{N}(H));$$
$$b_p^{(2)}(G \times_H X;\mathcal{N}(G)) \;=\; b_p^{(2)}(X;\mathcal{N}(H));$$

(vii) *Restriction to subgroups of finite index*
Let $H \subseteq G$ be a subgroup of finite index $[G:H]$. Let X be a G-space and let $\mathrm{res}_G^H X$ be the H-space obtained from X by restriction. Then

$$b_p^{(2)}(\mathrm{res}_G^H X;\mathcal{N}(H)) \;=\; [G:H] \cdot b_p^{(2)}(X;\mathcal{N}(G));$$

(viii) *Restriction with epimorphisms with finite kernel*
Let $p\colon G \to Q$ be an epimorphism of groups with finite kernel K. Let X be a Q-space. Let p^*X be the G-space obtained from X using p. Then

$$b_p^{(2)}(p^*X;\mathcal{N}(G)) \;=\; \frac{1}{|K|} \cdot b_p^{(2)}(X;\mathcal{N}(Q));$$

(ix) Zero-th homology and L^2-Betti number
Let X be a path-connected G-space. Then:

(a) There is an $\mathcal{N}(G)$-isomorphism $H_0^G(X;\mathcal{N}(G)) \xrightarrow{\cong} \mathcal{N}(G) \otimes_{\mathbb{C}G} \mathbb{C}$;

(b) $b_0^{(2)}(X;\mathcal{N}(G)) = |G|^{-1}$, where $|G|^{-1}$ is defined to be zero if the order $|G|$ of G is infinite;

(x) Euler-Poincaré formula
Let X be a free finite G-CW-complex. Let $\chi(G\backslash X)$ be the Euler characteristic of the finite CW-complex $G\backslash X$, i.e.

$$\chi(G\backslash X) \quad := \quad \sum_{p\geq 0}(-1)^p \cdot |I_p(G\backslash X)| \qquad \in \mathbb{Z},$$

where $|I_p(G\backslash X)|$ is the number of p-cells of $G\backslash X$. Then

$$\chi(G\backslash X) \quad = \quad \sum_{p\geq 0}(-1)^p \cdot b_p^{(2)}(X);$$

(xi) Morse inequalities
Let X be a free G-CW-complex of finite type. Then we get for $n \geq 0$

$$\sum_{p=0}^{n}(-1)^{n-p} \cdot b_p^{(2)}(X) \quad \leq \quad \sum_{p=0}^{n}(-1)^{n-p} \cdot |I_p(G\backslash X)|;$$

(xii) Poincaré duality
Let M be a cocompact free proper G-manifold of dimension n which is orientable. Then

$$b_p^{(2)}(M) \quad = \quad b_{n-p}^{(2)}(M, \partial M);$$

(xiii) Wedges
Let $X_1, X_2, \ldots, X_r$ be connected (pointed) CW-complexes of finite type and $X = \bigvee_{i=1}^r X_i$ be their wedge. Then

$$b_1^{(2)}(\widetilde{X}) - b_0^{(2)}(\widetilde{X}) \quad = \quad r - 1 + \sum_{j=1}^{r}\left(b_1^{(2)}(\widetilde{X_j}) - b_0^{(2)}(\widetilde{X_j})\right);$$

$$b_p^{(2)}(\widetilde{X}) \quad = \quad \sum_{j=1}^{r} b_p^{(2)}(\widetilde{X_j}) \qquad \text{for } 2 \leq p;$$

(xiv) Connected sums

Let M_1, M_2, ..., M_r be compact connected m-dimensional manifolds for $m \geq 3$. Let M be their connected sum $M_1 \# \ldots \# M_r$. Then

$$\begin{aligned} b_1^{(2)}(\widetilde{M}) - b_0^{(2)}(\widetilde{M}) &= r - 1 + \sum_{j=1}^{r} \left(b_1^{(2)}(\widetilde{M_j}) - b_0^{(2)}(\widetilde{M_j}) \right); \\ b_p^{(2)}(\widetilde{M}) &= \sum_{j=1}^{r} b_p^{(2)}(\widetilde{M_j}) \qquad \text{for } 2 \leq p \leq m-2. \end{aligned}$$

Example 2.8. If G is finite, then $b_p^{(2)}(X; \mathcal{N}(G))$ reduces to the classical Betti number $b_p(X)$ multiplied with the factor $|G|^{-1}$.

Remark 2.9 (Reading off L^2-Betti numbers from $H_p(X; \mathbb{C})$)**.** If $f \colon X \to Y$ is a G-map of free G-CW-complexes which induces isomorphisms $H_p^{\text{sing}}(f; \mathbb{C})$ for all $p \geq 0$, then Theorem 2.7 (i) implies

$$b_p^{(2)}(X; \mathcal{N}(G)) = b_p^{(2)}(Y; \mathcal{N}(G)).$$

This does not necessarily mean that one can read off $b_p^{(2)}(X; \mathcal{N}(G))$ from the singular homology $H_p(X; \mathbb{C})$ regarded as a $\mathbb{C}G$-module in general. In general there is for a free G-CW-complex X a spectral sequence converging to $H_{p+q}^G(X; \mathcal{N}(G))$, whose E^2-term is

$$E_{p,q}^2 = \operatorname{Tor}_p^{\mathbb{C}G}(H_q(X; \mathbb{C}), \mathcal{N}(G)).$$

There is no reason why the equality of the dimension of the E^2-term for two free G-CW-complexes X and Y implies that the dimension of $H_{p+q}^G(X; \mathcal{N}(G))$ and $H_{p+q}^G(Y; \mathcal{N}(G))$ agree. However, this is the case if the spectral sequence collapses from the dimension point of view. For instance, if we make the assumption $\dim_{\mathcal{N}(G)} \left(\operatorname{Tor}_p^{\mathbb{C}G}(M, \mathcal{N}(G)) \right) = 0$ for all $\mathbb{C}G$-modules M and $p \geq 2$, Additivity and Cofinality of $\dim_{\mathcal{N}(G)}$ (see Theorem 1.11) imply

$$\begin{aligned} & b_p^{(2)}(X; \mathcal{N}(G)) = \\ & \dim_{\mathcal{N}(G)} (\mathcal{N}(G) \otimes_{\mathbb{C}G} H_p(X; \mathbb{C})) + \dim_{\mathcal{N}(G)} \left(\operatorname{Tor}_1^{\mathbb{C}G}(H_{p-1}(X; \mathbb{C}), \mathcal{N}(G)) \right). \end{aligned}$$

The assumption above is satisfied if G is amenable (see Theorem 5.1) or G has cohomological dimension ≤ 1 over $\mathbb{C}$, for instance, if G is virtually free.

Remark 2.10 (L^2-Betti numbers ignore infinite isotropy)**.** Theorem 2.7 (iv) says that the L^2-Betti numbers do not see the part of a G-space X whose isotropy groups are infinite. In particular $b_p^{(2)}(X;\mathcal{N}(G)) = 0$ if X is a G-CW-complex whose isotropy groups are all infinite. This follows from the fact that for a subgroup $H \subseteq G$

$$\dim_{\mathcal{N}(G)}(\mathcal{N}(G) \otimes_{\mathbb{C}G} \mathbb{C}[G/H]) = \begin{cases} \frac{1}{|H|} & \text{if } |H| < \infty; \\ 0 & \text{if } |H| = \infty. \end{cases}$$

Remark 2.11 (L^2-Betti numbers often vanish)**.** An important phenomenon is that the L^2-Betti numbers of universal coverings of spaces and of groups tend to vanish more often than the classical Betti numbers. This allows to draw interesting conclusions as we will see later.

2.3 Comparison with Other Definitions

In this subsection we give a short overview of the previous definitions of L^2-Betti numbers. Originally they were defined in terms of heat kernels. Their analytic aspects are important, but we will only focus on their algebraic aspects in this survey article. So a reader may skip the brief explanations below.

The notion of L^2-Betti numbers is due to Atiyah [2]. He defined for a smooth Riemannian manifold with a free proper cocompact G-action by isometries its *analytic p-th L^2-Betti number* by the following expression in terms of the *heat kernel* $e^{-t\Delta_p}(x,y)$ of the p-th Laplacian Δ_p

$$b_p^{(2)}(M) = \lim_{t\to\infty} \int_{\mathcal{F}} \operatorname{tr}_{\mathbb{C}}(e^{-t\Delta_p}(x,x))\, dvol_x, \tag{2.12}$$

where $\mathcal{F}$ is a fundamental domain for the G-action and $\operatorname{tr}_{\mathbb{C}}$ denotes the trace of an endomorphism of a finite-dimensional vector space. The L^2-Betti numbers are invariants of the large times asymptotic of the heat kernel.

A *finitely generated Hilbert $\mathcal{N}(G)$-module* is a Hilbert space V together with a linear G-action by isometries such that there exists a linear isometric G-embedding into $l^2(G)^n$ for some $n \geq 0$. One can assign to it its *von Neumann dimension* by

$$\dim_{\mathcal{N}(G)}(V) := \operatorname{tr}_{\mathcal{N}(G)}(A) \quad \in [0,\infty),$$

where A is any idempotent matrix $A \in M(n,n;\mathcal{N}(G))$ such that the image of the G-equivariant operator $l^2(G)^n \to l^2(G)^n$ induced by A is isometrically linearly G-isomorphic to V.

The expression in (2.12) can be interpreted as the von Neumann dimension of the *space $\mathcal{H}^p_{(2)}(M)$ of square-integrable harmonic p-forms* on M, which is a finitely generated Hilbert $\mathcal{N}(G)$-module (see [2, Proposition 4.16 on page 63])

$$\lim_{t\to\infty} \int_{\mathcal{F}} \operatorname{tr}_{\mathbb{C}}(e^{-t\Delta_p}(x,x))\, dvol_x = \dim_{\mathcal{N}(G)}\left(\mathcal{H}^p_{(2)}(M)\right). \tag{2.13}$$

Given a cocompact free G-CW-complex X, one obtains a chain complex of finitely generated Hilbert $\mathcal{N}(G)$-modules $C_*^{(2)}(X) := C_*^c(X) \otimes_{\mathbb{Z}G} l^2(G)$. Its *reduced p-th L^2-homology* is the finitely generated Hilbert $\mathcal{N}(G)$-module

$$H_p^{(2)}(X; l^2(G)) = \ker(c_p^{(2)})/\overline{\operatorname{im}(c_{p+1}^{(2)})}. \tag{2.14}$$

Notice that we divide out the closure of the image of the $(p+1)$-th differential $c_{p+1}^{(2)}$ of $C_*^{(2)}(X)$ in order to ensure that we obtain a Hilbert space. Then by a result of Dodziuk [24] there is an isometric bijective G-operator

$$\mathcal{H}_{(2)}^p(M) \xrightarrow{\cong} H_p^{(2)}(K; l^2(G)), \tag{2.15}$$

where K is an equivariant triangulation of M. Finally one can show [74, Theorem 6.1]

$$b_p^{(2)}(K; \mathcal{N}(G)) = \dim_{\mathcal{N}(G)}\left(H_p^{(2)}(K; l^2(G))\right), \tag{2.16}$$

where $b_p^{(2)}(K; \mathcal{N}(G))$ is the p-th L^2-Betti number in the sense of Definition 2.1.

All in all we see that our Definition 2.1 of L^2-Betti numbers for arbitrary G-spaces extends the heat kernel definition of (2.12) for smooth Riemannian manifolds with a free proper cocompact G-action by isometries. More details of all these definitions and of their identifications can be found in [80, Chapter 1].

2.4 L^2-Euler Characteristic

In this section we introduce the notion of L^2-Euler characteristic.

If X is a G-CW-complex, denote by $I(X)$ the set of its equivariant cells. For a cell $c \in I(X)$ let (G_c) be the conjugacy class of subgroups of G given by its orbit type and let $\dim(c)$ be its dimension. Denote by $|G_c|^{-1}$ the inverse of the order of any representative of (G_c), where $|G_c|^{-1}$ is to be understood to be zero if the order is infinite.

Definition 2.17 (L^2-Euler characteristic). *Let G be a group and let X be a G-space. Define*

$$\begin{aligned}
h^{(2)}(X; \mathcal{N}(G)) &:= \textstyle\sum_{p \geq 0} b_p^{(2)}(X; \mathcal{N}(G)) \in [0, \infty];\\
\chi^{(2)}(X; \mathcal{N}(G)) &:= \textstyle\sum_{p \geq 0} (-1)^p \cdot b_p^{(2)}(X; \mathcal{N}(G)) \in \mathbb{R},\\
&\qquad\qquad \textit{if } h^{(2)}(X; \mathcal{N}(G)) < \infty;\\
m(X; G) &:= \textstyle\sum_{c \in I(X)} |G_c|^{-1} \in [0, \infty],\\
&\qquad\qquad \textit{if } X \textit{ is a } G\text{-}CW\textit{-complex};\\
h^{(2)}(G) &:= h^{(2)}(EG; \mathcal{N}(G)) \in [0, \infty];\\
\chi^{(2)}(G) &:= \chi^{(2)}(EG; \mathcal{N}(G)) \in \mathbb{R}, \quad \textit{if } h^{(2)}(G) < \infty.
\end{aligned}$$

We call $\chi^{(2)}(X; \mathcal{N}(G))$ and $\chi^{(2)}(G)$ the L^2-Euler characteristic *of X and G.*

The condition $h^{(2)}(X;\mathcal{N}(G)) < \infty$ ensures that the sum which appears in the definition of $\chi^{(2)}(X;\mathcal{N}(G))$ converges absolutely and that the following results are true. The reader should compare the next theorem with [15, Theorem 0.3 on page 191]. It essentially follows from Theorem 2.7. Details of its proof can be found in [80, Theorem 6.80].

Theorem 2.18 (L^2-Euler characteristic).

(i) *Generalized Euler-Poincaré formula*
Let X be a G-CW-complex with $m(X;G) < \infty$. Then

$$\begin{aligned} h^{(2)}(X;\mathcal{N}(G)) &< \infty; \\ \sum_{c\in I(X)} (-1)^{\dim(c)} \cdot |G_c|^{-1} &= \chi^{(2)}(X;\mathcal{N}(G)); \end{aligned}$$

(ii) *Sum formula*
Consider the following G-pushout

$$\begin{array}{ccc} X_0 & \xrightarrow{i_1} & X_1 \\ {\scriptstyle i_2}\downarrow & & \downarrow{\scriptstyle j_1} \\ X_2 & \xrightarrow[j_2]{} & X \end{array}$$

such that i_1 is a G-cofibration. Suppose that $h^{(2)}(X_i;\mathcal{N}(G)) < \infty$ for $i = 0, 1, 2$. Then

$$h^{(2)}(X;\mathcal{N}(G)) < \infty;$$

$$\chi^{(2)}(X;\mathcal{N}(G)) = \chi^{(2)}(X_1;\mathcal{N}(G)) + \chi^{(2)}(X_2;\mathcal{N}(G)) - \chi^{(2)}(X_0;\mathcal{N}(G));$$

(iii) *Comparison with the Borel construction*
Let X be a G-CW-complex. If for all $c \in I(X)$ the group G_c is finite or $b_p^{(2)}(G_c) = 0$ for all $p \geq 0$, then

$$\begin{aligned} b_p^{(2)}(X;\mathcal{N}(G)) &= b_p^{(2)}(EG \times X;\mathcal{N}(G)) \qquad \text{for } p \geq 0; \\ h^{(2)}(X;\mathcal{N}(G)) &= h^{(2)}(EG \times X;\mathcal{N}(G)); \\ \chi^{(2)}(X;\mathcal{N}(G)) &= \chi^{(2)}(EG \times X;\mathcal{N}(G)), \\ & \qquad \text{if } h^{(2)}(X;\mathcal{N}(G)) < \infty; \\ \sum_{c\in I(X)} (-1)^{\dim(c)} \cdot |G_c|^{-1} &= \chi^{(2)}(EG \times X;\mathcal{N}(G)), \text{ if } m(X;G) < \infty; \end{aligned}$$

(iv) *Invariance under non-equivariant $\mathbb{C}$-homology equivalences*
Suppose that $f\colon X \to Y$ is a G-equivariant map of G-CW-complexes

with $m(X;G) < \infty$ and $m(Y;G) < \infty$, such that the induced map $H_p(f;\mathbb{C})$ on homology with complex coefficients is bijective for all $p \geq 0$. Suppose that for all $c \in I(X)$ the group G_c is finite or $b_p^{(2)}(G_c) = 0$ for all $p \geq 0$, and analogously for all $d \in I(Y)$. Then

$$\begin{aligned}
\chi^{(2)}(X;\mathcal{N}(G)) &= \sum_{c \in I(X)} (-1)^{\dim(c)} \cdot |G_c|^{-1} \\
&= \sum_{d \in I(Y)} (-1)^{\dim(d)} \cdot |G_d|^{-1} \\
&= \chi^{(2)}(Y;\mathcal{N}(G));
\end{aligned}$$

(v) *Künneth formula*
Let X be a G-CW-complex and Y be an H-CW-complex. Then we get for the $G \times H$-CW-complex $X \times Y$

$$\begin{aligned}
m(X \times Y; G \times H) &= m(X;G) \cdot m(Y;H); \\
h^{(2)}(X \times Y; \mathcal{N}(G \times H)) &= h^{(2)}(X;\mathcal{N}(G)) \cdot h^{(2)}(Y;\mathcal{N}(H)); \\
\chi^{(2)}(X \times Y; \mathcal{N}(G \times H)) &= \chi^{(2)}(X;\mathcal{N}(G)) \cdot \chi^{(2)}(Y;\mathcal{N}(H)), \\
&\qquad \text{if } h^{(2)}(X;\mathcal{N}(G)), h^{(2)}(Y;\mathcal{N}(H)) < \infty,
\end{aligned}$$

where we use the convention that $0 \cdot \infty = 0$ and $r \cdot \infty = \infty$ for $r \in (0, \infty]$;

(vi) *Induction*
Let $H \subseteq G$ be a subgroup and let X be an H-space. Then

$$\begin{aligned}
m(G \times_H X; G) &= m(X;H); \\
h^{(2)}(G \times_H X; \mathcal{N}(G)) &= h^{(2)}(X;\mathcal{N}(H)); \\
\chi^{(2)}(G \times_H X; \mathcal{N}(G)) &= \chi^{(2)}(X;\mathcal{N}(H)), \qquad \text{if } h^{(2)}(X;\mathcal{N}(H)) < \infty;
\end{aligned}$$

(vii) *Restriction to subgroups of finite index*
Let $H \subseteq G$ be a subgroup of finite index $[G:H]$. Let X be a G-space and let $\operatorname{res}_G^H X$ be the H-space obtained from X by restriction. Then

$$\begin{aligned}
m(\operatorname{res}_G^H X; H) &= [G:H] \cdot m(X;G); \\
h^{(2)}(\operatorname{res}_G^H X; \mathcal{N}(H)) &= [G:H] \cdot h^{(2)}(X;\mathcal{N}(G)); \\
\chi^{(2)}(\operatorname{res}_G^H X; \mathcal{N}(H)) &= [G:H] \cdot \chi^{(2)}(X;\mathcal{N}(G)), \\
&\qquad \text{if } h^{(2)}(X;\mathcal{N}(G)) < \infty,
\end{aligned}$$

where $[G:H] \cdot \infty$ is understood to be ∞;

(viii) *Restriction with epimorphisms with finite kernel*

Let $p\colon G \to Q$ be an epimorphism of groups with finite kernel K. Let X be a Q-space. Let p^*X be the G-space obtained from X using p. Then

$$\begin{aligned} m(p^*X;G) &= |K|^{-1} \cdot m(X;Q); \\ h^{(2)}(p^*X;\mathcal{N}(G)) &= |K|^{-1} \cdot h^{(2)}(X;\mathcal{N}(Q)); \\ \chi^{(2)}(p^*X;\mathcal{N}(G)) &= |K|^{-1} \cdot \chi^{(2)}(X;\mathcal{N}(Q)), \text{ if } h^{(2)}(X;\mathcal{N}(Q)) < \infty. \end{aligned}$$

Remark 2.19 (L^2-Euler characteristic and virtual Euler characteristic)**.** The L^2-Euler characteristic generalizes the notion of the virtual Euler characteristic. Let X be a CW-complex which is *virtually homotopy finite*, i.e. there is a d-sheeted covering $p\colon \overline{X} \to X$ for some positive integer d such that $\overline{X}$ is homotopy equivalent to a finite CW-complex. Define the *virtual Euler characteristic* following Wall [105]

$$\chi_{\text{virt}}(X) := \frac{\chi(\overline{X})}{d}.$$

One easily checks that this is independent of the choice of $p\colon \overline{X} \to X$ since the classical Euler characteristic is multiplicative under finite coverings. Moreover, we conclude from Theorem 2.18 (i) and (vii) that for virtually homotopy finite X

$$\begin{aligned} m(\widetilde{X};\pi_1(X)) &< \infty; \\ \chi^{(2)}(\widetilde{X};\mathcal{N}(\pi_1(X))) &= \chi_{\text{virt}}(X). \end{aligned}$$

Remark 2.20 (L^2-Euler characteristic and orbifold Euler characteristic)**.** If X is a finite G-CW-complex, then $\sum_{c\in I(X)}(-1)^{\dim(c)} \cdot |G_c|^{-1}$ is also called *orbifold Euler characteristic* and agrees with the L^2-Euler characteristic by Theorem 2.18 (i).

3 Computations of L^2-Betti Numbers

In this section we state some cases where the L^2-Betti numbers $b_p^{(2)}(\widetilde{X})$ for certain compact manifolds or finite CW-complexes X can explicitly be computed. These computations give evidence for certain conjectures such as the Atiyah Conjecture 4.1 for $(G, d, \mathbb{Q})$ and the Singer Conjecture 9.1 which we will discuss later. Sometimes we will also make a few comments on their proofs in order to give some insight into the methods. Besides analytic methods, which will not be discussed, standard techniques from topology and algebra such as spectral sequences and Mayer-Vietoris sequences will play a role. With our algebraic setup and the nice properties of the dimension function such as Additivity and Cofinality these tools are directly available, whereas in the original settings, which we have briefly discussed in Subsection 2.3, these methods do not apply directly and, if at all, only after some considerable technical efforts.

3.1 Abelian Groups

Let X be a $\mathbb{Z}^n$-space. Then we get from (1.20)

$$b_p^{(2)}(X;\mathcal{N}(\mathbb{Z}^n)) \quad = \quad \dim_{\mathbb{C}[\mathbb{Z}^n]_{(0)}}\left(\mathbb{C}[\mathbb{Z}^n]_{(0)} \otimes_{\mathbb{C}[\mathbb{Z}^n]} H_p^{\mathrm{sing}}(X;\mathbb{C})\right). \quad (3.1)$$

Notice that $b_p^{(2)}(X;\mathcal{N}(\mathbb{Z}^n))$ is always an integer or ∞.

3.2 Finite Coverings

Let $p\colon X \to Y$ be a finite covering with d-sheets. Then we conclude from Theorem 2.7 (vii)

$$b_p^{(2)}(\widetilde{X}) \quad = \quad d \cdot b_p^{(2)}(\widetilde{Y}). \quad (3.2)$$

This implies for every connected CW-complex X which admits a selfcovering $X \to X$ with d-sheets for $d \geq 2$ that $b_p^{(2)}(\widetilde{X}) = 0$ for all $p \in \mathbb{Z}$. In particular

$$b_p^{(2)}(\widetilde{S^1}) \quad = \quad 0 \qquad \text{for all } p \in \mathbb{Z}. \quad (3.3)$$

3.3 Surfaces

Let F_g^d be the orientable closed surface of genus g with d embedded 2-disks removed. (As any non-orientable compact surface is finitely covered by an orientable surface, it suffices to handle the orientable case by (3.2).) From the value of the zero-th L^2-Betti number, the Euler-Poincaré formula and Poincaré duality (see Theorem 2.7 (ix), (x) and (xii)) and from the fact that a compact surface with boundary is homotopy equivalent to a bouquet of circles, we conclude

$$b_0^{(2)}(\widetilde{F_g^d}) \quad = \quad \begin{cases} 1 & \text{if } g = 0, d = 0, 1; \\ 0 & \text{otherwise}; \end{cases}$$

$$b_1^{(2)}(\widetilde{F_g^d}) \quad = \quad \begin{cases} 0 & \text{if } g = 0, d = 0, 1; \\ d + 2 \cdot (g-1) & \text{otherwise}; \end{cases}$$

$$b_2^{(2)}(\widetilde{F_g^d}) \quad = \quad \begin{cases} 1 & \text{if } g = 0, d = 0; \\ 0 & \text{otherwise}. \end{cases}$$

Of course $b_p^{(2)}(\widetilde{F_g^d}) = 0$ for $p \geq 3$.

3.4 Three-Dimensional Manifolds

In this subsection we state the values of the L^2-Betti numbers of compact orientable 3-manifolds.

We begin with collecting some basic notations and facts about 3-manifolds. In the sequel 3-manifold means connected compact orientable 3-manifold, possibly with boundary. A 3-manifold M is *prime* if for any decomposition of M as a connected sum $M_1 \# M_2$, M_1 or M_2 is homeomorphic to S^3. It is *irreducible* if every embedded 2-sphere bounds an embedded 3-disk. Every prime 3-manifold is either irreducible or is homeomorphic to $S^1 \times S^2$ [50, Lemma 3.13]. A 3-manifold M has a prime decomposition, i.e. one can write M as a connected sum

$$M = M_1 \# M_2 \# \ldots \# M_r,$$

where each M_j is prime, and this prime decomposition is unique up to renumbering and orientation preserving homeomorphism [50, Theorems 3.15, 3.21]. Recall that a connected CW-complex is called *aspherical* if $\pi_n(X) = 0$ for $n \geq 2$, or, equivalently, if $\widetilde{X}$ is contractible. Any aspherical 3-manifold is homotopy equivalent to an irreducible 3-manifold with infinite fundamental group or to a 3-disk. By the Sphere Theorem [50, Theorem 4.3], an irreducible 3-manifold is aspherical if and only if it is a 3-disk or has infinite fundamental group.

Let us say that a prime 3-manifold is *exceptional* if it is closed and no finite covering of it is homotopy equivalent to a Haken, Seifert or hyperbolic 3-manifold. No exceptional prime 3-manifolds are known. Both Thurston's Geometrization Conjecture and Waldhausen's Conjecture that any 3-manifold is finitely covered by a Haken manifold imply that there are none.

Details of the proof of the following theorem can be found in [69, Sections 5 and 6]. The proof is quite interesting since it uses both topological and analytic tools and relies on Thurston's Geometrization.

Theorem 3.4 (L^2-Betti numbers of 3-manifolds). *Let M be the connected sum $M_1 \# \ldots \# M_r$ of (compact connected orientable) prime 3-manifolds M_j which are non-exceptional. Assume that $\pi_1(M)$ is infinite. Then the L^2-Betti numbers of the universal covering $\widetilde{M}$ are given by*

$$\begin{aligned}
b_0^{(2)}(\widetilde{M}) &= 0; \\
b_1^{(2)}(\widetilde{M}) &= (r-1) - \sum_{j=1}^{r} \frac{1}{|\pi_1(M_j)|} + \left|\{C \in \pi_0(\partial M) \mid C \cong S^2\}\right| - \chi(M); \\
b_2^{(2)}(\widetilde{M}) &= (r-1) - \sum_{j=1}^{r} \frac{1}{|\pi_1(M_j)|} + \left|\{C \in \pi_0(\partial M) \mid C \cong S^2\}\right|; \\
b_3^{(2)}(\widetilde{M}) &= 0.
\end{aligned}$$

Notice that in the situation of Theorem 3.4 the p-th L^2-Betti number $b_p(\widetilde{M})$ is a rational number. It is an integer, if $\pi_1(M)$ is torsion-free, and vanishes, if M is aspherical.

3.5 Symmetric Spaces

Let L be a connected semisimple Lie group with finite center such that its Lie algebra has no compact ideal. Let $K \subseteq L$ be a maximal compact subgroup. Then the manifold $M := L/K$ equipped with a left L-invariant Riemannian metric is a symmetric space of non-compact type with $L = \mathrm{Isom}(M)^0$ and $K = \mathrm{Isom}(M)^0_x$, where $\mathrm{Isom}(M)^0$ is the identity component of the group of isometries $\mathrm{Isom}(M)$ and $\mathrm{Isom}(M)^0_x$ is the isotropy group of some point $x \in M$ under the $\mathrm{Isom}(M)^0$-action. Every symmetric space M of non-compact type can be written in this way. The space M is diffeomorphic to $\mathbb{R}^n$. Define its *fundamental rank*

$$\text{f-rk}(M) := \mathrm{rk}_{\mathbb{C}}(L) - \mathrm{rk}_{\mathbb{C}}(K),$$

where $\mathrm{rk}_{\mathbb{C}}(L)$ and $\mathrm{rk}_{\mathbb{C}}(K)$ denotes the so called complex rank of the Lie algebra of L and K respectively (see [62, page 128f]). For a compact Lie group K this is the same as the dimension of a maximal torus. The proof of the next result is due to Borel [6].

Theorem 3.5 (L^2-Betti numbers of symmetric spaces of non-compact type)**.** *Let M be a closed Riemannian manifold whose universal covering $\widetilde{M}$ is a symmetric space of non-compact type.*

Then $b_p^{(2)}(\widetilde{M}) \neq 0$ if and only if $\text{f-rk}(\widetilde{M}) = 0$ and $2p = \dim(M)$. If $\text{f-rk}(\widetilde{M}) = 0$, then $\dim(M)$ is even and for $2p = \dim(M)$ we get

$$0 \;<\; b_p^{(2)}(\widetilde{M}) \;=\; (-1)^p \cdot \chi(M).$$

This applies in particular to a hyperbolic manifold and thus we get the result of Dodziuk [25].

Theorem 3.6. *Let M be a hyperbolic closed Riemannian manifold of dimension n. Then*

$$b_p^{(2)}(\widetilde{M}) \begin{cases} = 0 & \text{if } 2p \neq n \\ > 0 & \text{if } 2p = n \end{cases}.$$

If n is even, then

$$(-1)^{n/2} \cdot \chi(M) \;>\; 0.$$

The strategy of the proof of Theorem 3.6 is the following. Because of the Euler-Poincaré formula (see Theorem 2.7 (x)) it suffices to show that $b_p^{(2)}(\widetilde{M}) = 0$ for $2p \neq n$ and $b_p^{(2)}(\widetilde{M}) > 0$ for $2p = n$. Because of the Hodge-deRham Theorem (see (2.15)) and the facts that the von Neumann dimension is faithful and $\widetilde{M}$ is isometrically diffeomorphic to the hyperbolic space $\mathbb{H}^n$, it remains to show that the space of harmonic L^2-integrable forms $\mathcal{H}^p_{(2)}(\mathbb{H}^n)$ is trivial for $2p \neq n$ and non-trivial for $2p = n$. Notice that this question is independent of M or the $\pi_1(M)$-action. Using the rotational symmetry of $\mathbb{H}^n$, this question is answered positively by Dodziuk [25].

More generally one has the following so called Proportionality Principle (see [80, Theorem 3.183].)

Theorem 3.7 (Proportionality Principle for L^2-Betti numbers). *Let M be a simply connected Riemannian manifold. Then there are constants $B_p^{(2)}(M)$ for $p \geq 0$ depending only on the Riemannian manifold M with the following property: For every discrete group G with a cocompact free proper action on M by isometries the following holds*

$$b_p^{(2)}(M;\mathcal{N}(G)) \quad = \quad B_p^{(2)}(M) \cdot \operatorname{vol}(G\backslash M).$$

3.6 Spaces with S^1-Action

The next two theorems are taken from [80, Corollary 1.43 and Theorem 6.65].

Theorem 3.8. **(L^2-Betti numbers and S^1-actions).** *Let X be a connected S^1-CW-complex. Suppose that for one orbit S^1/H (and hence for all orbits) the inclusion into X induces a map on π_1 with infinite image. (In particular the S^1-action has no fixed points.)*

Then we get

$$\begin{aligned} b_p^{(2)}(\widetilde{X}) &= 0 \quad \text{for } p \in \mathbb{Z}; \\ \chi(X) &= 0. \end{aligned}$$

Proof. We give an outline of the idea of the proof in the case where X is a cocompact S^1-CW-complex, because it is a very illuminating example. The proof in the general case is given in [80, Theorem 6.65]. It is useful to show the following slightly more general statement that for any finite S^1-CW-complex Y and S^1-map $f\colon Y \to X$ we get $b_p^{(2)}(f^*\widetilde{X};\mathcal{N}(\pi_1(X))) = 0$ for all $p \geq 0$, where $f^*\widetilde{X} \to Y$ is the pullback of the universal covering $\widetilde{X} \to X$ with f. We prove the latter statement by induction over the dimension and the number of S^1-equivariant cells in top dimension of Y. In the induction step we can assume that Y is an S^1-pushout

$$\begin{array}{ccc} S^1/H \times S^{n-1} & \xrightarrow{q} & Z \\ \downarrow & & \downarrow j \\ S^1/H \times D^n & \xrightarrow[Q]{} & Y \end{array}$$

for $n = \dim(Y)$. It induces a pushout of free finite $\pi_1(X)$-CW-complexes

$$\begin{array}{ccc} q^*j^*f^*\widetilde{X} & \longrightarrow & j^*f^*\widetilde{X} \\ \downarrow & & \downarrow \\ Q^*f^*\widetilde{X} & \longrightarrow & f^*\widetilde{X} \end{array}$$

The associated long exact Mayer-Vietoris sequence looks like

$$\begin{aligned} &\ldots H_p(q^*f^*\widetilde{X};\mathcal{N}(\pi_1(X))) \\ &\qquad \to H_p(Q^*f^*\widetilde{X};\mathcal{N}(\pi_1(X))) \oplus H_p(j^*f^*\widetilde{X};\mathcal{N}(\pi_1(X))) \\ &\qquad \to H_p(f^*\widetilde{X};\mathcal{N}(\pi_1(X))) \to H_{p-1}(q^*j^*f^*\widetilde{X};\mathcal{N}(\pi_1(X))) \\ &\qquad \to H_{p-1}(Q^*f^*\widetilde{X};\mathcal{N}(\pi_1(X))) \oplus H_{p-1}(j^*f^*\widetilde{X};\mathcal{N}(\pi_1(X))) \to \ldots \end{aligned}$$

Because of the Additivity of the dimension (see Theorem 1.11 (iii)b) it suffices to prove for all $p \in \mathbb{Z}$

$$\begin{aligned} \dim_{\mathcal{N}(\pi_1(X))}\left(H_p(j^*f^*\widetilde{X};\mathcal{N}(\pi_1(X)))\right) &= 0; \\ \dim_{\mathcal{N}(\pi_1(X))}\left(H_p(q^*f^*\widetilde{X};\mathcal{N}(\pi_1(X)))\right) &= 0; \\ \dim_{\mathcal{N}(\pi_1(X))}\left(H_p(Q^*f^*\widetilde{X};\mathcal{N}(\pi_1(X)))\right) &= 0. \end{aligned}$$

The induction hypothesis applies to $f \circ j\colon Z \to X$ and $f \circ j \circ q\colon S^1/H \times S^{n-1} \to X$. Hence it remains to show

$$\dim_{\mathcal{N}(\pi_1(X))}\left(H_p(Q^*f^*\widetilde{X};\mathcal{N}(\pi_1(X)))\right) = 0.$$

By elementary covering theory $Q^*f^*\widetilde{X}$ is $\pi_1(X)$-homeomorphic to $\pi_1(X) \times_j \widetilde{S^1/H} \times D^n$ for the injective group homomorphism $j\colon \pi_1(S^1/H) \to \pi_1(X)$ induced by $f \circ Q$. We conclude from the Künneth formula and the compatibility of dimension and induction (see Theorem 2.7 (v) and (vi))

$$\dim_{\mathcal{N}(\pi_1(X))}\left(H_p(Q^*f^*\widetilde{X};\mathcal{N}(\pi_1(X)))\right) = b_p^{(2)}(\widetilde{S^1/H}).$$

Since S^1/H is homeomorphic to S^1, we get $b_p^{(2)}(\widetilde{S^1/H}) = 0$ from (3.3). □

The next result is taken from [80, Corollary 1.43].

Theorem 3.9. *Let M be an aspherical closed manifold with non-trivial S^1-action. (Non-trivial means that $sx \neq x$ holds for at least one element $s \in S^1$ and one element $x \in M$). Then the action has no fixed points and the inclusion of any orbit into X induces an injection on the fundamental groups. All L^2-Betti numbers $b_p^{(2)}(\widetilde{M})$ are trivial and $\chi(M) = 0$.*

3.7 Mapping Tori

Let $f\colon X \to X$ be a selfmap. Its *mapping torus* T_f is obtained from the cylinder $X \times [0,1]$ by glueing the bottom to the top by the identification $(x,1) = (f(x),0)$. There is a canonical map $p\colon T_f \to S^1$ which sends (x,t) to $\exp(2\pi it)$. It induces a canonical epimorphism $\pi_1(T_f) \to \mathbb{Z} = \pi_1(S^1)$ if X is path-connected.

The following result is taken from [80, Theorem 6.63].

Theorem 3.10 (Vanishing of L^2-Betti numbers of mapping tori).
Let $f\colon X \to X$ be a cellular selfmap of a connected CW-complex X and let $\pi_1(T_f) \xrightarrow{\phi} G \xrightarrow{\psi} \mathbb{Z}$ be a factorization of the canonical epimorphism into epimorphisms ϕ and ψ. Suppose for given $p \geq 0$ that $b_p^{(2)}(G \times_{\phi \circ i} \widetilde{X}; \mathcal{N}(G)) < \infty$ and $b_{p-1}^{(2)}(G \times_{\phi \circ i} \widetilde{X}; \mathcal{N}(G)) < \infty$ holds, where $i\colon \pi_1(X) \to \pi_1(T_f)$ is the map induced by the obvious inclusion of X into T_f. Let $\overline{T_f}$ be the covering of T_f associated to ϕ, which is a free G-CW-complex. Then we get

$$b_p^{(2)}(\overline{T_f}; \mathcal{N}(G)) \quad = \quad 0.$$

Proof. We give the proof in the special case where X is a connected finite CW-complex and $\phi = \mathrm{id}$, i.e. we show for a connected finite CW-complex X that $b_p^{(2)}(\widetilde{T_f}) = 0$ for all $p \geq 0$. For each positive integer d there is a finite d-sheeted covering $\overline{T_f} \to T_f$ associated to the subgroup of index d in $\pi_1(T_f)$ which is the preimage of $d\mathbb{Z} \subseteq \mathbb{Z}$ under the canonical homomorphism $\pi_1(T_f) \to \mathbb{Z}$. There is a homotopy equivalence $T_{f^d} \to \overline{T_f}$. We conclude from (3.2) and homotopy invariance of L^2-Betti numbers (see Theorem 2.7 (i))

$$b_p^{(2)}(\widetilde{T_f}) \quad = \quad \frac{b_p^{(2)}(\widetilde{T_{f^d}})}{d}.$$

There is a CW-complex structure on T_{f^d} with $\beta_p(X) + \beta_{p-1}(X)$ p-cells, if $\beta_p(X)$ is the number of p-cells in X. We conclude from Additivity of the dimension function (see Theorem 1.11 (iii)b)

$$\begin{aligned} b_p^{(2)}(\widetilde{T_{f^d}}) \;\leq\; \dim_{\mathcal{N}(\pi_1(T_{f^d}))} & \left(\mathcal{N}(\pi_1(T_{f^d})) \otimes_{\mathbb{Z}\pi_1(T_{f^d})} C_p(\widetilde{T_{f^d}})\right) \\ & = \beta_p(X) + \beta_{p-1}(X). \end{aligned}$$

This implies for all positive integers d

$$0 \;\leq\; b_p^{(2)}(\widetilde{T_f}) \;\leq\; \frac{\beta_p(X) + \beta_{p-1}(X)}{d}.$$

Taking the limit for $d \to \infty$ implies $b_p^{(2)}(\widetilde{T_f}) = 0$. □

3.8 Fibrations

The next result is proved in [80, Lemma 6.6. and Theorem 6.67]. The proof is based on standard spectral sequence arguments and the fact that the dimension function is defined for arbitrary $\mathcal{N}(G)$-modules.

Theorem 3.11 (L^2-Betti numbers and fibrations).

(i) *Let* $F \xrightarrow{i} E \xrightarrow{p} B$ *be a fibration of connected CW-complexes. Consider a factorization* $p_*\colon \pi_1(E) \xrightarrow{\phi} G \xrightarrow{\psi} \pi_1(B)$ *of the map induced by* p *into epimorphisms* ϕ *and* ψ. *Let* $i_*\colon \pi_1(F) \to \pi_1(E)$ *be the homomorphism induced by the inclusion* i. *Suppose for a given integer* $d \geq 1$ *that* $b_p^{(2)}(G \times_{\phi \circ i_*} \widetilde{F}; \mathcal{N}(G)) = 0$ *for* $p \leq d-1$ *and* $b_d^{(2)}(G \times_{\phi \circ i_*} \widetilde{F}; \mathcal{N}(G)) < \infty$ *holds. Suppose that* $\pi_1(B)$ *contains an element of infinite order or finite subgroups of arbitrarily large order. Then* $b_p^{(2)}(G \times_\phi \widetilde{E}; \mathcal{N}(G)) = 0$ *for* $p \leq d$;

(ii) *Let* $F \xrightarrow{i} E \to B$ *be a fibration of connected CW-complexes. Consider a group homomorphism* $\phi\colon \pi_1(E) \to G$. *Let* $i_*\colon \pi_1(F) \to \pi_1(E)$ *be the homomorphism induced by the inclusion* i. *Suppose that for a given integer* $d \geq 0$ *the* L^2*-Betti number* $b_p^{(2)}(G \times_{\phi \circ i_*} \widetilde{F}; \mathcal{N}(G))$ *vanishes for all* $p \leq d$. *Then the* L^2*-Betti number* $b_p^{(2)}(G \times_\phi \widetilde{E}; \mathcal{N}(G))$ *vanishes for all* $p \leq d$.

4 The Atiyah Conjecture

In this section we discuss the Atiyah Conjecture

Conjecture 4.1 (Atiyah Conjecture). *Let* G *be a discrete group with an upper bound on the orders of its finite subgroups. Consider* $d \in \mathbb{Z}$, $d \geq 1$ *such that the order of every finite subgroup of* G *divides* d. *Let* F *be a field with* $\mathbb{Q} \subseteq F \subseteq \mathbb{C}$. *The* Atiyah Conjecture *for* (G, d, F) *says that for any finitely presented* FG*-module* M *we have*

$$d \cdot \dim_{\mathcal{N}(G)}(\mathcal{N}(G) \otimes_{FG} M) \in \mathbb{Z}.$$

4.1 Reformulations of the Atiyah Conjecture

We present equivalent reformulations of the Atiyah Conjecture 4.1.

Theorem 4.2 (Reformulations of the Atiyah Conjecture). *Let* G *be a discrete group. Suppose that there exists* $d \in \mathbb{Z}$, $d \geq 1$ *such that the order of every finite subgroup of* G *divides* d. *Let* F *be a field with* $\mathbb{Q} \subseteq F \subseteq \mathbb{C}$. *Then the following assertions are equivalent:*

(i) *The Atiyah Conjecture 4.1 is true for* (G, d, F), *i.e. for every finitely presented* FG*-module* M *we have*

$$d \cdot \dim_{\mathcal{N}(G)}(\mathcal{N}(G) \otimes_{FG} M) \in \mathbb{Z};$$

(ii) *For every* FG*-module* M *we have*

$$d \cdot \dim_{\mathcal{N}(G)}(\mathcal{N}(G) \otimes_{FG} M) \in \mathbb{Z} \amalg \{\infty\}.$$

Proof. See [80, Lemma 10.7 and Remark 10.11]. □

We mention that the Atiyah Conjecture 4.1 is true for (G, d, F) if and only if for any finitely generated subgroup $H \subseteq G$ the Atiyah Conjecture 4.1 is true for (H, d, F) (see [80, Lemma 10.4]).

The next result explains that the Atiyah Conjecture 4.1 for $(G, d, \mathbb{Q})$ for a finitely generated group G is a statement about the possible values of L^2-Betti numbers.

Theorem 4.3 (Reformulations of the Atiyah Conjecture for $F = \mathbb{Q}$)**.** *Let G be a finitely generated group with an upper bound $d \in \mathbb{Z}, d \geq 1$ on the orders of its finite subgroups. Then the following assertions are equivalent:*

(i) The Atiyah Conjecture 4.1 is true for $(G, d, \mathbb{Q})$;

(ii) For every free proper smooth cocompact G-manifold M without boundary and $p \in \mathbb{Z}$ we have

$$d \cdot b_p^{(2)}(M; \mathcal{N}(G)) \in \mathbb{Z};$$

(iii) For every finite free G-CW-complex X and $p \in \mathbb{Z}$ we have

$$d \cdot b_p^{(2)}(X; \mathcal{N}(G)) \in \mathbb{Z};$$

(iv) For every G-space X and $p \in \mathbb{Z}$ we have

$$d \cdot b_p^{(2)}(X; \mathcal{N}(G)) \in \mathbb{Z} \amalg \{\infty\}.$$

Proof. This follows from [80, Lemma 10.5] and Theorem 4.2. □

We mention that all the explicit computations presented in Section 3 are compatible with the Atiyah Conjecture 4.1.

4.2 The Ring Theoretic Version of the Atiyah Conjecture

In this subsection we consider the following *fundamental square of ring extensions*

$$\begin{array}{ccc} \mathbb{C}G & \xrightarrow{i} & \mathcal{N}(G) \\ {\scriptstyle j}\downarrow & & \downarrow{\scriptstyle k} \\ \mathcal{D}(G) & \xrightarrow[l]{} & \mathcal{U}(G) \end{array} \tag{4.4}$$

which we explain next.

As before $\mathbb{C}G$ is the complex group ring and $\mathcal{N}(G)$ is the group von Neumann algebra.

By $\mathcal{U}(G)$ we denote the algebra of affiliated operators. Instead of its functional analytic definition we describe it algebraically, namely, it is the Ore localization of $\mathcal{N}(G)$ with respect to the multiplicative subset of non-trivial zero-divisors in $\mathcal{N}(G)$. The proof that this multiplicative subset satisfies the Ore condition and basic definitions and properties of Ore localization and of $\mathcal{U}(G)$ can be found for instance in [80, Sections 8.1 and 8.2]. In particular $\mathcal{U}(G)$ is flat when regarded as an $\mathcal{N}(G)$-module. Moreover, the ring $\mathcal{U}(G)$ is a *von Neumann regular ring*, i.e. every finitely generated submodule of a projective module is a direct summand. This is a stronger condition than being semihereditary.

Given a finitely generated projective $\mathcal{U}(G)$-module Q, there is a finitely generated projective $\mathcal{N}(G)$-module P such that $\mathcal{U}(G) \otimes_{\mathcal{N}(G)} P$ and Q are $\mathcal{U}(G)$-isomorphic. If P_0 and P_1 are two finitely generated projective $\mathcal{N}(G)$-modules, then $P_0 \cong_{\mathcal{N}(G)} P_1 \Leftrightarrow \mathcal{U}(G) \otimes_{\mathcal{N}(G)} P_0 \cong_{\mathcal{U}(G)} \mathcal{U}(G) \otimes_{\mathcal{N}(G)} P_1$. This enables us to define a dimension function for $\dim_{\mathcal{U}(G)}$ with properties analogous to $\dim_{\mathcal{N}(G)}$ (see [80, Section 8.3], [98] or [99]).

Theorem 4.5. (Dimension function for arbitrary $\mathcal{U}(G)$-modules). *There exists precisely one dimension function*

$$\dim_{\mathcal{U}(G)} \colon \{\mathcal{U}(G)\text{-}modules\} \rightarrow [0, \infty]$$

which satisfies:

(i) *Extension Property*

If M is an $\mathcal{N}(G)$-module, then

$$\dim_{\mathcal{U}(G)} \big(\mathcal{U}(G) \otimes_{\mathcal{N}(G)} M\big) = \dim_{\mathcal{N}(G)}(M);$$

(ii) *Additivity*

If $0 \to M_0 \to M_1 \to M_2 \to 0$ is an exact sequence of $\mathcal{U}(G)$-modules, then

$$\dim_{\mathcal{U}(G)}(M_1) = \dim_{\mathcal{U}(G)}(M_0) + \dim_{\mathcal{U}(G)}(M_2);$$

(iii) *Cofinality*

Let $\{M_i \mid i \in I\}$ be a cofinal system of submodules of M. Then

$$\dim_{\mathcal{U}(G)}(M) = \sup\{\dim_{\mathcal{U}(G)}(M_i) \mid i \in I\};$$

(iv) *Continuity*

If $K \subseteq M$ is a submodule of the finitely generated $\mathcal{U}(G)$-module M, then

$$\dim_{\mathcal{U}(G)}(K) = \dim_{\mathcal{U}(G)}(\overline{K}).$$

Remark 4.6 (Comparing $\mathbb{Z} \subseteq \mathbb{Q}$ and $\mathcal{N}(G) \subseteq \mathcal{U}(G)$). Recall the Slogan 1.16 that the group von Neumann algebra $\mathcal{N}(G)$ behaves like the ring of integers $\mathbb{Z}$, provided one ignores the properties integral domain and Noetherian. This is supported by the construction and properties of $\mathcal{U}(G)$. Obviously $\mathcal{U}(G)$ plays the same role for $\mathcal{N}(G)$ as $\mathbb{Q}$ plays for $\mathbb{Z}$ as the definition of $\mathcal{U}(G)$ as the Ore localization of $\mathcal{N}(G)$ with respect to the multiplicative subset of non-zero-divisors and Theorem 4.5 show.

A subring $R \subseteq S$ is called *division closed* if each element in R, which is invertible in S, is already invertible in R. It is called *rationally closed* if each square matrix over R, which is invertible over S, is already invertible over R. Notice that the intersection of division closed subrings of S is again division closed, and analogously for rationally closed subrings. Hence the following definition makes sense.

Definition 4.7 (Division and rational closure). *Let S be a ring with subring $R \subseteq S$. The* division closure $\mathcal{D}(R \subseteq S)$ *or* rational closure $\mathcal{R}(R \subseteq S)$ *respectively is the smallest subring of S which contains R and is division closed or rationally closed respectively.*

The ring $\mathcal{D}(G)$ appearing in the fundamental square (4.4) is the rational closure of $\mathbb{C}G$ in $\mathcal{U}(G)$.

Conjecture 4.8 (Ring theoretic version of the Atiyah Conjecture). *Let G be a group for which there exists an upper bound on the orders of its finite subgroups. Then:*

(R) *The ring $\mathcal{D}(G)$ is semisimple;*

(K) *The composition*

$$\bigoplus_{H \subseteq G, |H| < \infty} K_0(\mathbb{C}H) \xrightarrow{a} K_0(\mathbb{C}G) \xrightarrow{j} K_0(\mathcal{D}(G))$$

is surjective, where a is induced by the various inclusions $H \to G$.

Lemma 4.9. *Let G be a group. Suppose that there exists $d \in \mathbb{Z}$, $d \geq 1$ such that the order of every finite subgroup of G divides d. If the group G satisfies the ring theoretic version of the Atiyah Conjecture 4.8, then the Atiyah Conjecture 4.1 for $(G, d, \mathbb{C})$ is true.*

Proof. Let M be a finitely presented $\mathbb{C}G$-module. Then $\mathcal{D}(G) \otimes_{\mathbb{C}G} M$ is a finitely generated projective $\mathcal{D}(G)$-module since $\mathcal{D}(G)$ is semisimple by assumption. We obtain a well-defined homomorphism of abelian groups

$$D\colon K_0(\mathcal{D}(G)) \to \mathbb{R}, \quad [P] \mapsto \dim_{\mathcal{U}(G)} \left(\mathcal{U}(G) \otimes_{\mathcal{D}(G)} P\right).$$

Because of the fundamental square (4.4) and Theorem 4.5 (i) we have

$$\dim_{\mathcal{N}(G)}(\mathcal{N}(G) \otimes_{\mathbb{C}G} M) \;=\; D([\mathcal{D}(G) \otimes_{\mathbb{C}G} M]).$$

Hence it suffices to show that $d \cdot \operatorname{im}(D)$ is contained in $\mathbb{Z}$. Because of assumption (**K**) it suffices to check for each finite subgroup $H \subseteq G$ and each finitely generated projective $\mathbb{C}H$-module P

$$d \cdot \dim_{\mathcal{U}(G)}(\mathcal{U}(G) \otimes_{\mathbb{C}G} \mathbb{C}G \otimes_{\mathbb{C}H} P) \;\in\; \mathbb{Z}.$$

Example 1.14 and Theorem 1.18 imply

$$\begin{aligned}
\dim_{\mathcal{U}(G)}(\mathcal{U}(G) \otimes_{\mathbb{C}G} \mathbb{C}G \otimes_{\mathbb{C}H} P) &= \dim_{\mathcal{N}(G)}(\mathcal{N}(G) \otimes_{\mathbb{C}G} \mathbb{C}G \otimes_{\mathbb{C}H} P) \\
&= \dim_{\mathcal{N}(G)}(\mathcal{N}(G) \otimes_{\mathcal{N}(H)} P) \\
&= \dim_{\mathcal{N}(H)}(P) \\
&= \frac{\dim_{\mathbb{C}}(P)}{|H|}.
\end{aligned}$$

Obviously $d \cdot \frac{\dim_{\mathbb{C}}(P)}{|H|} \in \mathbb{Z}$. □

4.3 The Atiyah Conjecture for Torsion-Free Groups

Remark 4.10 (The Atiyah Conjecture in the torsion-free case). Let G be a torsion-free group. Then we can choose $d = 1$ in the Atiyah Conjecture 4.1. The Atiyah Conjecture 4.1 for $(G, 1, F)$ says that $\dim_{\mathcal{N}(G)}(\mathcal{N}(G) \otimes_{FG} M) \in \mathbb{Z}$ holds for every finitely presented FG-module M and Theorem 4.2 says that then this holds automatically for all FG-modules M with $\dim_{\mathcal{N}(G)}(\mathcal{N}(G) \otimes_{FG} M) < \infty$. In the case, where $F = \mathbb{Q}$ and G is a torsion-free finitely generated group G, Theorem 4.3 implies that the Atiyah Conjecture 4.1 for $(G, 1, F)$ is equivalent to the statement that $b_p^{(2)}(X; \mathcal{N}(G)) \in \mathbb{Z}$ is true for all G-spaces X.

Remark 4.11 (The ring theoretic version of the Atiyah Conjecture in the torsion-free case). Let G be a torsion-free group. Then the ring theoretic version of the Atiyah Conjecture 4.8 reduces to the statement that $\mathcal{D}(G)$ is a skewfield. In this case we can assign to every $\mathcal{D}(G)$-module N its dimension $\dim_{\mathcal{D}(G)}(N) \in \mathbb{Z} \amalg \{\infty\}$ in the usual way and we get for every $\mathbb{C}G$-module M

$$\dim_{\mathcal{N}(G)}(\mathcal{N}(G) \otimes_{\mathbb{C}G} M) = \dim_{\mathcal{U}(G)}(\mathcal{U}(G) \otimes_{\mathbb{C}G} M) = \dim_{\mathcal{D}(G)}(\mathcal{D}(G) \otimes_{\mathbb{C}G} M).$$

Example 4.12 (The case $G = \mathbb{Z}^n$). In the case $G = \mathbb{Z}^n$ the fundamental square of ring extensions (4.4) can be identified with

$$\begin{array}{ccc}
\mathbb{C}[\mathbb{Z}^n] & \longrightarrow & L^\infty(T^n) \\
\downarrow & & \downarrow \\
\mathbb{C}[\mathbb{Z}^n]_{(0)} & \longrightarrow & MF(T^n)
\end{array}$$

where $MF(T^n)$ the ring of equivalence classes of measurable functions $T^n \to \mathbb{C}$. We have already proved

$$\dim_{\mathcal{N}(\mathbb{Z}^n)}(\mathcal{N}(\mathbb{Z}^n) \otimes_{\mathbb{C}[\mathbb{Z}^n]} M) = \dim_{\mathbb{C}[\mathbb{Z}^n]_{(0)}}(\mathbb{C}[\mathbb{Z}^n]_{(0)} \otimes_{\mathbb{C}[\mathbb{Z}^n]} M)$$

in Example 1.19.

4.4 The Atiyah Conjecture Implies the Kaplanski Conjecture

The following conjecture is a well-known conjecture about group rings.

Conjecture 4.13 (Kaplanski Conjecture). *Let F be a field and let G be a torsion-free group. Then FG contains no non-trivial zero-divisors.*

Theorem 4.14 (The Atiyah and the Kaplanski Conjecture). *Let G be a torsion-free group and let F be a field with $\mathbb{Q} \subseteq F \subseteq \mathbb{C}$. Then the Atiyah Conjecture 4.1 for $(G, 1, F)$ implies the Kaplanski Conjecture 4.13 for F and G.*

Proof. Let $u \in FG$ be a zero-divisor. Then the kernel of the $\mathcal{N}(G)$-map $r_u \colon \mathcal{N}(G) \to \mathcal{N}(G)$ given by right multiplication with u is non-trivial. Since $\mathcal{N}(G)$ is semihereditary, the image of r_u is projective. Hence both $\ker(r_u)$ and $\mathcal{N}(G)/\ker(r_u)$ are finitely generated projective $\mathcal{N}(G)$-modules. Additivity of $\dim_{\mathcal{N}(G)}$ implies

$$0 < \dim_{\mathcal{N}(G)}(\ker(r_u)) \leq \dim_{\mathcal{N}(G)}(\mathcal{N}(G)) = 1.$$

We conclude from Remark 4.10 that $\dim_{\mathcal{N}(G)}(\ker(r_u))$ is an integer. Additivity of $\dim_{\mathcal{N}(G)}$ implies

$$\dim_{\mathcal{N}(G)}(\mathcal{N}(G)/\ker(r_u)) = 0.$$

We conclude $\mathcal{N}(G)/\ker(r_u) = 0$ and hence $u = 0$. □

4.5 The Status of the Atiyah Conjecture

Let $l^\infty(G, \mathbb{R})$ be the space of equivalence classes of bounded functions from G to $\mathbb{R}$ with the supremum norm. Denote by 1 the constant function with value 1.

Definition 4.15 (Amenable group). *A group G is called* amenable, *if there is a (left) G-invariant linear operator $\mu \colon l^\infty(G, \mathbb{R}) \to \mathbb{R}$ with $\mu(1) = 1$, which satisfies for all $f \in l^\infty(G, \mathbb{R})$*

$$\inf\{f(g) \mid g \in G\} \leq \mu(f) \leq \sup\{f(g) \mid g \in G\}.$$

The latter condition is equivalent to the condition that μ is bounded and $\mu(f) \geq 0$ if $f(g) \geq 0$ for all $g \in G$.

Definition 4.16 (Elementary amenable group). *The* class of elementary amenable groups $\mathcal{EAM}$ *is defined as the smallest class of groups which has the following properties:*

(i) It contains all finite and all abelian groups;

(ii) It is closed under taking subgroups;

(iii) It is closed under taking quotient groups;

(iv) It is closed under extensions, i.e. if $1 \to H \to G \to K \to 1$ *is an exact sequence of groups and* H *and* K *belong to* $\mathcal{EAM}$, *then also* $G \in \mathcal{EAM}$;

(v) It is closed under directed unions, *i.e. if* $\{G_i \mid i \in I\}$ *is a directed system of subgroups such that* $G = \bigcup_{i \in I} G_i$ *and each* G_i *belongs to* $\mathcal{EAM}$, *then* $G \in \mathcal{EAM}$. *(Directed means that for two indices* i *and* j *there is a third index* k *with* $G_i, G_j \subseteq G_k$.*)*

The class of amenable groups satisfies all the conditions appearing in Definition 4.16. Hence every elementary amenable group is amenable. The converse is not true.

Definition 4.17 (Linnell's class of groups $\mathcal{C}$). *Let* $\mathcal{C}$ *be the smallest class of groups, which contains all free groups and is closed under directed unions and extensions with elementary amenable quotients.*

The next result is due to Linnell [65].

Theorem 4.18 (Linnell's Theorem). *Let* G *be a group in* $\mathcal{C}$. *Suppose that there exists* $d \in \mathbb{Z}$, $d \geq 1$ *such that the order of every finite subgroup of* G *divides* d. *Then the ring theoretic version of the Atiyah Conjecture 4.8 for* G *and hence the Atiyah Conjecture 4.1 for* $(G, d, \mathbb{C})$ *are true.*

The next definition and the next theorem are due to Schick [101].

Definition 4.19. *Let* $\mathcal{D}$ *be the smallest non-empty class of groups such that*

(i) If $p\colon G \to A$ *is an epimorphism of a torsion-free group* G *onto an elementary amenable group* A *and if* $p^{-1}(B) \in \mathcal{D}$ *for every finite group* $B \subseteq A$, *then* $G \in \mathcal{D}$;

(ii) $\mathcal{D}$ *is closed under taking subgroups;*

(iii) $\mathcal{D}$ *is closed under colimits and inverse limits over directed systems.*

Theorem 4.20. *(i) If the group* G *belongs to* $\mathcal{D}$, *then* G *is torsion-free and the Atiyah Conjecture 4.1 for* $(G, 1, \mathbb{Q})$ *is true for* G;

(ii) The class $\mathcal{D}$ *is closed under direct sums, direct products and free products. Every residually torsion-free elementary amenable group belongs to* $\mathcal{D}$.

More information about the status of the Atiyah Conjecture 4.1 can be found for instance in [80, Subsection 10.1.3].

4.6 Groups Without Bound on the Order of Its Finite Subgroups

Given a group G, let $\mathcal{FIN}(G)$ be the set of finite subgroups of G. Denote by

$$\frac{1}{|\mathcal{FIN}(G)|}\mathbb{Z} \subseteq \mathbb{Q} \tag{4.21}$$

the additive subgroup of $\mathbb{R}$ generated by the set of rational numbers $\{\frac{1}{|H|} \mid H \in \mathcal{FIN}(G)\}$.

There is the following formulation of the Atiyah Conjecture for arbitrary groups in the literature.

Conjecture 4.22 (Atiyah Conjecture for arbitrary groups G). *A group G satisfies the* Atiyah Conjecture *if for every finitely presented $\mathbb{C}G$-module M we have*

$$\dim_{\mathcal{N}(G)}(\mathcal{N}(G) \otimes_{\mathbb{C}G} M) \in \frac{1}{|\mathcal{FIN}(G)|}\mathbb{Z}.$$

There do exist counterexamples to this conjecture. The *lamplighter group* L is defined by the semidirect product

$$L := \left(\bigoplus_{n \in \mathbb{Z}} \mathbb{Z}/2\right) \rtimes \mathbb{Z}$$

with respect to the shift automorphism of $\bigoplus_{n\in\mathbb{Z}} \mathbb{Z}/2$, which sends $(x_n)_{n\in\mathbb{Z}}$ to $(x_{n-1})_{n\in\mathbb{Z}}$. Let $e_0 \in \bigoplus_{n\in\mathbb{Z}} \mathbb{Z}/2$ be the element whose entries are all zero except the entry at 0. Denote by $t \in \mathbb{Z}$ the standard generator of $\mathbb{Z}$ which we will also view as an element of L. Then $\{e_0t, t\}$ is a set of generators for L. The associated *Markov operator* $M\colon l^2(G) \to l^2(G)$ is given by right multiplication with $\frac{1}{4} \cdot (e_0t + t + (e_0t)^{-1} + t^{-1})$. It is related to the Laplace operator $\Delta_0\colon l^2(G) \to l^2(G)$ of the Cayley graph of G by $\Delta_0 = 4 \cdot \mathrm{id} - 4 \cdot M$. The following result is a special case of the main result in the paper of Grigorchuk and Żuk [41, Theorem 1 and Corollary 3] (see also [40]). An elementary proof can be found in [22].

Theorem 4.23 (Counterexample to the Atiyah Conjecture for arbitrary groups). *The von Neumann dimension of the kernel of the Markov operator M of the lamplighter group L associated to the set of generators $\{e_0t, t\}$ is $1/3$. In particular L does not satisfy the Atiyah Conjecture 4.22.*

To the author's knowledge there is no example of a group G for which there is a finitely presented $\mathbb{C}G$-module M such that $\dim_{\mathcal{N}(G)}(\mathcal{N}(G) \otimes_{\mathbb{Z}G} M)$ is irrational.

Let $A = \bigoplus_{n \in \mathbb{Z}} \mathbb{Z}/2$. Because this group is locally finite, it satisfies the Atiyah Conjecture for arbitrary groups 4.22, i.e. $\dim_{\mathcal{N}(G)}(\mathcal{N}(G) \otimes_{\mathbb{C}A} M) \in \mathbb{Z}[1/2]$ for every finitely presented $\mathbb{C}A$-module M. On the other hand, each non-negative real number r can be realized as $\dim_{\mathcal{N}(G)}(\mathcal{N}(G) \otimes_{\mathbb{C}A} M)$ for a finitely generated $\mathbb{C}A$-module (see [80, Example 10.13]). Notice that there is no upper bound on the orders of finite subgroups of A, so that this is no contradiction to Theorem 4.2.

5 Flatness Properties of the Group von Neumann Algebra

The proof of next result can be found in [77, Theorem 5.1] or [80, Theorem 6.37].

Theorem 5.1. (Dimension-flatness of $\mathcal{N}(G)$ over $\mathbb{C}G$ for amenable G). *Let G be amenable and M be a $\mathbb{C}G$-module. Then*

$$\dim_{\mathcal{N}(G)}\left(\mathrm{Tor}_p^{\mathbb{C}G}(\mathcal{N}(G), M)\right) = 0 \qquad \text{for } p \geq 1,$$

where we consider $\mathcal{N}(G)$ as an $\mathcal{N}(G)$-$\mathbb{C}G$-bimodule.

It implies using an easy spectral sequence argument

Theorem 5.2 (L^2-Betti numbers and homology in the amenable case). *Let G be an amenable group and X be a G-space. Then*

(i) $b_p^{(2)}(X; \mathcal{N}(G)) = \dim_{\mathcal{N}(G)}\left(\mathcal{N}(G) \otimes_{\mathbb{C}G} H_p^{\mathrm{sing}}(X; \mathbb{C})\right)$;

(ii) Suppose that X is a G-CW-complex with $m(X; G) < \infty$. Then

$$\chi^{(2)}(X) = \sum_{c \in I(X)} (-1)^{\dim(c)} \cdot |G_c|^{-1} = \sum_{p \geq 0} (-1)^p \cdot \dim_{\mathcal{N}(G)}\left(\mathcal{N}(G) \otimes_{\mathbb{C}G} H_p(X; \mathbb{C})\right).$$

Further applications of Theorem 5.1 will be discussed in Section 6 and Section 7.

Conjecture 5.3. (Amenability and dimension-flatness of $\mathcal{N}(G)$ over $\mathbb{C}G$). *A group G is amenable if and only if for every $\mathbb{C}G$-module M*

$$\dim_{\mathcal{N}(G)}\left(\mathrm{Tor}_p^{\mathbb{C}G}(\mathcal{N}(G), M)\right) = 0 \qquad \text{for } p \geq 1$$

holds.

Remark 5.4 (Evidence for Conjecture 5.3). Theorem 5.1 proves the "only if"-statement of Conjecture 5.3. Some evidence for the "if"-statement of Conjecture 5.3 comes from the following fact. Notice that a group which contains a non-abelian free group as a subgroup, cannot be amenable.

Suppose that G contains a free group $\mathbb{Z}*\mathbb{Z}$ of rank 2 as a subgroup. Notice that $S^1 \vee S^1$ is a model for $B(\mathbb{Z}*\mathbb{Z})$. Its cellular $\mathbb{C}[\mathbb{Z}*\mathbb{Z}]$-chain complex yields an exact sequence $0 \to \mathbb{C}[\mathbb{Z}*\mathbb{Z}]^2 \to \mathbb{C}[\mathbb{Z}*\mathbb{Z}] \to \mathbb{C} \to 0$, where $\mathbb{C}$ is equipped with the trivial $\mathbb{Z}*\mathbb{Z}$-action. One easily checks $b_1^{(2)}(\widetilde{S^1 \vee S^1}) = -\chi(S^1 \vee S^1) = 1$. This implies

$$\dim_{\mathcal{N}(\mathbb{Z}*\mathbb{Z})}\left(\operatorname{Tor}_1^{\mathbb{C}[\mathbb{Z}*\mathbb{Z}]}(\mathcal{N}(\mathbb{Z}*\mathbb{Z}), \mathbb{C})\right) = 1.$$

We conclude from Theorem 1.18 (i)

$$\mathcal{N}(G) \otimes_{\mathcal{N}(\mathbb{Z}*\mathbb{Z})} \operatorname{Tor}_1^{\mathbb{C}[\mathbb{Z}*\mathbb{Z}]}(\mathcal{N}(\mathbb{Z}*\mathbb{Z}), \mathbb{C}) = \operatorname{Tor}_1^{\mathbb{C}G}(\mathcal{N}(G), \mathbb{C}G \otimes_{\mathbb{C}[\mathbb{Z}*\mathbb{Z}]} \mathbb{C}).$$

Theorem 1.18 (ii) implies

$$\dim_{\mathcal{N}(G)}\left(\operatorname{Tor}_1^{\mathbb{C}G}(\mathcal{N}(G), \mathbb{C}G \otimes_{\mathbb{C}[\mathbb{Z}*\mathbb{Z}]} \mathbb{C})\right) = 1.$$

One may ask for which groups the von Neumann algebra $\mathcal{N}(G)$ is flat as a $\mathbb{C}G$-module. This is true if G is virtually cyclic, i.e. G is finite or contains $\mathbb{Z}$ as a normal subgroup of finite index. There is some evidence for the following conjecture (see [77, Remark 5.15]).

Conjecture 5.5 (Flatness of $\mathcal{N}(G)$ over $\mathbb{C}G$). *The group von Neumann algebra $\mathcal{N}(G)$ is flat over $\mathbb{C}G$ if and only if G is virtually cyclic.*

6 Applications to Group Theory

Recall the Definition 2.1 of the L^2-Betti numbers of a group G by $b_p^{(2)}(G) := b_p^{(2)}(EG; \mathcal{N}(G))$. In this section we present tools for and examples of computations of the L^2-Betti numbers and discuss applications to group theory. We will explain in Remark 7.8 that for a torsion-free group with a model of finite type for BG the knowledge of $b_p^{(2)}(G; \mathcal{N}(G))$ is the same as the knowledge of the reduced L^2-homology $H_p^{(2)}(EG, l^2(G))$, or, equivalently, of $\mathbf{P}H_p^G(EG; \mathcal{N}(G))$ if G satisfies the Atiyah Conjecture 4.1 for $(G, 1, \mathbb{Q})$.

6.1 L^2-Betti Numbers of Groups

Theorem 2.7 implies:

Theorem 6.1 (L^2-Betti numbers and Betti numbers of groups). *In the sequel we use the conventions $0 \cdot \infty = 0$, $r \cdot \infty = \infty$ for $r \in (0, \infty]$ and $r + \infty = \infty$ for $r \in [0, \infty]$ and put $|G|^{-1} = 0$ for $|G| = \infty$. Let $G_1, G_2, \ldots$ be a sequence of non-trivial groups.*

(i) *Free amalgamated products*
For $r \in \{2, 3, \ldots\} \amalg \{\infty\}$ we get

$$\begin{aligned}
b_0^{(2)}(*_{i=1}^r G_i) &= 0; \\
b_1^{(2)}(*_{i=1}^r G_i) &= \begin{cases} r - 1 + \sum_{i=1}^r \left(b_1^{(2)}(G_i) - \frac{1}{|G_i|} \right) & \text{, if } r < \infty; \\ \infty & \text{, if } r = \infty; \end{cases} \\
b_p^{(2)}(*_{i=1}^r G_i) &= \sum_{i=1}^r b_p^{(2)}(G_i) \qquad \text{for } p \geq 2; \\
b_p(*_{i=1}^r G_i) &= \sum_{i=1}^r b_p(G_i) \qquad \text{for } p \geq 1;
\end{aligned}$$

(ii) *Künneth formula*

$$\begin{aligned}
b_p^{(2)}(G_1 \times G_2) &= \sum_{i=0}^p b_i^{(2)}(G_1) \cdot b_{p-i}^{(2)}(G_2); \\
b_p(G_1 \times G_2) &= \sum_{i=0}^p b_i(G_1) \cdot b_{p-i}(G_2);
\end{aligned}$$

(iii) *Restriction to subgroups of finite index*
For a subgroup $H \subseteq G$ of finite index $[G : H]$ we get

$$b_p^{(2)}(H) = [G : H] \cdot b_p^{(2)}(G);$$

(iv) *Extensions with finite kernel*
Let $1 \to H \to G \to Q \to 1$ be an extension of groups with finite H. Then

$$b_p^{(2)}(Q) = |H| \cdot b_p^{(2)}(G);$$

(v) *Zero-th L^2-Betti number*
We have $b_0^{(2)}(G) = 0$ for $|G| = \infty$ and $b_0^{(2)}(G) = |G|^{-1}$ for $|G| < \infty$.

Example 6.2 (Independence of L^2-Betti numbers and Betti numbers). Given an integer $l \geq 1$ and a sequence $r_1, r_2, \ldots, r_l$ of non-negative rational numbers, we can construct a group G such that BG is of finite type and

$$\begin{aligned} b_p^{(2)}(G) &= \begin{cases} r_p & \text{for } 1 \leq p \leq l; \\ 0 & \text{for } l+1 \leq p; \end{cases} \\ b_p(G) &= 0 \qquad \text{for } p \geq 1, \end{aligned}$$

holds as follows.

For integers $m \geq 0$, $n \geq 1$ and $i \geq 1$ define

$$G_i(m,n) = \mathbb{Z}/n \times \left(*_{k=1}^{2m+2}\mathbb{Z}/2\right) \times \left(\prod_{j=1}^{i-1} *_{l=1}^{4}\mathbb{Z}/2\right)$$

One easily checks using Theorem 6.1.

$$\begin{aligned} b_i^{(2)}(G_i(m,n)) &= \frac{m}{n}; \\ b_p^{(2)}(G_i(m,n)) &= 0 \qquad \text{for } p \neq i; \\ b_p(G_i(m,n)) &= 0 \qquad \text{for } p \geq 1. \end{aligned}$$

Define the desired group G as follows. For $l = 1$ put $G = G_1(m,n)$ if $r_1 = m/n$. It remains to treat the case $l \geq 2$. Choose integers $n \geq 1$ and $k \geq l$ with $r_1 = \frac{k-2}{n}$. Fix for $i = 2, 3, \ldots, k$ integers $m_i \geq 0$ and $n_i \geq 1$ such that $\frac{m_i}{n \cdot n_i} = r_i$ holds for $1 \leq i \leq l$ and $m_i = 0$ holds for $i > l$. Put

$$G \;=\; \mathbb{Z}/n \times *_{i=2}^{k} G_i(m_i, n_i).$$

One easily checks using Theorem 6.1 that G has the prescribed L^2-Betti numbers and Betti numbers and a model for BG of finite type.

On the other hand we can construct for any sequence n_1, n_2, ... of non-negative integers a CW-complex X of finite type such that $b_p(X) = n_p$ and $b_p^{(2)}(\widetilde{X}) = 0$ holds for $p \geq 1$, namely take

$$X = B(\mathbb{Z}/2 * \mathbb{Z}/2) \times \bigvee_{p=1}^{\infty} \left(\bigvee_{i=1}^{n_p} S^p\right).$$

This example shows by considering the $(l+1)$-skeleton that for a finite connected CW-complex X the only general relation between the L^2-Betti numbers $b_p^{(2)}(\widetilde{X})$ of its universal covering $\widetilde{X}$ and the Betti numbers $b_p(X)$ of X is given by the Euler-Poincaré formula (see Theorem 2.7 (x))

$$\sum_{p \geq 0} (-1)^p \cdot b_p^{(2)}(\widetilde{X}) \;=\; \chi(X) \;=\; \sum_{p \geq 0} (-1)^p \cdot b_p(X).$$

6.2 Vanishing of L^2-Betti Numbers of Groups

Let d be a non-negative integer or $d = \infty$. In this subsection we want to investigate the class of groups

$$\mathcal{B}_d \quad := \quad \{G \mid b_p^{(2)}(G) = 0 \text{ for } 0 \leq p \leq d\}. \tag{6.3}$$

Notice that $\mathcal{B}_0$ is the class of infinite groups by Theorem 6.1 (v).

Theorem 6.4. *Let d be a non-negative integer or $d = \infty$. Then:*

(i) The class $\mathcal{B}_\infty$ contains all infinite amenable groups;

(ii) If G contains a normal subgroup H with $H \in \mathcal{B}_d$, then $G \in \mathcal{B}_d$;

(iii) If G is the union of a directed system of subgroups $\{G_i \mid i \in I\}$ such that each G_i belongs to $\mathcal{B}_d$, then $G \in \mathcal{B}_d$;

(iv) Suppose that there are groups G_1 and G_2 and group homomorphisms $\phi_i \colon G_0 \to G_i$ for $i = 1, 2$ such that ϕ_1 and ϕ_2 are injective, G_0 belongs to $\mathcal{B}_{d-1}$, G_1 and G_2 belong to $\mathcal{B}_d$ and G is the amalgamated product $G_1 *_{G_0} G_2$ with respect to ϕ_1 and ϕ_2. Then G belongs to $\mathcal{B}_d$;

(v) Let $1 \to H \to G \to K \to 1$ be an exact sequence of groups such that $b_p^{(2)}(H)$ is finite for all $p \leq d$. Suppose that K is infinite amenable or suppose that BK has finite d-skeleton and there is an injective endomorphism $j \colon K \to K$ whose image has finite index, but is not equal to K. Then $G \in \mathcal{B}_d$;

(vi) Let $1 \to H \to G \to K \to 1$ be an exact sequence of groups such that $H \in \mathcal{B}_{d-1}$, $b_d^{(2)}(H) < \infty$ and K contains an element of infinite order or finite subgroups of arbitrary large order. Then $G \in \mathcal{B}_d$;

(vii) Let $1 \to H \to G \to K \to 1$ be an exact sequence of infinite countable groups such that $b_1^{(2)}(H) < \infty$. Then $G \in \mathcal{B}_1$.

Proof. (i) We get $b_p^{(2)}(G) = 0$ for $p = 0$ from Theorem 6.1 (v). The case $p \geq 1$ follows from Theorem 5.2 (i) since $H_p^{\text{sing}}(EG; \mathbb{C}) = 0$ for $p \geq 1$.

(ii) Apply Theorem 3.11 (ii) to the fibration $BH \to BG \to B(G/H)$.

(iii) The proof is based on a colimit argument. See [80, Theorem 7.2 (3)].

(iv) The proof is based on a Mayer-Vietoris argument. See [80, Theorem 7.2 (4)].

(v) See [80, Theorem 7.2 (5)].

(vi) This follows from Theorem 3.11 (i) applied to the fibration $BH \to BG \to B(G/H)$.

(vii) This is proved by Gaboriau [38, Theorem 6.8]. □

More information about the vanishing of the first L^2-Betti number can be found for instance in [4]. Obviously the following is true

Lemma 6.5. *If G belongs to $\mathcal{B}_\infty$, then $\chi^{(2)}(G) = 0$.*

Remark 6.6 (The Theorem of Cheeger and Gromov). We rediscover from Theorem 6.4 the result of Cheeger and Gromov [15] that all the L^2-Betti numbers of an infinite amenable group G vanish. A detailed comparison of our approach and the approach by Cheeger and Gromov to L^2-Betti numbers can be found in [80, Remark 6.76].

Remark 6.7 (Advantage of the general definition of L^2-Betti numbers). Recall that we have given criterions for $G \in \mathcal{B}_\infty$ in Theorem 6.4. Now it becomes clear why it is worth while to extend the classical notion of the Euler characteristic $\chi(G) := \chi(BG)$ for groups G with finite BG to arbitrary groups. For instance it may very well happen for a group G with finite BG that G contains a normal group H which is not even finitely generated and has in particular no finite model for BH and which belongs to $\mathcal{B}_\infty$ (for instance, H is amenable). Then the classical Euler characteristic is not defined any more for H, but we can still conclude that the classical Euler characteristic of G vanishes by Remark 2.19, Theorem 6.4 and Lemma 6.5.

6.3 L^2-Betti Numbers of Some Specific Groups

Example 6.8 (Thompson's group). Next we explain the following observation about *Thompson's group F*. It is the group of orientation preserving dyadic PL-automorphisms of $[0,1]$, where dyadic means that all slopes are integral powers of 2 and the break points are contained in $\mathbb{Z}[1/2]$. It has the presentation

$$F = \langle x_0, x_1, x_2, \ldots \mid x_i^{-1} x_n x_i = x_{n+1} \text{ for } i < n \rangle.$$

This group has some very interesting properties. Its classifying space BF is of finite type [8] but is not homotopy equivalent to a finite dimensional CW-complex since F contains $\mathbb{Z}^n$ as a subgroup for all $n \geq 0$ [8, Proposition 1.8]. It is not elementary amenable and does not contain a subgroup which is free on two generators [7], [10]. Hence it is a very interesting question whether F is amenable or not. We conclude from Theorem 6.4 (i) that a necessary condition for F to be amenable is that $b_p^{(2)}(F)$ vanishes for all $p \geq 0$. By [80, Theorem 7.10] this condition is satisfied.

Example 6.9 (Artin groups). Davis and Leary [19] compute for every Artin group A the reduced L^2-cohomology and thus the L^2-Betti numbers of the universal covering $\widetilde{S_A}$ of its *Salvetti complex* S_A. The Salvetti complex S_A is a CW-complex which is conjectured to be a model for the classifying space BA of A. This conjecture is known to be true in many cases and implies that the L^2-Betti numbers of A are given by the L^2-Betti numbers of $\widetilde{S_A}$.

Example 6.10 (Right angled Coxeter groups). The L^2-homology and the L^2-Betti numbers of right angled Coxeter groups are treated by Davis and Okun [20]. More details will be given in Remark 9.6.

Example 6.11 (Fundamental groups of surfaces and 3-manifolds). Let G be the fundamental group of a compact orientable surface F_g^d of genus g with d boundary components. Suppose that G is non-trivial which is equivalent to the condition that $d \geq 1$ or $g \geq 1$. Then F_g^d is a model for BG and we have computed $b_p^{(2)}(G) = b_p^{(2)}(\widetilde{F_g^d})$ in Subsection 3.3.

Let G be the fundamental group of a compact orientable 3-manifold M. The case $|G| < \infty$ is clear, since then the universal covering is homotopy equivalent to a sphere or contractible. So let us assume $|G| = \infty$. Under the condition that M in non-exceptional, we have computed $b_p^{(2)}(\widetilde{M})$ in Theorem 3.4. If M is prime, then either $M = S^1 \times S^2$ and $G = \mathbb{Z}$ and $b_p^{(2)}(G) = 0$ for all $p \geq 0$ or M is irreducible, in which case M is aspherical and $b_p^{(2)}(G) = b_p^{(2)}(\widetilde{M})$.

Suppose that M is not prime. Then still $b_1^{(2)}(G) = b_1^{(2)}(\widetilde{M})$ by Theorem 2.7 (i)a since the classifying map $M \to BG$ is 2-connected. Suppose the prime decomposition of M looks like $M = \#_{i=1}^r M_i$. Then $G = *_{i=1}^r G_i$ for $G_i = \pi_1(M_i)$. We know $b_p^{(2)}(G_i)$ for each i if each M_i is non-exceptional and we get $b_p^{(2)}(G) = \sum_{i=1}^r b_p^{(2)}(G_i)$ for $p \geq 2$ from Theorem 6.1 (i).

Example 6.12 (One relator groups). Let $G = \langle g_1, g_2, \dots g_s \mid R \rangle$ be a torsion-free one relator group for $s \in \{2, 3 \dots\} \amalg \{\infty\}$ and one non-trivial relation R. Then

$$b_p^{(2)}(G) = \begin{cases} 0 & \text{if } p \neq 1; \\ s-2 & \text{if } p = 1 \text{ and } s < \infty; \\ \infty & \text{if } p = 1 \text{ and } s = \infty. \end{cases}$$

We only treat the case $s < \infty$, the general case is obtained from it by taking the free amalgamated product with a free group. Because the 2-dimensional CW-complex X associated to the given presentation is a model for BG (see [85, chapter III §§9 -11]) and satisfies $\chi(X) = s - 2$, it suffices to prove $b_2^{(2)}(G) = 0$. We sketch the argument of Dicks and Linnell for this claim. Howie [56] has shown that such a group G is locally indicable and hence left-orderable. A result of Linnell [64, Theorem 2] for left-orderable groups says that an element $\alpha \in \mathbb{C}G$ with $\alpha \neq 0$ is a non-zero-divisor in $\mathcal{U}(G)$. This implies that the second differential $c_2^{\mathcal{U}(G)}$ in the chain complex $\mathcal{U}(G) \otimes_{\mathbb{C}G} C_*(EG)$ is injective. Since $\mathcal{U}(G)$ is flat over $\mathcal{N}(G)$, we get from Theorem 4.5

$$\begin{aligned} b_2^{(2)}(G) &= \dim_{\mathcal{N}(G)} \left(H_p^G(EG; \mathcal{N}(G))\right) = \dim_{\mathcal{U}(G)} \left(H_p^G(EG; \mathcal{U}(G))\right) \\ &= \dim_{\mathcal{U}(G)} \left(\ker\left(c_2^{\mathcal{U}(G)}\right)\right) = 0. \end{aligned}$$

Linnell has an extensions of this argument to non-torsion-free one-relator groups G with $s \geq 2$ generators. (The case $s = 1$ is obvious.) Such a group contains a cyclic subgroup $\mathbb{Z}/k$ such that any finite subgroup is subconjugated to $\mathbb{Z}/k$ and then

$$b_p^{(2)}(G) = \begin{cases} 0 & \text{if } p \neq 1; \\ s - 1 - \frac{1}{k} & \text{if } p = 1 \text{ and } s < \infty; \\ \infty & \text{if } p = 1 \text{ and } s = \infty. \end{cases}$$

Example 6.13 (Lattices). Let L be a connected semisimple Lie group with finite center such that its Lie algebra has no compact ideal. Let $G \subseteq L$ be a lattice, i.e. a discrete subgroup of finite covolume. We want to compute its L^2-Betti numbers. There is a subgroup $G_0 \subseteq G$ of finite index which is torsion-free. Since $b_p^{(2)}(G) = [G : G_0] \cdot b_p^{(2)}(G_0)$, it suffices to treat the case $G = G_0$, i.e. $G \subseteq L$ is a torsion-free lattice.

Let $K \subseteq L$ be a maximal compact subgroup. Put $M = G\backslash L/K$. Then the space $L/K = \widetilde{M}$ is a symmetric space of non-compact type. We have already mentioned in Theorem 3.5 that the work of Borel [6] implies for cocompact G that $b_p^{(2)}(G) = b_p^{(2)}(\widetilde{M}) \neq 0$ if and only if $\text{f-rk}(\widetilde{M}) = 0$ and $2p = \dim(M)$. This is actually true without the condition "cocompact", because the condition "finite covolume" is enough.

Next we deal with the general case of a connected Lie group L. Let $\text{Rad}(L)$ be its radical. One can choose a compact normal subgroup $K \subseteq L$ such that $R = \text{Rad}(L) \times K$ is a normal subgroup of L and the quotient $L_1 = L/R$ is a semisimple Lie group such that its Lie algebra has no compact ideal. Then $G_1 = L/L \cap R$ is a lattice in L_1 and $G \cap R$ is a lattice in R. The group $G \cap R$ is a normal amenable subgroup of G. If $G \cap R$ is infinite, we get $b_p^{(2)}(G) = 0$ for all $p \geq 0$ from Theorem 6.4. If $G \cap R$ is finite, we get $b_p^{(2)}(G) = |G \cap R|^{-1} \cdot b_p^{(2)}(G_1)$ for all $p \geq 0$ from Theorem 6.1 (iv). If the center of L_1 is infinite, the center of G_1 must also be infinite and hence $b_p^{(2)}(G_1) = 0$ for all $p \geq 0$ by Theorem 6.4. Suppose that the center of L_1 is finite. Then we know already how to compute the L^2-Betti numbers of G_1 from the explanation above.

Given a lattice G in a connected Lie group, $b_1^{(2)}(G) > 0$ is true if and only if G is commensurable with a torsion-free lattice in $PSL_2(\mathbb{R})$, or, equivalently commensurable with a surface group for genus ≥ 2 or a finitely generated non-abelian free group (see Eckmann [27] or Lott [68, Theorem 2]).

6.4 Deficiency and L^2-Betti Numbers of Groups

Let G be a finitely presented group. Define its *deficiency* $\text{def}(G)$ to be the maximum $g(P) - r(P)$, where P runs over all presentations P of G and $g(P)$ is the number of generators and $r(P)$ is the number of relations of a presentation P.

Next we reprove the well-known fact that the maximum appearing in the definition of the deficiency does exist.

Lemma 6.14. *Let G be a group with finite presentation*

$$P = \langle s_1, s_2, \dots, s_g \mid R_1, R_2, \dots, R_r \rangle$$

Let $\phi\colon G \to K$ be any group homomorphism. Then

$$\begin{aligned} g(P) - r(P) \;\le\; 1 - b_0^{(2)}(K \times_\phi EG; \mathcal{N}(K)) + b_1^{(2)}(K \times_\phi EG; \mathcal{N}(K)) \\ - b_2^{(2)}(K \times_\phi EG; \mathcal{N}(K)). \end{aligned}$$

Proof. Given a presentation P with g generators and r relations, let X be the associated finite 2-dimensional CW-complex. It has one 0-cell, g 1-cells, one for each generator, and r 2-cells, one for each relation. There is an obvious isomorphism from $\pi_1(X)$ to G so that we can choose a map $f\colon X \to BG$ which induces an isomorphism on the fundamental groups. It induces a 2-connected K-equivariant map $\overline{f}\colon K \times_\phi \widetilde{X} \to K \times_\phi \widetilde{EG}$. We conclude from Theorem 2.7 (i)a

$$\begin{aligned} b_p^{(2)}(K \times_\phi \widetilde{X}; \mathcal{N}(K)) &= b_p^{(2)}(K \times_\phi EG; \mathcal{N}(K)) \qquad \text{for } p = 0, 1; \\ b_2^{(2)}(K \times_\phi \widetilde{X}; \mathcal{N}(K)) &\ge b_2^{(2)}(K \times_\phi EG; \mathcal{N}(K)). \end{aligned}$$

We conclude from the L^2-Euler-Poincaré formula (see Theorem 2.18 (i))

$$\begin{aligned} g - r &= 1 - \chi^{(2)}(K \times_\phi \widetilde{X}; \mathcal{N}(K)) \\ &= 1 - b_0^{(2)}(K \times_\phi \widetilde{X}; \mathcal{N}(K)) + b_1^{(2)}(K \times_\phi \widetilde{X}; \mathcal{N}(K)) \\ &\qquad - b_2^{(2)}(K \times_\phi \widetilde{X}; \mathcal{N}(K)) \\ &\le 1 - b_0^{(2)}(K \times_\phi EG; \mathcal{N}(K)) + b_1^{(2)}(K \times_\phi EG; \mathcal{N}(K)) \\ &\qquad - b_2^{(2)}(K \times_\phi EG; \mathcal{N}(K)). \quad \square \end{aligned}$$

Example 6.15 (Deficiency of some groups)**.** Sometimes the deficiency is realized by the "obvious" presentation. For instance the deficiency of a free group $\langle s_1, s_2, \dots, s_g \mid \emptyset \rangle$ on g letters is indeed g. The cyclic group $\mathbb{Z}/n$ of order n has the presentation $\langle t \mid t^n = 1 \rangle$ and its deficiency is 0. The group $\mathbb{Z}/n \times \mathbb{Z}/n$ has the presentation $\langle s, t \mid s^n, t^n, [s,t] \rangle$ and its deficiency is -1.

Remark 6.16 (Non-additivity of the deficiency)**.** The deficiency is not additive under free products by the following example which is a special case of a more general example due to Hog, Lustig and Metzler [55, Theorem 3 on page 162]. The group $(\mathbb{Z}/2 \times \mathbb{Z}/2) * (\mathbb{Z}/3 \times \mathbb{Z}/3)$ has the obvious presentation

$$\langle s_0, t_0, s_1, t_1 \mid s_0^2 = t_0^2 = [s_0, t_0] = s_1^3 = t_1^3 = [s_1, t_1] = 1 \rangle$$

One may think that its deficiency is -2. However, it turns out that its deficiency is -1, realized by the following presentation

$$\langle s_0, t_0, s_1, t_1 \mid s_0^2 = 1, [s_0, t_0] = t_0^2, s_1^3 = 1, [s_1, t_1] = t_1^3, t_0^2 = t_1^3 \rangle.$$

This shows that it is important to get upper bounds on the deficiency of groups. Writing down presentations gives lower bounds, but it is not clear whether a given presentation realizes the deficiency.

Lemma 6.17. *Let G be a finitely presented group and let $\phi\colon G \to K$ be a homomorphism such that $b_1^{(2)}(K \times_\phi EG; \mathcal{N}(K)) = 0$. Then*

(i) $\operatorname{def}(G) \leq 1$*;*

(ii) Let M be a closed oriented 4-manifold with G as fundamental group. Then

$$|\operatorname{sign}(M)| \leq \chi(M).$$

Proof. (i) This follows directly from Lemma 6.14.
(ii) This is a consequence of the L^2-Signature Theorem due to Atiyah [2]. Details of the proof can be found in [80, Lemma 7.22]. □

Theorem 6.18. *Let $1 \to H \xrightarrow{i} G \xrightarrow{q} K \to 1$ be an exact sequence of infinite groups. Suppose that G is finitely presented and one of the following conditions is satisfied:*

(i) $b_1^{(2)}(H) < \infty$*;*

(ii) The classical first Betti number of H satisfies $b_1(H) < \infty$ and K belongs to $\mathcal{B}_1$.

Then

(i) $\operatorname{def}(G) \leq 1$*;*

(ii) Let M be a closed oriented 4-manifold with G as fundamental group. Then

$$|\operatorname{sign}(M)| \leq \chi(M).$$

Proof. If condition (i) is satisfied, then $b_p^{(2)}(G) = 0$ for $p = 0, 1$ by Theorem 6.4 (vii), and the claim follows from Lemma 6.17.

Suppose that condition (ii) is satisfied. There is a spectral sequence converging to $H_{p+q}^K(K \times_q EG; \mathcal{N}(K))$ with E^2-term

$$E_{p,q}^2 = \operatorname{Tor}_p^{\mathbb{C}K}(H_q(BH; \mathbb{C}), \mathcal{N}(K))$$

[108, Theorem 5.6.4 on page 143]. Since $H_q(BH; \mathbb{C})$ is $\mathbb{C}$ with the trivial K-action for $q = 0$ and finite dimensional as complex vector space by assumption for $q=1$, we conclude $\dim_{\mathcal{N}(K)}(E^2_{p,q})=0$ for $p+q=1$ from the assumption $b_1^{(2)}(K)=0$. This implies $b_1^{(2)}(K\times_q EG;\mathcal{N}(K))=0$ and the claim follows from Lemma 6.17. □

Theorem 6.18 generalizes results in [29], [58], where also some additional information is given. Furthermore see [49], [63]. We mention the result of Hitchin [54] that a connected closed oriented smooth 4-manifold which admits an Einstein metric satisfies the stronger inequality $|\operatorname{sign}(M)|\le\frac{2}{3}\cdot\chi(M)$.

Finally we mention the following result of Lott [68, Theorem 2] (see also [28]) which generalizes a result of Lubotzky [70]. The statement we present here is a slight improvement of Lott's result due to Hillman [53].

Theorem 6.19 (Lattices of positive deficiency). *Let L be a connected Lie group. Let G be a lattice in L. If $\operatorname{def}(G)>0$, then one of the following assertions holds:*

(i) G is a lattice in $PSL_2(\mathbb{C})$;

(ii) $\operatorname{def}(G)=1$. Moreover, either G is isomorphic to a torsion-free non-uniform lattice in $\mathbb{R}\times PSL_2(\mathbb{R})$ or $PSL_2(\mathbb{C})$, or G is $\mathbb{Z}$ or $\mathbb{Z}^2$.

7 G- and K-Theory

In this section we discuss the projective class group $K_0(\mathcal{N}(G))$ of a group von Neumann algebra. We present applications of its computation to $G_0(\mathbb{C}G)$ and the Whitehead group $\operatorname{Wh}(G)$ of a group G.

7.1 The K_0-group of a Group von Neumann Algebra

In this subsection we want to investigate the projective class group of a group von Neumann algebra.

Definition 7.1 (Definition of $K_0(R)$ and $G_0(R)$). *Let R be an (associative) ring (with unit). Define its* projective class group *$K_0(R)$ to be the abelian group whose generators are isomorphism classes $[P]$ of finitely generated projective R-modules P and whose relations are $[P_0]+[P_2]=[P_1]$ for any exact sequence $0\to P_0\to P_1\to P_2\to 0$ of finitely generated projective R-modules.*

Define the Grothendieck group of finitely generated modules *$G_0(R)$ analogously but replace finitely generated projective with finitely generated.*

The group K_0 is known for any von Neumann algebra (see for instance [80, Subsection 9.2.1]. For simplicity we only treat the von Neumann algebra $\mathcal{N}(G)$ of a group here.

The next result is taken from [60, Theorem 7.1.12 on page 462, Proposition 7.4.5 on page 483, Theorem 8.2.8 on page 517, Proposition 8.3.10 on page 525, Theorem 8.4.3 on page 532].

Theorem 7.2 (The universal trace). *There is a map*

$$\mathrm{tr}^u_{\mathcal{N}(G)} : \mathcal{N}(G) \to \mathcal{Z}(\mathcal{N}(G))$$

into the center $\mathcal{Z}(\mathcal{N}(G))$ of $\mathcal{N}(G)$ called the center valued trace *or* universal trace *of $\mathcal{N}(G)$, which is uniquely determined by the following two properties:*

(i) *$\mathrm{tr}^u_{\mathcal{N}(G)}$ is a trace with values in the center, i.e. $\mathrm{tr}^u_{\mathcal{N}(G)}$ is $\mathbb{C}$-linear, for $a \in \mathcal{N}(G)$ with $a \geq 0$ we have $\mathrm{tr}^u_{\mathcal{N}(G)}(a) \geq 0$ and $\mathrm{tr}^u_{\mathcal{N}(G)}(ab) = \mathrm{tr}^u_{\mathcal{N}(G)}(ba)$ for all $a, b \in \mathcal{N}(G)$;*

(ii) *$\mathrm{tr}^u_{\mathcal{N}(G)}(a) = a$ for all $a \in \mathcal{Z}(\mathcal{N}(G))$.*

The map $\mathrm{tr}^u_{\mathcal{N}(G)}$ has the following further properties:

(iii) *$\mathrm{tr}^u_{\mathcal{N}(G)}$ is faithful, i.e. $\mathrm{tr}^u_{\mathcal{N}(G)}(a) = 0 \Leftrightarrow a = 0$ for $a \in \mathcal{N}(G), a \geq 0$;*

(iv) *$\mathrm{tr}^u_{\mathcal{N}(G)}$ is normal, i.e. for a monotone increasing net $\{a_i \mid i \in I\}$ of positive elements a_i with supremum a we have $\mathrm{tr}^u_{\mathcal{N}(G)}(a) = \sup\{\mathrm{tr}\,(a_i) \mid i \in I\}$, or, equivalently, $\mathrm{tr}^u_{\mathcal{N}(G)}$ is continuous with respect to the ultra-weak topology on $\mathcal{N}(G)$;*

(v) *$||\mathrm{tr}^u_{\mathcal{N}(G)}(a)|| \leq ||a||$ for $a \in \mathcal{N}(G)$;*

(vi) *$\mathrm{tr}^u_{\mathcal{N}(G)}(ab) = a\,\mathrm{tr}^u_{\mathcal{N}(G)}(b)$ for all $a \in \mathcal{Z}(\mathcal{N}(G))$ and $b \in \mathcal{N}(G)$;*

(vii) *Let p and q be projections in $\mathcal{N}(G)$. Then $p \sim q$, i.e. $p = uu^*$ and $q = u^*u$ for some element $u \in \mathcal{N}(G)$, if and only if $\mathrm{tr}^u_{\mathcal{N}(G)}(p) = \mathrm{tr}^u_{\mathcal{N}(G)}(q)$;*

(viii) *Any linear functional $f: \mathcal{N}(G) \to \mathbb{C}$ which is continuous with respect to the norm topology on $\mathcal{N}(G)$ and which is central, i.e. $f(ab) = f(ba)$ for all $a, b \in \mathcal{N}(G)$ factorizes as*

$$\mathcal{N}(G) \xrightarrow{\mathrm{tr}^u_{\mathcal{N}(G)}} \mathcal{Z}(\mathcal{N}(G)) \xrightarrow{f|_{\mathcal{Z}(\mathcal{N}(G))}} \mathbb{C}.$$

Definition 7.3 (Center valued dimension). *For a finitely generated projective $\mathcal{N}(G)$-module P define its* center valued von Neumann dimension *by*

$$\dim^u_{\mathcal{N}(G)}(P) := \sum_{i=1}^{n} \mathrm{tr}^u_{\mathcal{N}(G)}(a_{i,i}) \in \mathcal{Z}(\mathcal{N}(G))^{\mathbb{Z}/2} = \{a \in \mathcal{Z}(\mathcal{N}(G)) \mid a = a^*\}$$

for any matrix $A = (a_{i,j})_{i,j} \in M_n(\mathcal{N}(G))$ with $A^2 = A$ such that the image of the map $r_A : \mathcal{N}(G)^n \to \mathcal{N}(G)^n$ induced by right multiplication with A is $\mathcal{N}(G)$-isomorphic to P.

There is a classification of von Neumann algebras into certain types. We only need to know what the type of a group von Neumann algebra is.

Lemma 7.4. *Let G be a discrete group. Let G_f be the normal subgroup of G consisting of elements $g \in G$ whose centralizer has finite index (or, equivalently, whose conjugacy class (g) consists of finitely many elements). Then:*

(i) The group von Neumann algebra $\mathcal{N}(G)$ is of type I_f if and only if G is virtually abelian;

(ii) The group von Neumann algebra $\mathcal{N}(G)$ is of type II_1 if and only if the index of G_f in G is infinite;

(iii) Suppose that G is finitely generated. Then $\mathcal{N}(G)$ is of type I_f if G is virtually abelian, and of type II_1 if G is not virtually abelian;

(iv) The group von Neumann algebra $\mathcal{N}(G)$ is a factor, *i.e. its center consists of $\{r \cdot 1_{\mathcal{N}(G)} \mid r \in \mathbb{C}\}$, if and only if G_f is the trivial group.*

Proof. (i) This is proved in [61], [103].

(ii) This is proved in [61],[88].

(iii) This follows from assertions (i) and (ii) since for finitely generated G the group G_f has finite index in G if and only if G is virtually abelian.

(iv) This follows from [23, Proposition 5 in III.7.6 on page 319]. □

The next result follows from [60, Theorem 8.4.3 on page 532, Theorem 8.4.4 on page 533].

Theorem 7.5 (K_0 of finite von Neumann algebras). *Let G be a group.*

(i) The following statements are equivalent for two finitely generated projective $\mathcal{N}(G)$-modules P and Q:

(a) P and Q are $\mathcal{N}(G)$-isomorphic;

(b) P and Q are stably $\mathcal{N}(G)$-isomorphic, i.e. $P \oplus V$ and $Q \oplus V$ are $\mathcal{N}(G)$-isomorphic for some finitely generated projective $\mathcal{N}(G)$-module V;

(c) $\dim^u_{\mathcal{N}(G)}(P) = \dim^u_{\mathcal{N}(G)}(Q)$;

(d) $[P] = [Q]$ in $K_0(\mathcal{N}(G))$;

(ii) The center valued dimension induces an injection

$$\dim^u_{\mathcal{N}(G)}\colon K_0(\mathcal{N}(G)) \to \mathcal{Z}(\mathcal{N}(G))^{\mathbb{Z}/2} = \{a \in \mathcal{Z}(\mathcal{N}(G)) \mid a = a^*\},$$

where the group structure on $\mathcal{Z}(\mathcal{N}(G))^{\mathbb{Z}/2}$ comes from addition. If $\mathcal{N}(G)$ is of type II_1, this map is an isomorphism. □

Remark 7.6 (Group von Neumann algebras and representation theory). Theorem 7.5 shows that the group von Neumann algebra is the right generalization of the complex group ring from finite groups to infinite groups if one is concerned with representation theory of finite groups. Namely, let G be a finite group. Recall that a finite dimensional complex G-representation V is the same as a finitely generated $\mathbb{C}G$-module and that $K_0(\mathbb{C}G)$ is the same as the complex representation ring. Moreover, two finite dimensional G-representations V and W are linearly G-isomorphic if and only if they have the same character. Recall that the character is a class function. One easily checks that the complex vector space of class functions on a finite group G is the same as the center $\mathcal{Z}(\mathbb{C}G)$ and that the character of V contains the same information as $\dim^u_{\mathcal{N}(G)}(V)$.

Remark 7.7 (Factors). Suppose that $\mathcal{N}(G)$ is a factor, i.e. its center consists of $\{r \cdot 1_{\mathcal{N}(G)} \mid r \in \mathbb{C}\}$. By Lemma 7.4 (iv) this is the case if and only if G_f is the trivial group. Then $\dim_{\mathcal{N}(G)} = \dim^u_{\mathcal{N}(G)}$ and two finitely generated projective $\mathcal{N}(G)$-modules P and Q are $\mathcal{N}(G)$-isomorphic if and only if $\dim_{\mathcal{N}(G)}(P) = \dim_{\mathcal{N}(G)}(Q)$ holds. This has the consequence that for a free G-CW-complex X of finite type the p-th L^2-Betti number determines the isomorphism type of $\mathbf{P}H_p^G(X;\mathcal{N}(G))$. In particular we must have $\mathbf{P}H_p^G(X;\mathcal{N}(G)) \cong_{\mathcal{N}(G)} \mathcal{N}(G)^n$ provided that $n = b_p^{(2)}(X;\mathcal{N}(G))$ is an integer. If one prefers to work with reduced L^2-homology, this is equivalent to the statement that $H_p^{(2)}(X;l^2(G))$ is isometrically G-linearly isomorphic to $l^2(G)^n$ provided that $n = b_p^{(2)}(X;\mathcal{N}(G))$ is an integer.

Remark 7.8 (The reduced L^2-cohomology of torsion-free groups). Let G be a torsion-free group. Suppose that it satisfies the Atiyah Conjecture 4.1 for $(G, 1, \mathbb{Q})$. Suppose that there is a model for BG of finite type. Then we get for all p that $\mathbf{P}H_p^G(EG;\mathcal{N}(G)) \cong_{\mathcal{N}(G)} \mathcal{N}(G)^n$, or, equivalently, that $H_p^{(2)}(X;l^2(G))$ is isometrically G-linearly isomorphic to $l^2(G)^n$ if the integer n is given by $n = b_p^{(2)}(X;\mathcal{N}(G))$. This claim is proved in [80, solution to Exercise 10.11 on page 546].

We mention that the inclusion $i\colon \mathcal{N}(G) \to \mathcal{U}(G)$ induces an isomorphism

$$K_0(\mathcal{N}(G)) \xrightarrow{\cong} K_0(\mathcal{U}(G)).$$

The Farrell-Jones Conjecture for $K_0(\mathbb{C}G)$, the Bass Conjecture and the passage in K_0 from $\mathbb{Z}G$ to $\mathbb{C}G$ and to $\mathcal{N}(G)$ is discussed in [80, Section 9.5.2] and [81].

7.2 The K_1- and L-groups of a Group von Neumann Algebra

A complete calculation of the K_1-group and of the L-groups of any von Neumann algebra and of the associated algebra of affiliated operators can be found in [80, Section 9.3 and Section 9.4], [83] and [99].

7.3 Applications to G-theory of Group Rings

Theorem 7.9 (Detecting $G_0(\mathbb{C}G)$ by $K_0(\mathcal{N}(G))$ for amenable groups). *If G is amenable, the map*

$$l\colon G_0(\mathbb{C}G) \to K_0(\mathcal{N}(G)), \qquad [M] \mapsto [\mathbf{P}\mathcal{N}(G) \otimes_{\mathbb{C}G} M]$$

is a well-defined homomorphism. If $f\colon K_0(\mathbb{C}G) \to G_0(\mathbb{C}G)$ is the forgetful map sending $[P]$ to $[P]$ and $i_*\colon K_0(\mathbb{C}G) \to K_0(\mathcal{N}(G))$ is induced by the inclusion $i\colon \mathbb{C}G \to \mathcal{N}(G)$, then the composition $l \circ f$ agrees with i_*.

Proof. This is essentially a consequence of the dimension-flatness of $\mathcal{N}(G)$ over $\mathbb{C}G$ (see Theorem 5.1). Details of the proof can be found in [80, Theorem 9.64]. □

Now one can combine Theorem 7.5 and Theorem 7.9 to detect elements in $G_0(\mathbb{C}G)$ for amenable G. In particular one can show

$$\dim_{\mathbb{Q}}\left(\mathbb{Q} \otimes_{\mathbb{Z}} G_0(\mathbb{C}G)\right) \quad \geq \quad |\operatorname{con}(G)_{f,cf}|, \tag{7.10}$$

where $\operatorname{con}(G)_{f,cf}$ is the set of conjugacy classes (g) of elements $g \in G$ such that g has finite order and (g) contains only finitely many elements. Notice that $\operatorname{con}(G)_{f,cf}$ contains at least one element, namely the unit element e.

Remark 7.11 (The non-vanishing of $[RG]$ in $G_0(RG)$ for amenable groups). A direct consequence of Theorem 7.9 is that for an amenable group G the class $[\mathbb{C}G]$ in $G_0(\mathbb{C}G)$ generates an infinite cyclic subgroup. Namely, the dimension induces a well-defined homomorphism

$$\dim_{\mathcal{N}(G)}\colon G_0(\mathbb{C}G) \to \mathbb{R}, \quad [M] \;\mapsto\; \dim_{\mathcal{N}(G)}\left(\mathcal{N}(G) \otimes_{\mathbb{C}G} M\right),$$

which sends $[\mathbb{C}G]$ to 1. This result has been extended by Elek [32] to finitely generated amenable groups and arbitrary fields F, i.e. there is a well-defined homomorphism $G_0(FG) \to \mathbb{R}$, which sends $[FG]$ to 1 and is given by a certain rank function on finitely generated FG-modules.

The class $[RG]$ in $K_0(RG)$ is never zero for a commutative integral domain R with quotient field $R_{(0)}$. The augmentation $RG \to R$ and the map $K_0(R) \to \mathbb{Z}$, $\;[P] \mapsto \dim_{R_{(0)}}(R_{(0)} \otimes_R P)$ together induce a homomorphism $K_0(RG) \to \mathbb{Z}$ which sends $[RG]$ to 1. A decisive difference between $K_0(RG)$ and $G_0(RG)$ is that $[RG] = 0$ is possible in $G_0(RG)$ as the following example shows.

Example 7.12 (The vanishing of $[RG]$ in $G_0(RG)$ for groups G containing $\mathbb{Z}*\mathbb{Z}$)**.** We abbreviate $F_2 = \mathbb{Z}*\mathbb{Z}$. Suppose that G contains F_2 as a subgroup. Let R be a ring. Then

$$[RG] \;=\; 0 \quad \in G_0(RG)$$

holds by the following argument. Induction with the inclusion $F_2 \to G$ induces a homomorphism $G_0(RF_2) \to G_0(RG)$ which sends $[RF_2]$ to $[RG]$. Hence it suffices to show $[RF_2] = 0$ in $G_0(RF_2)$. The cellular chain complex of the universal covering of $S^1 \vee S^1$ yields an exact sequence of RF_2-modules $0 \to (RF_2)^2 \to RF_2 \to R \to 0$, where R is equipped with the trivial F_2-action. This implies $[RF_2] = -[R]$ in $G_0(RF_2)$. Hence it suffices to show $[R] = 0$ in $G_0(RF_2)$. Choose an epimorphism $f\colon F_2 \to \mathbb{Z}$. Restriction with f defines a homomorphism $G_0(R\mathbb{Z}) \to G_0(RF_2)$. It sends the class of R viewed as trivial $R\mathbb{Z}$-module to the class of R viewed as trivial RF_2-module. Hence it remains to show $[R] = 0$ in $G_0(R\mathbb{Z})$. This follows from the exact sequence $0 \to R\mathbb{Z} \xrightarrow{s-1} R\mathbb{Z} \to R \to 0$ for s a generator of $\mathbb{Z}$ which comes from the cellular $R\mathbb{Z}$-chain complex of $\widetilde{S^1}$.

Remark 7.11 and Example 7.12 give some evidence for

Conjecture 7.13. (Amenability and the regular representation in G-theory). *Let R be a commutative integral domain. Then a group G is amenable if and only if $[RG] \neq 0$ in $G_0(RG)$.*

Remark 7.14 (The Atiyah Conjecture for amenable groups and $G_0(\mathbb{C}G)$)**.** Assume that G is amenable and that there is an upper bound on the orders of finite subgroups of G. Then the Atiyah Conjecture 4.1 for $(G, d, \mathbb{C})$ is true if and only if the image of the map

$$\dim_{\mathcal{N}(G)}\colon G_0(\mathbb{C}G) \to \mathbb{R}, \qquad [M] \mapsto \dim_{\mathcal{N}(G)}(\mathcal{N}(G) \otimes_{\mathbb{C}G} M)$$

is contained in $\{r \in \mathbb{R} \mid d \cdot r \in \mathbb{Z}\}$.

Example 7.15 ($K_0(\mathbb{C}G) \to G_0(\mathbb{C}G)$ is not necessarily surjective)**.** Let $A = \bigoplus_{n\in\mathbb{Z}} \mathbb{Z}/2$. This abelian group is locally finite. Hence the map

$$\bigoplus_{H\subseteq A, |H|<\infty} K_0(\mathbb{C}H) \to K_0(\mathbb{C}A)$$

is surjective and the image of

$$\dim_{\mathcal{N}(G)}\colon K_0(\mathbb{C}A) \to \mathbb{R}, \qquad [P] \mapsto \dim_{\mathcal{N}(A)}(\mathcal{N}(A) \otimes_{\mathbb{C}A} P)$$

is $\mathbb{Z}[1/2]$. On the other hand the argument in [80, Example 10.13] shows that the map

$$\dim_{\mathcal{N}(G)}\colon G_0(\mathbb{C}A) \to \mathbb{R}, \qquad [M] \mapsto \dim_{\mathcal{N}(A)}(\mathcal{N}(A) \otimes_{\mathbb{C}A} M)$$

is surjective. In particular the obvious map $K_0(\mathbb{C}A) \to G_0(\mathbb{C}A)$ is not surjective.

7.4 Applications to the Whitehead Group

The *Whitehead group* $\mathrm{Wh}(G)$ of a group G is the quotient of $K_1(\mathbb{Z}G)$ by the subgroup which consists of elements given by units of the shape $\pm g \in \mathbb{Z}G$ for $g \in G$. Let $i\colon H \to G$ be the inclusion of a normal subgroup $H \subseteq G$. It induces a homomorphism $i_0\colon \mathrm{Wh}(H) \to \mathrm{Wh}(G)$. The conjugation action of G on H and on G induces a G-action on $\mathrm{Wh}(H)$ and on $\mathrm{Wh}(G)$ which turns out to be trivial on $\mathrm{Wh}(G)$. Hence i_0 induces homomorphisms

$$i_1 : \mathbb{Z} \otimes_{\mathbb{Z}G} \mathrm{Wh}(H) \to \mathrm{Wh}(G); \tag{7.16}$$
$$i_2 : \mathrm{Wh}(H)^G \to \mathrm{Wh}(G). \tag{7.17}$$

Theorem 7.18 (Detecting elements in $\mathrm{Wh}(G)$). *Let $i\colon H \to G$ be the inclusion of a normal finite subgroup H into an arbitrary group G. Then the maps i_1 and i_2 defined in* (7.16) *and* (7.17) *have finite kernel.*

Proof. See [80, Theorem 9.38]. □

We emphasize that Theorem 7.18 above holds for all groups G. It seems to be related to the Farrell-Jones Isomorphism Conjecture.

8 Measurable Group Theory

In this section we want to discuss an interesting relation between L^2-Betti numbers and measurable group theory. We begin with formulating the main result.

Definition 8.1 (Measure equivalence). *Two countable groups G and H are called* measure equivalent *if there exist commuting measure-preserving free actions of G and H on some standard Borel space (Ω, μ) with non-zero Borel measure μ such that the actions of both G and H admit measure fundamental domains X and Y of finite measure.*

The triple (Ω, X, Y) is called a measure coupling *of G and H. The index of (Ω, X, Y) is the quotient $\frac{\mu(X)}{\mu(Y)}$.*

Here are some explanations. A *Polish space* is a separable topological space which is metrizable by a complete metric. A *measurable space* $\Omega = (\Omega, \mathcal{A})$ is a set Ω together with a σ-algebra $\mathcal{A}$. It is called a *standard Borel space* if it is isomorphic to a Polish space with its Borel σ-algebra. (The Polish space is not part of the structure, only its existence is required.) More information about this notion of measure equivalence can be found for instance in [36], [37] and [46, 0.5E].

The following result is due to Gaboriau [38, Theorem 6.3]. We will discuss its applications and sketch the proof based on homological algebra and the dimension function due to R. Sauer [100].

Theorem 8.2 (Measure equivalence and L^2-Betti numbers). *Let G and H be two countable groups which are measure equivalent. If $C > 0$ is the index of a measure coupling, then we get for all $p \geq 0$*

$$b_p^{(2)}(G) = C \cdot b_p^{(2)}(H).$$

The general strategy of the proof of Theorem 8.2 is as follows. In the first step one introduces the notion of a standard action $G \curvearrowright X$ and of a weak orbit equivalence of standard actions of index C and shows that two groups G and H are measure equivalent of index C if and only if there exist standard actions $G \curvearrowright X$ and $H \curvearrowright Y$ which are weakly orbit equivalent with index C. In the second step one assigns to a standard action $G \curvearrowright X$ L^2-Betti numbers $b_p^{(2)}(G \curvearrowright X)$, which involve only data that is invariant under orbit equivalence. Hence $b_p^{(2)}(G \curvearrowright X)$ itself depends only on the orbit equivalence class of $G \curvearrowright X$. In order to deal with weak orbit equivalences, one has to investigate the behaviour of the L^2-Betti numbers of $b_p^{(2)}(G \curvearrowright X)$ under restriction. Finally one proves that the L^2-Betti numbers of a standard action $G \curvearrowright X$ agree with the L^2-Betti numbers of G itself.

A version of Theorem 8.2 for L^2-torsion is presented in Conjecture 11.30.

8.1 Measure Equivalence and Quasi-Isometry

Remark 8.3 (Measure equivalence is the measure theoretic version of quasi-isometry). The notion of measure equivalence can be viewed as the measure theoretic analogue of the metric notion of quasi-isometric groups. Namely, two finitely generated groups G_0 and G_1 are *quasi-isometric* if and only if there exist commuting proper (continuous) actions of G_0 and G_1 on some locally compact space such that each action has a cocompact fundamental domain [46, 0.2 C_2' on page 6].

Example 8.4 (Infinite amenable groups). Every countable infinite amenable group is measure equivalent to $\mathbb{Z}$ (see [94]). Since obviously all the L^2-Betti numbers of $\mathbb{Z}$ vanish, Theorem 8.2 implies the result of Cheeger and Gromov that all the L^2-Betti numbers of an infinite amenable group vanish.

Remark 8.5 (L^2-Betti numbers and quasi-isometry). If the finitely generated groups G_0 and G_1 are quasi-isometric and there exist finite models for BG_0 and BG_1, then $b_p^{(2)}(G_0) = 0 \Leftrightarrow b_p^{(2)}(G_1) = 0$ holds (see [46, page 224], [95]). But in general it is not true that there is a constant $C > 0$ such that $b_p^{(2)}(G_0) = C \cdot b_p^{(2)}(G_1)$ holds for all $p \geq 0$ (cf. [39, page 7], [46, page 233], [109]).

Remark 8.6 (Measure equivalence versus quasi-isometry). If F_g denotes the free group on g generators, then define $G_n := (F_3 \times F_3) * F_n$ for $n \geq 2$. The groups G_m and G_n are quasi-isometric for $m, n \geq 2$ (see [21, page 105 in IV-B.46], [109, Theorem 1.5]) and have finite models for their classifying spaces. One easily checks using Theorem 6.1 that $b_1^{(2)}(G_n) = n$ and $b_2^{(2)}(G_n) = 4$.

Theorem 8.2 due to Gaboriau implies that G_m and G_n are measure equivalent if and only if $m = n$ holds. Hence there are finitely presented groups which are quasi-isometric but not measure equivalent.

The converse is also true. The groups $\mathbb{Z}^n$ and $\mathbb{Z}^m$ are infinite amenable and hence measure equivalent. But they are not quasi-isometric for different m and n since n is the growth rate of $\mathbb{Z}^n$ and the growth rate is a quasi-isometry invariant.

Notice that Theorem 8.2 implies that the sign of the Euler characteristic of a group G is an invariant under measure equivalence, which is not true for quasi-isometry by the example of the groups G_n above.

Let the two groups G and H act on the same metric space X properly and cocompactly by isometries. If X is second countable and proper, then G and H are measure equivalent. [100, Theorem 2.36]. If X is a geodesic and proper, then G and H are quasi-isometric.

Remark 8.7 (Kazhdan's property (T)). Kazhdan's property (T) is an invariant under measure equivalence [36, Theorem 8.2]. There exist quasi-isometric finitely generated groups G_0 and G_1 such that G_0 has Kazhdan's property (T) and G_1 not (see [39, page 7]). Hence G_0 and G_1 are quasi-isometric but not measure equivalent.

The rest of this section is devoted to an outline of the proof of Theorem 8.2 due to R. Sauer [100] which is simpler and more algebraic than the original one of Gaboriau [38] and may have the potential to apply also to L^2-torsion.

8.2 Discrete Measured Groupoids

A *groupoid* is a small category in which all morphisms are isomorphisms. We will identify a groupoid $\underline{G}$ with its set of morphisms. Then the set of objects $\underline{G}^0$ can be considered as a subset of $\underline{G}$ via the identity morphisms. There are four canonical maps,

$$\begin{array}{llcll} \text{source map} & s\colon \underline{G} & \to & \underline{G}^0, & (f\colon x \to y) \mapsto x; \\ \text{target map} & t\colon \underline{G} & \to & \underline{G}^0, & (f\colon x \to y) \mapsto y; \\ \text{inverse map} & i\colon \underline{G} & \to & \underline{G}, & f \mapsto f^{-1}; \\ \text{composition} & \circ\colon \underline{G}^2 & \to & \underline{G}, & (f,g) \mapsto f \circ g, \end{array}$$

where $\underline{G}^2$ is $\{(f,g) \in \underline{G} \times \underline{G} \mid s(f) = t(g)\}$. We will often abbreviate $f \circ g$ by fg.

A *discrete measurable groupoid* is a groupoid $\underline{G}$ equipped with the structure of a standard Borel space such that the inverse map and the composition are measurable maps and $s^{-1}(x)$ is countable for all objects $x \in \underline{G}^0$. Then

$\underline{G}^0 \subseteq \underline{G}$ is a Borel subset, the source and the target maps are measurable and $t^{-1}(x)$ is countable for all objects $x \in \underline{G}^0$.

Let μ be a probability measure on $\underline{G}^0$. Then for each measurable subset $A \subseteq \underline{G}$ the function

$$\underline{G}^0 \to \mathbb{C}, \quad x \mapsto |s^{-1}(x) \cap A|$$

is measurable and we obtain a σ-finite measure μ_s on $\underline{G}$ by

$$\mu_s(A) := \int_{\underline{G}^0} |s^{-1}(x) \cap A| \, d\mu(x).$$

It is called the left counting measure of μ. The right counting measure μ_t is defined analogously replacing the source map s by the target map t. We call μ invariant if $\mu_s = \mu_t$, or, equivalently, if $i_*\mu_s = \mu_s$. A discrete measurable groupoid $\underline{G}$ together with an invariant measure μ on $\underline{G}^0$ is called a *discrete measured groupoid.* Given a Borel subset $A \subseteq \underline{G}^0$ with $\mu(A) > 0$, there is the *restricted discrete measured groupoid* $\underline{G}|_A = s^{-1}(A) \cap t^{-1}(A)$, which is equipped with the *normalized measure* $\frac{1}{\mu(A)} \cdot \mu|_A$.

An *isomorphism of discrete measured groupoids* $f \colon \underline{G} \to \underline{H}$ is an isomorphisms of groupoids which preserves the measures. Given measurable subsets $A \subseteq \underline{G}^0$ and $B \subseteq \underline{H}^0$ such that $t(s^{-1}(A))$ and $t(s^{-1}(B))$ have full measure in $\underline{G}^0$ and $\underline{H}^0$ respectively, we call an isomorphism of discrete measured groupoids $f \colon \underline{G}_A \to \underline{H}_B$ a *weak isomorphism of discrete measured groupoids.*

Example 8.8 (Orbit equivalence relation)**.** Consider the countable group G with an action $G \curvearrowright X$ on a standard Borel space X with probability measure μ by μ-preserving isomorphisms. The *orbit equivalence relation*

$$\mathcal{R}(G \curvearrowright X) := \{(x, gx) \mid x \in X, g \in G\} \subseteq X \times X$$

becomes a discrete measured groupoid by the obvious groupoid structure and measure.

An action $G \curvearrowright X$ of a countable group G is called *standard* if X is a standard Borel measure space with a probability measure μ, the action is by μ-preserving Borel isomorphisms and the action is *essentially free*, i.e. the stabilizer of almost every $x \in X$ is trivial. Every countable group G admits a standard action, which is given by the shift action on $\prod_{g \in G} [0, 1]$. Notice that this G-action is not free but essentially free.

Two standard actions $G \curvearrowright X$ and $H \curvearrowright Y$ are *weakly orbit equivalent* if there are Borel subsets $A \subseteq X$ and $B \subseteq Y$, which meet almost every orbit and have positive measure in X and Y respectively, and a Borel isomorphism $f \colon A \to B$, which preserves the normalized measures on A and B and satisfies

$$f(G \cdot x \cap A) = H \cdot f(x) \cap B$$

for almost all $x \in A$. If A has full measure in X and B has full measure in Y, then the two standard actions are called *orbit equivalent.* The map f is called a *weak orbit equivalence* or *orbit equivalence* respectively. The *index of a weak orbit equivalence* of f is the quotient $\frac{\mu(A)}{\mu(B)}$. The next result is due to Furman [37, Theorem 3.3].

Theorem 8.9 (Measure equivalence and weak orbit equivalence). *Two countable groups are measure equivalent with respect to a measure coupling of index $C > 0$ if and only if there exist standard actions of G and H which are weakly orbit equivalent with index C.*

8.3 Groupoid Rings

Let $\underline{G}$ be a discrete measured groupoid with invariant measure μ on $\underline{G}^0$. For a function $\phi\colon \underline{G} \to \mathbb{C}$ and $x \in \underline{G}^0$ put

$$\begin{aligned} S(\phi)(x) &:= |\{g \in \underline{G} \mid \phi(g) \neq 0, s(g) = x\} \quad \in \{0,1,2\ldots\} \amalg \{\infty\}; \\ T(\phi)(x) &:= |\{g \in \underline{G} \mid \phi(g) \neq 0, t(g) = x\} \quad \in \{0,1,2\ldots\} \amalg \{\infty\}. \end{aligned}$$

Let $\mu_{\underline{G}} = \mu_s = \mu_t$ be the measure on $\underline{G}$ induced by μ. Let $L^\infty(\underline{G}) = L^\infty(\underline{G}; \mu_{\underline{G}})$ be the $\mathbb{C}$-algebra of equivalence classes of essentially bounded measurable functions $\underline{G} \to \mathbb{C}$. Define $L^\infty(\underline{G}^0) = L^\infty(\underline{G}^0; \mu)$ analogously. Define the *groupoid ring* of $\underline{G}$ as the subset

$$\mathbb{C}\underline{G} := \{\phi \in L^\infty(\underline{G}) \mid S(\phi) \text{ and } T(\phi) \text{ are essentially bounded on } \underline{G}\}. \tag{8.10}$$

The addition comes from the pointwise addition in $L^\infty(\underline{G})$. Multiplication comes from the convolution product

$$(\phi \cdot \psi)(g) = \sum_{\substack{g_1, g_2 \in \underline{G} \\ g_2 \circ g_1 = g}} \phi(g_1) \cdot \psi(g_2).$$

An involution of rings on $\mathbb{C}\underline{G}$ is defined by $(\phi^*)(g) := \overline{\phi(i(g))}$. Define the augmentation homomorphism $\epsilon\colon \mathbb{C}\underline{G} \to L^\infty(\underline{G}^0)$ by sending ϕ to $\epsilon(\phi)\colon \underline{G}^0 \to \mathbb{C}$, $x \mapsto \sum_{g \in s^{-1}(x)} \phi(g)$. Notice that ϵ is in general not a ring homomorphism, it is only compatible with the additive structure. It becomes a homomorphism of $\mathbb{C}\underline{G}$-modules if we equip $L^\infty(\underline{G}^0)$ with the following $\mathbb{C}\underline{G}$-module structure

$$\phi \cdot f := \epsilon(\phi \cdot j(f)) \quad \text{for } \phi \in \mathbb{C}\underline{G}, f \in L^\infty(\underline{G}^0),$$

where $j\colon L^\infty(\underline{G}^0) \to \mathbb{C}\underline{G}$ is the inclusion of rings, which is given by extending a function on $\underline{G}^0$ to $\underline{G}$ by putting it to be zero outside $\underline{G}^0$.

Given a group G and a ring R together with a homomorphism $c\colon G \to \operatorname{aut}(R)$, define the *crossed product ring* $R *_c G$ as the free R-module with G as R-basis and the multiplication given by

$$\left(\sum_{g \in G} r_g \cdot g\right) \cdot \left(\sum_{g \in G} s_g \cdot g\right) = \sum_{g \in G} \left(\sum_{\substack{g_1, g_2 \in G, \\ g = g_1 g_2}} r_{g_1} \cdot c(g_1)(s_{g_2}) \right) \cdot g.$$

Given a standard action $G \curvearrowright X$, let $L^\infty(X) * G$ be the crossed product ring $L^\infty(X) *_c G$ with respect to the group homomorphism $c\colon G \to \operatorname{aut}(L^\infty(X))$ which sends g to the automorphism given by composition with $l_{g^{-1}}\colon X \to X, \quad x \mapsto g^{-1}x$. We obtain an injective ring homomorphism

$$k\colon L^\infty(X) * G \to \mathbb{C}\mathcal{R}(G \curvearrowright X)$$

which sends $\sum_{g \in G} f_g \cdot g$ to the function $(gx, x) \mapsto f_g(gx)$. In the sequel we will regard $L^\infty(X) * G$ as a subring of $\mathbb{C}\mathcal{R}(G \curvearrowright X)$ using k.

Next we briefly explain how one can associate to the groupoid ring $\mathbb{C}\underline{G}$ of a discrete measured groupoid $\underline{G}$ a von Neumann algebra $\mathcal{N}(\underline{G})$, which is finite, or, equivalently, which possesses a faithful finite normal trace. One can define on $\mathbb{C}\underline{G}$ an inner product

$$\langle \phi, \psi \rangle = \int_{\underline{G}} \phi(g) \cdot \overline{\psi(g)}\, d\mu_{\underline{G}}.$$

Then $\mathbb{C}\underline{G}$ as a $\mathbb{C}$-algebra with involution and the scalar product above satisfies the axioms of a *Hilbert algebra* A, i.e. we have $\langle y, x \rangle = \langle x^*, y^* \rangle$ for $x, y \in A$, $\langle xy, z \rangle = \langle y, x^*z \rangle$ for $x, y, z \in A$ and the map $A \to A, \ y \mapsto yx$ is continuous for all $x \in A$. Let H_A be the Hilbert space completion of A with respect to the given inner product. Define the von Neumann algebra $\mathcal{N}(A)$ associated to A by the $\mathbb{C}$-algebra with involution $\mathcal{B}(H_A)^A$ which consists of all bounded left A-invariant operators $H_A \to H_A$. The standard trace is given by

$$\operatorname{tr}_{\mathcal{N}(A)}\colon \mathcal{N}(A) \to \mathbb{C}, \quad f \mapsto \langle f(1_A), 1_A \rangle.$$

We do get a dimension function as in Theorem 1.11 for $\mathcal{N}(A)$.

Our main example will be $\mathcal{N}(G \curvearrowright X) := \mathcal{N}(\mathbb{C}\mathcal{R}(G \curvearrowright X))$ for a standard action $G \curvearrowright X$ of G.

If G is a countable group and $\underline{G} = G$ is the associated discrete measured groupoid with one object, then $\mathbb{C}\underline{G} = \mathbb{C}G$, $l^2(G) = H_{\mathbb{C}\underline{G}}$ and the definition of $\mathcal{N}(\underline{G})$ and $\operatorname{tr}_{\mathcal{N}(\underline{G})}$ above agrees with the previous Definition 1.1 of $\mathcal{N}(G)$ and $\operatorname{tr}_{\mathcal{N}(G)}$.

Remark 8.11 (Summary and Relevance of the algebraic structures associated to a standard action)**.** Let $G \curvearrowright X$ be a standard action. We have the following commutative diagram of inclusions of rings

$$\begin{array}{ccccccc}
\mathbb{C} & \longrightarrow & \mathbb{C}G & \xrightarrow{=} & \mathbb{C}G & \longrightarrow & \mathcal{N}(G) \\
\downarrow & & \downarrow & & \downarrow & & \downarrow \\
L^\infty(X) & \longrightarrow & L^\infty(X) * G & \longrightarrow & \mathbb{C}\mathcal{R}(G \curvearrowright X) & \longrightarrow & \mathcal{N}(G \curvearrowright X)
\end{array}$$

There is a $\mathbb{C}\underline{G}$-module structure on $L^\infty(\mathcal{R}(G \curvearrowright X)^0) = L^\infty(X)$. Its restriction to $L^\infty(X) * G \subseteq \mathbb{C}\mathcal{R}(G \curvearrowright X)$ is the obvious $L^\infty(X) * G$-module structure on $L^\infty(X)$.

The following observation will be crucial. Given two standard actions $G \curvearrowright X$ and $G \curvearrowright Y$, an orbit equivalence f from $G \curvearrowright X$ to $H \curvearrowright Y$ induces isomorphisms of rings, all denoted by f_*, such that the following diagram with inclusions as horizontal maps commutes

$$\begin{array}{ccccc}
L^\infty(X) & \longrightarrow & \mathbb{C}\mathcal{R}(G \curvearrowright X) & \longrightarrow & \mathcal{N}(G \curvearrowright X) \\
{\scriptstyle f_*}\downarrow{\scriptstyle \cong} & & {\scriptstyle f_*}\downarrow{\scriptstyle \cong} & & {\scriptstyle f_*}\downarrow{\scriptstyle \cong} \\
L^\infty(Y) & \longrightarrow & \mathbb{C}\mathcal{R}(H \curvearrowright Y) & \longrightarrow & \mathcal{N}(G \curvearrowright Y)
\end{array}$$

It is *not* true that f induces a ring map $L^\infty(X) * G \to L^\infty(Y) * H$, since we only require that f maps orbits to orbits but nothing is demanded about equivariance of f with respect to some homomorphism of groups from $G \to H$. The crossed product ring $L^\infty(X) * G$ contains too much information about the group G itself. Hence we shall only involve $L^\infty(X)$, $\mathbb{C}\mathcal{R}(G \curvearrowright X)$, and $\mathcal{N}(G \curvearrowright X)$ in any algebraic construction which is designed to be invariant under orbit equivalence.

8.4 L^2-Betti Numbers of Standard Actions

Definition 8.12. *Let $\underline{G}$ be a discrete measured groupoid. Define its* p-th L^2-Betti number *by*

$$b_p^{(2)}(\underline{G}) = \dim_{\mathcal{N}(\underline{G})}\left(\operatorname{Tor}_p^{\mathbb{C}\underline{G}}\left(\mathcal{N}(\underline{G}), L^\infty(\underline{G}^0)\right)\right).$$

Given a standard action $G \curvearrowright X$, define its p-th L^2-Betti number *as the p-th L^2-Betti number of the associated orbit equivalence relation $\mathcal{R}(G \curvearrowright X)$, i.e.*

$$b_p^{(2)}(G \curvearrowright X) = \dim_{\mathcal{N}(G \curvearrowright X)}\left(\operatorname{Tor}_p^{\mathbb{C}\mathcal{R}(G \curvearrowright X)}\left(\mathcal{N}(G \curvearrowright X), L^\infty(X)\right)\right).$$

Notice that Theorem 8.2 is true if we can prove the following three lemmas.

Lemma 8.13. *If two standard actions $G \curvearrowright X$ and $H \curvearrowright Y$ are orbit equivalent, then they have the same L^2-Betti numbers.*

Lemma 8.14. *Let $\underline{G}$ be a discrete measured groupoid. Let $A \subseteq \underline{G}$ be a Borel subset such that $t(s^{-1}(A))$ has full measure in $\underline{G}^0$. Then we get for all $p \geq 0$*

$$b_p^{(2)}(\underline{G}) \;=\; \mu(A) \cdot b_p^{(2)}(\underline{G}|_A).$$

Lemma 8.15. *Let $G \curvearrowright X$ be a standard action. Then we get for all $p \geq 0$*

$$b_p^{(2)}(G \curvearrowright X) \;=\; b_p^{(2)}(G).$$

Lemma 8.13 follows directly from Remark 8.11. The hard part of the proof of Theorem 8.2 is indeed the proof of the remaining two Lemmas 8.14 and 8.15. This is essentially done by developing some homological algebra over finite von Neumann algebras taking the dimension for arbitrary modules into account.

8.5 Invariance of L^2-Betti Numbers under Orbit Equivalence

As an illustration we sketch the proof of Lemma 8.15. It follows from the following chain of equalities which we explain briefly below.

$$\begin{aligned}
b_p^{(2)}(G) = \cdots & \\
\cdots\; & \dim_{\mathcal{N}(G)}\left(\operatorname{Tor}_p^{\mathbb{C}G}(\mathcal{N}(G), \mathbb{C})\right) && (8.16)\\
= \; & \dim_{\mathcal{N}(G \curvearrowright X)}\left(\mathcal{N}(G \curvearrowright X) \otimes_{\mathcal{N}(G)} \operatorname{Tor}_p^{\mathbb{C}G}(\mathcal{N}(G), \mathbb{C})\right) && (8.17)\\
= \; & \dim_{\mathcal{N}(G \curvearrowright X)}\left(\operatorname{Tor}_p^{\mathbb{C}G}(\mathcal{N}(G \curvearrowright X), \mathbb{C})\right) && (8.18)\\
= \; & \dim_{\mathcal{N}(G \curvearrowright X)}\left(\operatorname{Tor}_p^{L^\infty(X)*G}(\mathcal{N}(G \curvearrowright X), L^\infty(X) * G \otimes_{\mathbb{C}G} \mathbb{C})\right) && (8.19)\\
= \; & \dim_{\mathcal{N}(G \curvearrowright X)}\left(\operatorname{Tor}_p^{L^\infty(X)*G}(\mathcal{N}(G \curvearrowright X), L^\infty(X))\right) && (8.20)\\
= \; & \dim_{\mathcal{N}(G \curvearrowright X)}\left(\operatorname{Tor}_p^{\mathbb{C}\mathcal{R}(G \curvearrowright X)}(\mathcal{N}(G \curvearrowright X), L^\infty(X))\right) && (8.21)\\
= \; & b_p^{(2)}(G \curvearrowright X). && (8.22)
\end{aligned}$$

Equations (8.16) and (8.22) are true by definition. The inclusion of von Neumann algebras $\mathcal{N}(G) \to \mathcal{N}(G \curvearrowright X)$ preserves the traces. This implies that the functor $\mathcal{N}(G \curvearrowright X) \otimes_{\mathcal{N}(G)} -$ from $\mathcal{N}(G)$-modules to $\mathcal{N}(G \curvearrowright X)$-modules is faithfully flat and preserves dimensions. The proof of this fact is completely analogous to the proof of Theorem 1.18. This shows (8.17) and (8.18). For every $\mathbb{C}G$-module M there is a natural $L^\infty(X) * G$-isomorphism

$$L^\infty(X) * G \otimes_{\mathbb{C}G} M \xrightarrow{\cong} L^\infty(X) \otimes_{\mathbb{C}} M.$$

This shows that $L^\infty(X) * G$ is flat as $\mathbb{C}G$-module and that (8.19) and (8.20) are true. The hard part is now to prove (8.21), which is the decisive step, since here one eliminates $L^\infty(X) * G$ from the picture and stays with terms which depend only on the orbit equivalence class of $G \curvearrowright X$. Its proof involves homological algebra and dimension theory. It is not true that the relevant Tor-terms are isomorphic, they only have the same dimension.

This finishes the outline of the proof of Lemma 8.15 and of Theorem 8.2. The complete proof can be found in [100].

9 The Singer Conjecture

In this section we briefly discuss the following conjecture.

Conjecture 9.1 (Singer Conjecture). *If M is an aspherical closed manifold, then*

$$b_p^{(2)}(\widetilde{M}) = 0 \qquad \text{if } 2p \neq \dim(M).$$

If M is a closed connected Riemannian manifold with negative sectional curvature, then

$$b_p^{(2)}(\widetilde{M}) \begin{cases} = 0 & \text{if } 2p \neq \dim(M); \\ > 0 & \text{if } 2p = \dim(M). \end{cases}$$

We mention that all the explicit computations presented in Section 3 are compatible with the Singer Conjecture 9.1. A version of the Singer Conjecture for L^2-torsion will be presented in Conjecture 11.28.

9.1 The Singer Conjecture and the Hopf Conjecture

Because of the Euler-Poincaré formula $\chi(M) = \sum_{p\geq 0}(-1)^p \cdot b_p^{(2)}(\widetilde{M})$ (see Theorem 2.7 (x)) the Singer Conjecture 9.1 implies the following conjecture in case M is aspherical or has negative sectional curvature.

Conjecture 9.2 (Hopf Conjecture). *If M is an aspherical closed manifold of even dimension, then*

$$(-1)^{\dim(M)/2} \cdot \chi(M) \geq 0.$$

If M is a closed Riemannian manifold of even dimension with sectional curvature $\sec(M)$, *then*

$$\begin{array}{rclcrcl} (-1)^{\dim(M)/2} \cdot \chi(M) & > & 0 & \text{if} & \sec(M) & < & 0; \\ (-1)^{\dim(M)/2} \cdot \chi(M) & \geq & 0 & \text{if} & \sec(M) & \leq & 0; \\ \chi(M) & = & 0 & \text{if} & \sec(M) & = & 0; \\ \chi(M) & \geq & 0 & \text{if} & \sec(M) & \geq & 0; \\ \chi(M) & > & 0 & \text{if} & \sec(M) & > & 0. \end{array}$$

In original versions of the Singer Conjecture 9.1 and the Hopf Conjecture 9.2 the statements for aspherical manifolds did not appear. Every Riemannian manifold with non-positive sectional curvature is aspherical by Hadamard's Theorem.

9.2 Pinching Conditions

The following two results are taken from the paper by Jost and Xin [59, Theorem 2.1 and Theorem 2.3].

Theorem 9.3. *Let M be a closed connected Riemannian manifold of dimension* $\dim(M) \geq 3$. *Suppose that there are real numbers $a > 0$ and $b > 0$ such that the sectional curvature satisfies $-a^2 \leq \sec(M) \leq 0$ and the Ricci curvature is bounded from above by $-b^2$. If the non-negative integer p satisfies $2p \neq \dim(M)$ and $2pa \leq b$, then*

$$b_p^{(2)}(\widetilde{M}) \;=\; 0.$$

Theorem 9.4. *Let M be a closed connected Riemannian manifold of dimension* $\dim(M) \geq 4$. *Suppose that there are real numbers $a > 0$ and $b > 0$ such that the sectional curvature satisfies $-a^2 \leq \sec(M) \leq -b^2$. If the non-negative integer p satisfies $2p \neq \dim(M)$ and $(2p-1) \cdot a \leq (\dim(M) - 2) \cdot b$, then*

$$b_p^{(2)}(\widetilde{M}) \;=\; 0.$$

The next result is a consequence of a result of Ballmann and Brüning [3, Theorem B on page 594].

Theorem 9.5. *Let M be a closed connected Riemannian manifold. Suppose that there are real numbers $a > 0$ and $b > 0$ such that the sectional curvature satisfies $-a^2 \leq \sec(M) \leq -b^2$. If the non-negative integer p satisfies $2p <$ $\dim(M) - 1$ and $p \cdot a < (\dim(M) - 1 - p) \cdot b$, then*

$$b_p^{(2)}(\widetilde{M}) \;=\; 0.$$

Theorem 9.4 and Theorem 9.5 are improvements of the older results by Donnelly and Xavier [26].

Remark 9.6 (Right angled Coxeter groups and Coxeter complexes)**.** Next we mention the work of Davis and Okun [20]. A simplicial complex L is called a *flag complex* if each finite non-empty set of vertices which pairwise are connected by edges spans a simplex of L. To such a flag complex they associate a right-angled Coxeter group W_L defined by the following presentation [20, Definition 5.1]. Generators are the vertices v of L. Each generator v satisfies $v^2 = 1$. If two vertices v and w span an edge, there is the relation $(vw)^2 = 1$. Given a finite flag complex L, Davis and Okun associate to it a finite proper W_L-CW-complex Σ_L, which turns out to be a model for the classifying space of the family of finite subgroups $E_{\mathcal{FIN}}(W_L)$ [20, 6.1, 6.1.1 and 6.1.2]. Equipped with a specific metric, Σ_L turns out to be non-positive curved in a combinatorial sense, namely, it is a CAT(0)-space [20, 6.5.3]. If L is a generalized rational homology $(n-1)$-sphere, i.e. a homology $(n-1)$-manifold with the same rational homology as S^{n-1}, then Σ_L is a polyhedral homology n-manifold with rational coefficients [20, 7.4]. So Σ_L is a reminiscence of the universal covering of a closed n-dimensional manifold with non-positive sectional curvature and fundamental group W_L. In view of the Singer Conjecture 9.1 the conjecture makes sense that $b_p^{(2)}(\Sigma_L; \mathcal{N}(W_L)) = 0$ for $2p \neq n$ provided that the underlying topological space of L is S^{n-1} (or, more generally, that it is a homology $(n-1)$-sphere) [20, Conjecture 0.4 and 8.1]. Davis and Okun show that the conjecture is true in dimension $n \leq 4$ and that it is true in dimension $(n+1)$ if it holds in dimension n and n is odd [20, Theorem 9.3.1 and Theorem 10.4.1].

9.3 The Singer Conjecture and Kähler Manifolds

Definition 9.7. *Let (M, g) be a connected Riemannian manifold. A $(p-1)$-form $\eta \in \Omega^{p-1}(M)$ is* bounded *if $||\eta||_\infty := \sup\{||\eta||_x \mid x \in M\} < \infty$ holds, where $||\eta||_x$ is the norm on $\mathrm{Alt}^{p-1}(T_x M)$ induced by g_x. A p-form $\omega \in \Omega^p(M)$ is called d(bounded) if $\omega = d(\eta)$ holds for some bounded $(p-1)$-form $\eta \in \Omega^{p-1}(M)$. A p-form $\omega \in \Omega^p(M)$ is called $\widetilde{d}$(bounded) if its lift $\widetilde{\omega} \in \Omega^p(\widetilde{M})$ to the universal covering $\widetilde{M}$ is d(bounded).*

The next definition is taken from [45, 0.3 on page 265].

Definition 9.8 (Kähler hyperbolic manifold). *A* Kähler hyperbolic manifold *is a closed connected Kähler manifold (M, h) whose fundamental form ω is $\widetilde{d}$(bounded).*

Example 9.9 (Examples of Kähler hyperbolic manifolds). The following list of examples of Kähler hyperbolic manifolds is taken from [45, Example 0.3]:

(i) M is a closed Kähler manifold which is homotopy equivalent to a Riemannian manifold with negative sectional curvature;

(ii) M is a closed Kähler manifold such that $\pi_1(M)$ is word-hyperbolic in the sense of [44] and $\pi_2(M) = 0$;

(iii) $\widetilde{M}$ is a symmetric Hermitian space of non-compact type;

(iv) M is a complex submanifold of a Kähler hyperbolic manifold;

(v) M is a product of two Kähler hyperbolic manifolds.

The following result is due to Gromov [44, Theorem 1.2.B and Theorem 1.4.A on page 274]. A detailed discussion of the proof and the consequences of this theorem can also be found in [80, Chapter 11].

Theorem 9.10. **(L^2-Betti numbers of Kähler hyperbolic manifolds).** *Let M be a Kähler hyperbolic manifold of complex dimension m and real dimension $n = 2m$. Then*

$$\begin{aligned} b_p^{(2)}(\widetilde{M}) &= 0 \qquad \text{if } p \neq m; \\ b_m^{(2)}(\widetilde{M}) &> 0; \\ (-1)^m \cdot \chi(M) &> 0. \end{aligned}$$

10 The Approximation Conjecture

This section is devoted to the following conjecture.

Conjecture 10.1 (Approximation Conjecture). *A group G satisfies the* Approximation Conjecture *if the following holds:*

Let $\{G_i \mid i \in I\}$ be an inverse system of normal subgroups of G directed by inclusion over the directed set I. Suppose that $\bigcap_{i \in I} G_i = \{1\}$. Let X be a G-CW-complex of finite type. Then $G_i \backslash X$ is a G/G_i-CW-complex of finite type and

$$b_p^{(2)}(X; \mathcal{N}(G)) = \lim_{i \in I} b_p^{(2)}(G_i \backslash X; \mathcal{N}(G/G_i)).$$

Remark 10.2 (The Approximation Conjecture for subgroups of finite index). Let us consider the special case where the inverse system $\{G_i \mid i \in I\}$ is given by a nested sequence of normal subgroups of finite index

$$G = G_0 \supset G_1 \supset G_2 \supset G_3 \supset \ldots.$$

Notice that then $b_p^{(2)}(G_i \backslash X; \mathcal{N}(G/G_i)) = \frac{b_p(G_i \backslash X)}{[G:G_i]}$, where $b_p(G_i \backslash X)$ is the classical p-th Betti number of the finite CW-complex $G_i \backslash X$. In this special case Conjecture 10.1 was formulated by Gromov [46, pages 20, 231] and proved in [72, Theorem 0.1]. Thus we get an asymptotic relation between the L^2-Betti numbers and Betti numbers, namely

$$b_p^{(2)}(X; \mathcal{N}(G)) = \lim_{i \to \infty} \frac{b_p(G_i \backslash X)}{[G : G_i]},$$

although the Betti numbers of a connected finite CW-complex Y and the L^2-Betti numbers of its universal covering $\widetilde{Y}$ have nothing in common except the fact that their alternating sum equals $\chi(Y)$ (see Example 6.2).

Interesting variations of this result for not necessarily normal subgroups of finite index and Betti-numbers with coefficients in representations can be found in the paper by Farber [34].

Definition 10.3. *Let $\mathcal{G}$ be the smallest class of groups which contains the trivial group and is closed under the following operations:*

(i) Amenable quotient
Let $H \subseteq G$ be a (not necessarily normal) subgroup. Suppose that $H \in \mathcal{G}$ and the quotient G/H is an amenable discrete homogeneous space. (For the precise definition of amenable discrete homogeneous space see for instance [80, Definition 13.8]. If $H \subseteq G$ is normal and G/H is amenable, then G/H is an amenable discrete homogeneous space.)

Then $G \in \mathcal{G}$;

(ii) Colimits
If $G = \operatorname{colim}_{i \in I} G_i$ is the colimit of the directed system $\{G_i \mid i \in I\}$ of groups indexed by the directed set I and each G_i belongs to $\mathcal{G}$, then G belongs to $\mathcal{G}$;

(iii) Inverse limits
If $G = \lim_{i \in I} G_i$ is the limit of the inverse system $\{G_i \mid i \in I\}$ of groups indexed by the directed set I and each G_i belongs to $\mathcal{G}$, then G belongs to $\mathcal{G}$;

(iv) Subgroups
If H is isomorphic to a subgroup of the group G with $G \in \mathcal{G}$, then $H \in \mathcal{G}$;

(v) Quotients with finite kernel
Let $1 \to K \to G \to Q \to 1$ be an exact sequence of groups. If K is finite and G belongs to $\mathcal{G}$, then Q belongs to $\mathcal{G}$.

Next we provide some information about the class $\mathcal{G}$. Notice that in the original definition of $\mathcal{G}$ due to Schick [102, Definition 1.12] the resulting class is slightly smaller: there it is required that the class contains the trivial subgroup and is closed under operations (i), (ii), (iii) and (iv), but not necessarily under operation (v). The proof of the next lemma can be found in [80, Lemma 13.11].

Lemma 10.4. *(i) A group G belongs to $\mathcal{G}$ if and only if every finitely generated subgroup of G belongs to $\mathcal{G}$;*

(ii) The class $\mathcal{G}$ is residually closed, i.e. if there is a nested sequence of normal subgroups $G = G_0 \supset G_1 \supset G_2 \supset \dots$ such that $\bigcap_{i \geq 0} G_i = \{1\}$ and each quotient G/G_i belongs to $\mathcal{G}$, then G belongs to $\mathcal{G}$;

(iii) Any residually amenable and in particular any residually finite group belongs to $\mathcal{G}$;

(iv) Suppose that G belongs to $\mathcal{G}$ and $f \colon G \to G$ is an endomorphism. Define the "mapping torus group" G_f to be the quotient of $G * \mathbb{Z}$ obtained by introducing the relations $t^{-1}gt = f(g)$ for $g \in G$ and $t \in \mathbb{Z}$ a fixed generator. Then G_f belongs to $\mathcal{G}$;

(v) *Let $\{G_j \mid j \in J\}$ be a set of groups with $G_j \in \mathcal{G}$. Then the direct sum $\bigoplus_{j \in J} G_j$ and the direct product $\prod_{j \in J} G_j$ belong to $\mathcal{G}$.*

The proof of the next result can be found in [80, Theorem 13.3]. It is a mild generalization of the results of Schick [101] and [102], where the original proof of the Approximation Conjecture for subgroups of finite index was generalized to the much more general setting above and then applied to the Atiyah Conjecture. The connection between the Approximation Conjecture and the Atiyah Conjecture for torsion-free groups comes from the obvious fact that a convergent series of integers has an integer as limit.

Theorem 10.5 (Status of the Approximation Conjecture). *Every group G which belongs to the class $\mathcal{G}$ (see Definition 10.3) satisfies the Approximation Conjecture 10.1.*

11 L^2-Torsion

Recall that L^2-Betti numbers are modelled on Betti numbers. Analogously one can generalize the classical notion of Reidemeister torsion to an L^2-setting, which will lead to the notion of L^2-torsion. The L^2-torsion may be viewed as a secondary L^2-Betti number just as the Reidemeister torsion can be viewed as a secondary Betti number. Namely, the Reidemeister torsion is only defined if all the Betti numbers (with coefficients in a suitable representation) vanish, and similarly the L^2-torsion is defined only if the L^2-Betti numbers vanish. Both invariants give valuable information about the spaces in question.

11.1 The Fuglede-Kadison Determinant

In this subsection we briefly explain the notion of the Fuglede-Kadison determinant. We have extended the notion of the (classical) dimension of a finite dimensional complex vector space to the von Neumann dimension of a finitely generated projective $\mathcal{N}(G)$-module (and later even to arbitrary $\mathcal{N}(G)$-modules). Similarly we want to generalize the classical determinant of an endomorphism of a finite dimensional complex vector space to the Fuglede-Kadison determinant of an $\mathcal{N}(G)$-endomorphism $f\colon P \to P$ of a finitely generated projective $\mathcal{N}(G)$-module P and of an $\mathcal{N}(G)$-map $f\colon \mathcal{N}(G)^m \to \mathcal{N}(G)^n$ of based finitely generated $\mathcal{N}(G)$-modules. This is necessary since for the definition of Reidemeister torsion one needs determinants and hence for the definition of L^2-torsion one has to develop an appropriate L^2-analogue.

Definition 11.1 (Spectral density function). *Let $f\colon \mathcal{N}(G)^m \to \mathcal{N}(G)^n$ be an $\mathcal{N}(G)$-homomorphism. Let $\nu(f)\colon l^2(G)^m \to l^2(G)^n$ be the associated bounded G-equivariant operator (see Remark 1.7). Denote by $\{E_\lambda^{f^*f} \mid \lambda \in \mathbb{R}\}$ the* *right-continuous family of spectral projections of the positive operator* $\nu(f^*f)$. *Define the* spectral density function of f *by*

$$F_f\colon \mathbb{R} \to [0,\infty) \quad \lambda \mapsto \dim_{\mathcal{N}(G)}\left(\operatorname{im}(E^{f^*f}_{\lambda^2})\right).$$

The spectral density function is monotone and right-continuous. It takes values in $[0,m]$. Here and in the sequel $|x|$ denotes the norm of an element x of a Hilbert space and $||T||$ the operator norm of a bounded operator T. Since $\nu(f)$ and $\nu(f^*f)$ have the same kernel, $\dim_{\mathcal{N}(G)}(\ker(f)) = F_f(0)$.

Example 11.2 (Spectral density function for finite G)**.** Suppose that G is finite. Then $\mathbb{C}G = \mathcal{N}(G) = l^2(G)$ and $\nu(f) = f$. Let $0 \leq \lambda_0 < \ldots < \lambda_r$ be the eigenvalues of f^*f and μ_i be the multiplicity of λ_i, i.e. the dimension of the eigenspace of λ_i. Then the spectral density function is a right continuous step function which is zero for $\lambda < 0$ and has a step of height $\frac{\mu_i}{|G|}$ at each $\sqrt{\lambda_i}$.

Example 11.3 (Spectral density function for $G = \mathbb{Z}^n$)**.** Let $G = \mathbb{Z}^n$. We use the identification $\mathcal{N}(\mathbb{Z}^n) = L^\infty(T^n)$ of Example 1.4. For $f \in L^\infty(T^n)$ the spectral density function F_{M_f} of $M_f\colon L^2(T^n) \to L^2(T^n)$, $g \mapsto g \cdot f$ sends λ to the volume of the set $\{z \in T^n \mid |f(z)| \leq \lambda\}$.

Definition 11.4. (Fuglede-Kadison determinant). *Let* $f\colon \mathcal{N}(G)^m \to \mathcal{N}(G)^n$ *be an* $\mathcal{N}(G)$*-map. Let* $F_f(\lambda)$ *be the spectral density function of Definition 11.1 which is a monotone non-decreasing right-continuous function. Let* dF *be the unique measure on the Borel* σ*-algebra on* $\mathbb{R}$ *which satisfies* $dF(]a,b]) = F(b) - F(a)$ *for* $a < b$. *Then define the* Fuglede-Kadison determinant

$$\det_{\mathcal{N}(G)}(f) \in [0,\infty)$$

by the positive real number

$$\det_{\mathcal{N}(G)}(f) \;=\; \exp\left(\int_{0+}^{\infty} \ln(\lambda)\; dF\right)$$

if the Lebesgue integral $\int_{0+}^{\infty} \ln(\lambda)\; dF$ *converges to a real number and by* 0 *otherwise.*

Notice that in the definition above we do not require $m = n$ or that f is injective or f is surjective.

Example 11.5 (Fuglede-Kadison determinant for finite G)**.** To illustrate this definition, we look at the example where G is finite. We essentially get the classical determinant $\det_{\mathbb{C}}$. Namely, we have computed the spectral density function for finite G in Example 11.2. Let λ_1, λ_2, ..., λ_r be the non-zero eigenvalues of f^*f with multiplicity μ_i. Then one obtains, if $\overline{f^*f}$ is the automorphism of the orthogonal complement of the kernel of f^*f induced by f^*f,

$$\det_{\mathcal{N}(G)}(f) = \exp\left(\sum_{i=1}^{r} \frac{\mu_i}{|G|} \cdot \ln(\sqrt{\lambda_i})\right) = \prod_{i=1}^{r} \lambda_i^{\frac{\mu_i}{2\cdot|G|}} = \left(\det_{\mathbb{C}}\left(\overline{f^*f}\right)\right)^{\frac{1}{2\cdot|G|}}.$$

If $f\colon \mathbb{C}G^m \to \mathbb{C}G^m$ is an automorphism, we get

$$\det_{\mathcal{N}(G)}(f) \;=\; |\det_{\mathbb{C}}(f)|^{\frac{1}{|G|}}.$$

Example 11.6 (Fuglede-Kadison determinant over $\mathcal{N}(\mathbb{Z}^n)$)**.** Let $G = \mathbb{Z}^n$. We use the identification $\mathcal{N}(\mathbb{Z}^n) = L^\infty(T^n)$ of Example 1.4. For $f \in L^\infty(T^n)$ we conclude from Example 11.3

$$\det_{\mathcal{N}(\mathbb{Z}^n)}\left(M_f\colon L^2(T^n) \to L^2(T^n)\right) = \exp\left(\int_{T^n} \ln(|f(z)|) \cdot \chi_{\{u\in S^1 \mid f(u)\neq 0\}}\, dvol_z\right)$$

using the convention $\exp(-\infty) = 0$.

Here are some basic properties of this notion. A morphism $f\colon \mathcal{N}(G)^m \to \mathcal{N}(G)^n$ has dense image if the closure $\overline{\operatorname{im}(f)}$ of its image in $\mathcal{N}(G)^n$ in the sense of Definition 1.10 is $\mathcal{N}(G)^n$. The adjoint A^* of a matrix $A = (a_{i,j}) \in M(m,n;\mathcal{N}(G))$ is the matrix in $M(n,m;\mathcal{N}(G))$ given by $(a^*_{j,i})$, where the map $*\colon \mathcal{N}(G) \to \mathcal{N}(G)$ sends an operator a to its adjoint a^*. The adjoint $f^*\colon \mathcal{N}(G)^n \to \mathcal{N}(G)^m$ of $f\colon \mathcal{N}(G)^m \to \mathcal{N}(G)^n$ is given by the matrix A^* if f is given by the matrix A. The proof of the next result can be found in [80, Theorem 3.14].

Theorem 11.7 (Fuglede-Kadison determinant)**.**

(i) Composition

Let $f\colon \mathcal{N}(G)^l \to \mathcal{N}(G)^m$ and $g\colon \mathcal{N}(G)^m \to \mathcal{N}(G)^n$ be $\mathcal{N}(G)$-homomorphisms such that f has dense image and g is injective. Then

$$\det_{\mathcal{N}(G)}(g \circ f) \;=\; \det_{\mathcal{N}(G)}(g) \cdot \det_{\mathcal{N}(G)}(f);$$

(ii) Additivity

Let the maps $f_1\colon \mathcal{N}(G)^{m_1} \to \mathcal{N}(G)^{n_1}$, $f_2\colon \mathcal{N}(G)^{m_2} \to \mathcal{N}(G)^{n_2}$ and $f_3\colon \mathcal{N}(G)^{m_2} \to \mathcal{N}(G)^{n_1}$ be $\mathcal{N}(G)$-homomorphisms such that f_1 has dense image and f_2 is injective. Then

$$\det_{\mathcal{N}(G)}\begin{pmatrix} f_1 & f_3 \\ 0 & f_2 \end{pmatrix} \;=\; \det_{\mathcal{N}(G)}(f_1) \cdot \det_{\mathcal{N}(G)}(f_2);$$

(iii) Invariance under adjoint map

Let $f \colon \mathcal{N}(G)^m \to \mathcal{N}(G)^n$ be an $\mathcal{N}(G)$-homomorphism. Then

$$\det_{\mathcal{N}(G)}(f) = \det_{\mathcal{N}(G)}(f^*);$$

(iv) Induction
Let $f \colon \mathcal{N}(H)^m \to \mathcal{N}(H)^n$ be an $\mathcal{N}(H)$-homomorphism, and let $i \colon H \to G$ be an injective group homomorphism. Then

$$\det_{\mathcal{N}(G)}(i_* f) \;=\; \det_{\mathcal{N}(H)}(f).$$

Definition 11.8. (Fuglede-Kadison determinant of $\mathcal{N}(G)$-endomorphisms of finitely generated projective modules). *Let $f \colon P \to P$ be an endomorphism of a finitely generated projective $\mathcal{N}(G)$-module P. Choose a finitely generated projective $\mathcal{N}(G)$-module Q and an $\mathcal{N}(G)$-isomorphism $u \colon \mathcal{N}(G)^n \xrightarrow{\cong} P \oplus Q$. Define the* Fuglede-Kadison determinant

$$\det_{\mathcal{N}(G)}(f) \;\in\; [0, \infty)$$

by the Fuglede-Kadison determinant in the sense of Definition 11.4

$$\det_{\mathcal{N}(G)}\left(u^{-1} \circ (f \oplus \mathrm{id}_Q) \circ u\right).$$

This definition is independent of the choices of Q and u by Theorem 11.7. Notice that in Definition 11.8 no $\mathcal{N}(G)$-basis appear but that it works only for endomorphisms, whereas in Definition 11.4 we work with finitely generated free based modules but do not require that the source and target of f are isomorphic. There is an obvious analogue of Theorem 11.7 for the Fuglede-Kadison determinant of endomorphisms of finitely generated projective $\mathcal{N}(G)$-modules.

11.2 The Determinant Conjecture

It will be important for applications to geometry to study the Fuglede-Kadison determinant of $\mathcal{N}(G)$-maps $f \colon \mathcal{N}(G)^m \to \mathcal{N}(G)^n$ which come by induction from $\mathbb{Z}G$-maps or $\mathbb{C}G$-maps. The following example is taken from [80, Example 3.22].

Example 11.9 (Fuglede-Kadison determinant of maps coming from elements in $\mathbb{C}[\mathbb{Z}]$). Consider a non-trivial element $p \in \mathbb{C}[\mathbb{Z}] = \mathbb{C}[z, z^{-1}]$. We can write

$$p(z) \;=\; C \cdot z^n \cdot \prod_{k=1}^{l} (z - a_k)$$

for non-zero complex numbers $C, a_1, \ldots, a_l$ and non-negative integers n, l. Let $r_p \colon \mathcal{N}(\mathbb{Z}) \to \mathcal{N}(\mathbb{Z})$ be the $\mathcal{N}(\mathbb{Z})$-map given by right multiplication with p. Then

$$\det_{\mathcal{N}(\mathbb{Z})}(r_p) = |C| \cdot \prod_{\substack{1 \le k \le l, \\ |a_k| > 1}} |a_k|.$$

Definition 11.10 (Determinant class). *A group G is of* $\det \geq 1$-class *if for each $A \in M(m, n; \mathbb{Z}G)$ the Fuglede-Kadison determinant (see Definition 11.4) of the morphism $r_A \colon \mathcal{N}(G)^m \to \mathcal{N}(G)^n$ given by right multiplication with A satisfies*

$$\det_{\mathcal{N}(G)}(r_A) \geq 1.$$

Conjecture 11.11 (Determinant Conjecture). *Every group G is of* $\det \geq 1$*-class.*

The proof of the next result can be found in [80, Theorem 13.3]. It is a mild generalization of the results of Schick [101] and [102].

Theorem 11.12 (Status of the Determinant Conjecture). *Every group G which belongs to the class $\mathcal{G}$ (see Definition 10.3) satisfies the Determinant Conjecture 11.11.*

One easily checks that the Fuglede-Kadison determinant defines a homomorphism of abelian groups

$$\Phi^G : \mathrm{Wh}(G) \to (0, \infty) = \{r \in \mathbb{R} \mid r > 0\} \tag{11.13}$$

with respect to the group structure given by multiplication of positive real numbers on the target. We mention the following conjecture.

Conjecture 11.14 (Triviality of the map induced by the Fuglede-Kadison determinant on $\mathrm{Wh}(G)$). *The map $\Phi^G \colon \mathrm{Wh}(G) \to (0, \infty)$ is trivial.*

Lemma 11.15. *(i) If G satisfies the Determinant Conjecture 11.11, then G satisfies Conjecture 11.14;*

(ii) The Approximation Conjecture 10.1 for G and the inverse system $\{G_i \mid i \in I\}$ is true if each group G_i is of $\det \geq 1$*-class.*

Proof. See [80, Theorem 13.3 (1) and Lemma 13.6]. □

11.3 Definition and Basic Properties of L^2-Torsion

We will consider L^2-torsion only for universal coverings and in the L^2-acyclic case. A more general setting is treated in [80, Section 3.4].

Definition 11.16 (det-L^2-acyclic). *Let X be a finite connected CW-complex with fundamental group $\pi = \pi_1(X)$. Let $C_*^{\mathcal{N}}(\widetilde{X})$ be the $\mathcal{N}(\pi)$-chain complex $\mathcal{N}(G) \otimes_{\mathbb{Z}G} C_*(\widetilde{X})$ with p-th differential $c_p^{\mathcal{N}} = \mathrm{id}_{\mathcal{N}(G)} \otimes_{\mathbb{Z}} c_p$. We say that X is* det-L^2-acyclic *if for each p we get for the Fuglede-Kadison determinant of $c_p^{\mathcal{N}}$ and for the p-th L^2-Betti number of $\widetilde{X}$*

$$\begin{aligned} \det_{\mathcal{N}(\pi)}\left(c_p^{\mathcal{N}}\right) &> 0; \\ b_p^{(2)}(\widetilde{X}) &= 0. \end{aligned}$$

If X is det-L^2*-acyclic, we define the* L^2-torsion *of $\widetilde{X}$ by*

$$\rho^{(2)}(\widetilde{X}) = -\sum_{p \geq 0} (-1)^p \cdot \ln\left(\det_{\mathcal{N}(G)}\left(c_p^{\mathcal{N}}\right)\right) \quad \in \mathbb{R}.$$

If X is a finite CW-complex, we call it det-L^2*-acyclic if each component C is* det-L^2*-acyclic. In this case we define*

$$\rho^{(2)}(\widetilde{X}) := \sum_{C \in \pi_0(X)} \rho^{(2)}(\widetilde{C}).$$

The condition that $\widetilde{X}$ is L^2-acyclic is not needed for the definition of L^2-torsion, but is necessary to ensure the basic and useful properties which we will discuss below.

Remark 11.17 (L^2-torsion in terms of the Laplacian). One can express the L^2-torsion also in terms of the Laplacian which is closer to the notions of analytic torsion and analytic L^2-torsion. After a choice of cellular $\mathbb{Z}\pi$-basis, every $\mathcal{N}(\pi)$-chain module $C_p^{\mathcal{N}}(\widetilde{X})$ looks like $\mathcal{N}(\pi)^{n_p}$ for appropriate non-negative integers n_p. Hence we can assign to $c_p^{\mathcal{N}} \colon C_p^{\mathcal{N}}(\widetilde{X}) \to C_{p-1}^{\mathcal{N}}(\widetilde{X})$ its adjoint $\left(c_p^{\mathcal{N}}\right)^* : C_{p-1}^{\mathcal{N}}(\widetilde{X}) \to C_p^{\mathcal{N}}(\widetilde{X})$ which is given by the matrix $(a_{j,i}^*)$ if $c_p^{(2)}$ is given by the matrix $(a_{i,j})$. Define the p-th Laplace homomorphism $\Delta_p \colon C_p^{\mathcal{N}}(\widetilde{X}) \to C_p^{\mathcal{N}}(\widetilde{X})$ to be the $\mathcal{N}(\pi)$-homomorphism $\left(c_p^{\mathcal{N}}\right)^* \circ c_p^{\mathcal{N}} + c_{p+1}^{\mathcal{N}} \circ \left(c_{p+1}^{\mathcal{N}}\right)^*$. Then $\widetilde{X}$ is det-L^2-acyclic if and only if Δ_p is injective and has dense image, i.e. the closure of its image in $C_p^{\mathcal{N}}(\widetilde{X})$ is $C_p^{\mathcal{N}}(\widetilde{X})$, and $\det_{\mathcal{N}(\pi)}(\Delta_p) > 0$. In this case we get

$$\rho^{(2)}(\widetilde{X}) = -\frac{1}{2} \cdot \sum_{p \geq 0} (-1)^p \cdot p \cdot \ln\left(\det_{\mathcal{N}(\pi)}(\Delta_p)\right).$$

This follows from [80, Lemma 3.30].

The next theorem presents the basic properties of $\rho^{(2)}(\widetilde{X})$ and is proved in [80, Theorem 3.96]. Notice the formal analogy between the behaviour of $\rho^{(2)}(\widetilde{X})$ and the classical Euler characteristic $\chi(X)$.

Theorem 11.18. (Cellular L^2-torsion for universal coverings).

(i) Homotopy invariance

Let $f\colon X \to Y$ be a homotopy equivalence of finite CW-complexes. Let $\tau(f) \in \mathrm{Wh}(\pi_1(Y))$ be its Whitehead torsion (see [17]). Suppose that $\widetilde{X}$ or $\widetilde{Y}$ is det-L^2*-acyclic. Then both $\widetilde{X}$ and $\widetilde{Y}$ are* det-L^2*-acyclic and*

$$\rho^{(2)}(\widetilde{Y}) - \rho^{(2)}(\widetilde{X}) \;=\; \Phi^{\pi_1(Y)}(\tau(f)),$$

where $\Phi^{\pi_1(Y)}\colon \mathrm{Wh}(\pi_1(Y)) = \bigoplus_{C \in \pi_0(Y)} \mathrm{Wh}(\pi_1(C)) \to \mathbb{R}$ is the sum of the maps $\Phi^{\pi_1(C)}$ of (11.13);

(ii) Sum formula

Consider the pushout of finite CW-complexes such that j_1 is an inclusion of CW-complexes, j_2 is cellular and X inherits its CW-complex structure from X_0, X_1 and X_2

$$\begin{array}{ccc} X_0 & \xrightarrow{j_1} & X_1 \\ {\scriptstyle j_2}\downarrow & & \downarrow{\scriptstyle i_1} \\ X_2 & \xrightarrow[i_2]{} & X \end{array}$$

Assume $\widetilde{X_0}$, $\widetilde{X_1}$ and $\widetilde{X_2}$ are det-L^2*-acyclic and that for $k = 0,1,2$ the map $\pi_1(i_k)\colon \pi_1(X_k) \to \pi_1(X)$ induced by the obvious map $i_k\colon X_k \to X$ is injective for all base points in X_k.*

Then $\widetilde{X}$ is det-L^2*-acyclic and we get*

$$\rho^{(2)}(\widetilde{X}) \;=\; \rho^{(2)}(\widetilde{X_1}) + \rho^{(2)}(\widetilde{X_2}) - \rho^{(2)}(\widetilde{X_0});$$

(iii) Poincaré duality

Let M be a closed manifold of even dimension. Equip it with some CW-complex structure. Suppose that $\widetilde{M}$ is det-L^2*-acyclic. Then*

$$\rho^{(2)}(\widetilde{M}) \;=\; 0;$$

(iv) Product formula

Let X and Y be finite CW-complexes. Suppose that $\widetilde{X}$ is det-L^2*-acyclic. Then $\widetilde{X \times Y}$ is* det-L^2*-acyclic and*

$$\rho^{(2)}(\widetilde{X \times Y}) \;=\; \chi(Y) \cdot \rho^{(2)}(\widetilde{X});$$

(v) *Multiplicativity*

Let $X \to Y$ be a finite covering of finite CW-complexes with d sheets. Then $\widetilde{X}$ is det-L^2*-acyclic if and only if $\widetilde{Y}$ is* det-L^2*-acyclic and in this case*

$$\rho^{(2)}(\widetilde{X}) = d \cdot \rho^{(2)}(\widetilde{Y}).$$

The next three results are taken from [80, Corollary 3.103, Theorem 3.105 and Theorem 3.111]. There is also a more general version of Theorem 11.19 for fibrations (see [80, Theorem 3.100]).

Theorem 11.19 (L^2-torsion and fiber bundles). *Suppose that $F \to E \xrightarrow{p} B$ is a (locally trivial) fiber bundle of finite CW-complexes with B connected. Suppose that for one (and hence all) $b \in B$ the inclusion of the fiber F_b into E induces an injection on the fundamental groups for all base points in F_b and $\widetilde{F_b}$ is* det-L^2*-acyclic. Then $\widetilde{E}$ is* det-L^2*-acyclic and*

$$\rho^{(2)}(\widetilde{E}) = \chi(B) \cdot \rho^{(2)}(\widetilde{F}).$$

Theorem 11.20 (L^2-torsion and S^1-actions). *Let X be a connected S^1-CW-complex of finite type. Suppose that for one orbit S^1/H (and hence for all orbits) the inclusion into X induces a map on π_1 with infinite image. (In particular the S^1-action has no fixed points.) Then $\widetilde{X}$ is* det-L^2*-acyclic and*

$$\rho^{(2)}(\widetilde{X}) = 0.$$

Theorem 11.21 (L^2-torsion on aspherical closed S^1-manifolds). *Let M be an aspherical closed manifold with non-trivial S^1-action. Then the action has no fixed points and the inclusion of any orbit into M induces an injection on the fundamental groups. Moreover, $\widetilde{M}$ is* det-L^2*-acyclic and*

$$\rho^{(2)}(\widetilde{M}) = 0.$$

The assertion for the L^2-torsion in the theorem below is the main result of [106] (see also [107]). Its proof is based on localization techniques.

Theorem 11.22 (L^2-torsion and aspherical CW-complexes). *Let X be an aspherical finite CW-complex. Suppose that its fundamental group $\pi_1(X)$ contains an elementary amenable infinite normal subgroup H and $\pi_1(X)$ is of* det ≥ 1*-class. Then $\widetilde{X}$ is* det-L^2*-acyclic and*

$$\rho^{(2)}(\widetilde{X}) = 0.$$

Remark 11.23 (Homotopy invariance of L^2-torsion). Notice that Conjecture 11.14 implies because of Theorem 11.18 (i) the homotopy invariance of the L^2-torsion. i.e. for two homotopy equivalent det-L^2-acyclic finite CW-complexes X and Y we have $\rho^{(2)}(\widetilde{X}) = \rho^{(2)}(\widetilde{Y})$.

11.4 Computations of L^2-Torsion

Remark 11.24 (Analytic L^2-torsion). It is important to know for the following specific calculations that there is an analytic version of L^2-torsion in terms of the heat kernel due to Lott [66] and Mathai [86] and that a deep result of Burghelea, Friedlander, Kappeler and McDonald [9] says that the analytic one agrees with the one presented here.

The following result is due to Hess and Schick [51].

Theorem 11.25 (Analytic L^2-torsion of hyperbolic manifolds). *Let $d = 2n + 1$ be an odd integer. To d one can associate an explicit real number $C_d > 0$ with the following property:*

For every closed hyperbolic d-dimensional manifold M we have

$$\rho^{(2)}(\widetilde{M}) \;=\; (-1)^n \cdot C_d \cdot \operatorname{vol}(M),$$

where $\operatorname{vol}(M)$ *is the volume of M.*

The existence of a real number C_d with $\rho^{(2)}(\widetilde{M}) \;=\; (-1)^n \cdot C_d \cdot \operatorname{vol}(M)$ follows from the version of the Proportionality Principle for L^2-Betti numbers (see Theorem 3.7) for L^2-torsion (see [80, Theorem 3.183]). The point is that this number C_d is given explicitly. For instance $C_3 = \frac{1}{6\pi}$ and $C_5 = \frac{31}{45\pi^2}$. For each odd d there exists a rational number r_d such that $C_d = \pi^{-n} \cdot r_d$ holds. The proof of this result is based on calculations involving the heat kernel on hyperbolic space.

Remark 11.26 (L^2-torsion of symmetric spaces of non-compact type). More generally, the L^2-torsion $\rho^{(2)}(\widetilde{M})$ for an aspherical closed manifold M whose universal covering $\widetilde{M}$ is a symmetric space is computed by Olbricht [93].

The following result is proved in [84, Theorem 0.6].

Theorem 11.27 (L^2-torsion of 3-manifolds). *Let M be a compact connected orientable prime 3-manifold with infinite fundamental group such that the boundary of M is empty or a disjoint union of incompressible tori. Suppose that M satisfies Thurston's Geometrization Conjecture, i.e. there is a geometric toral splitting along disjoint incompressible 2-sided tori in M whose pieces are Seifert manifolds or hyperbolic manifolds. Let M_1, M_2, ..., M_r be the hyperbolic pieces. They all have finite volume [90, Theorem B on page 52]. Then $\widetilde{M}$ is* det-L^2*-acyclic and*

$$\rho^{(2)}(\widetilde{M}) \;=\; -\frac{1}{6\pi} \cdot \sum_{i=1}^{r} \operatorname{vol}(M_i).$$

In particular, $\rho^{(2)}(\widetilde{M})$ is 0 if and only if there are no hyperbolic pieces.

11.5 Some Open Conjectures about L^2-Torsion

All the computations and results above give evidence and are compatible with the following conjectures about L^2-torsion taken from [80, Theorem 11.3].

Conjecture 11.28 (L^2-torsion for aspherical manifolds). *If M is an aspherical closed manifold of odd dimension, then $\widetilde{M}$ is* det-L^2*-acyclic and*

$$(-1)^{\frac{\dim(M)-1}{2}} \cdot \rho^{(2)}(\widetilde{M}) \ \geq \ 0.$$

If M is a closed connected Riemannian manifold of odd dimension with negative sectional curvature, then $\widetilde{M}$ is det-L^2*-acyclic and*

$$(-1)^{\frac{\dim(M)-1}{2}} \cdot \rho^{(2)}(\widetilde{M}) \ > \ 0.$$

If M is an aspherical closed manifold whose fundamental group contains an amenable infinite normal subgroup, then $\widetilde{M}$ is det-L^2*-acyclic and*

$$\rho^{(2)}(\widetilde{M}) \ = \ 0.$$

Consider a closed orientable manifold M of dimension n. Let $[M;\mathbb{R}]$ be the image of the fundamental class $[M] \in H_n^{\mathrm{sing}}(M;\mathbb{Z})$ under the change of coefficient map $H_n^{\mathrm{sing}}(M;\mathbb{Z}) \to H_n^{\mathrm{sing}}(M;\mathbb{R})$. Define the L^1-norm on $C_n^{\mathrm{sing}}(M;\mathbb{R})$ by sending $\sum_{i=1}^{s} r_i \cdot [\sigma_i \colon \Delta_n \to M]$ to $\sum_{i=1}^{s} |r_i|$. It induces a seminorm on $H_n(M;\mathbb{R})$. Define the *simplicial volume* $||M|| \in \mathbb{R}$ to be the seminorm of $[M;\mathbb{R}]$. More information about the simplicial volume can be found for instance in [42], [47] and [57], and in [80, Chapter 14], where also the following conjecture is discussed.

Conjecture 11.29 (Simplicial volume and L^2-invariants). *Let M be an aspherical closed orientable manifold of dimension ≥ 1. Suppose that its simplicial volume $||M||$ vanishes. Then $\widetilde{M}$ is* det-L^2*-acyclic and*

$$\rho^{(2)}(\widetilde{M}) \ = \ 0.$$

The simplicial volume is a special invariant concerning bounded cohomology. The point of this conjecture is that it suggests a connection between bounded cohomology and L^2-invariants such as L^2-cohomology and L^2-torsion.

We have already seen that L^2-Betti numbers are up to scaling invariant under measure equivalence. The next conjecture is interesting because it would give a sharper invariant in case all the L^2-Betti numbers vanish, namely the vanishing of the L^2-torsion.

Conjecture 11.30 (Measure equivalence and L^2-torsion). *Let G_i for $i = 0, 1$ be a group such that there is a finite CW-model for BG_i and EG_i is* det-L^2*-acyclic. Suppose that G_0 and G_1 are measure equivalent. Then*

$$\rho^{(2)}(EG_0;\mathcal{N}(G_0)) = 0 \ \Leftrightarrow \ \rho^{(2)}(EG_1;\mathcal{N}(G_1)) = 0.$$

11.6 L^2-Torsion of Group Automorphisms

In this section we explain that for a group automorphism $f\colon G \to G$ the L^2-torsion applied to the $(G \rtimes_f \mathbb{Z})$-CW-complex $E(G \rtimes_f \mathbb{Z})$ gives an interesting new invariant, provided that G is of det ≥ 1-class and satisfies certain finiteness assumptions. It seems to be worthwhile to investigate it further. The following definition and theorem are taken from [80, Definition 7.26 and Theorem 7.27].

Definition 11.31 (L^2-torsion of group automorphisms). *Let $f\colon G \to G$ be a group automorphism. Suppose that there is a finite CW-model for BG and G is of* det ≥ 1*-class. Define the* L^2-torsion *of f by*

$$\rho^{(2)}(f\colon G \to G) \;:=\rho^{(2)}(\widetilde{B(G \rtimes_f \mathbb{Z})}) \qquad \in \mathbb{R}.$$

Next we present the basic properties of this invariant. Notice that its behaviour is similar to the Euler characteristic $\chi(G) := \chi(BG)$.

Theorem 11.32. *Suppose that all groups appearing below have finite CW-models for their classifying spaces and are of* det ≥ 1*-class.*

(i) Suppose that G is the amalgamated product $G_1 *_{G_0} G_2$ for subgroups $G_i \subseteq G$ and the automorphism $f\colon G \to G$ is the amalgamated product $f_1 *_{f_0} f_2$ for automorphisms $f_i\colon G_i \to G_i$. Then

$$\rho^{(2)}(f) \;=\; \rho^{(2)}(f_1) + \rho^{(2)}(f_2) - \rho^{(2)}(f_0);$$

(ii) Let $f\colon G \to H$ and $g\colon H \to G$ be isomorphisms of groups. Then

$$\rho^{(2)}(f \circ g) \;=\; \rho^{(2)}(g \circ f).$$

In particular $\rho^{(2)}(f)$ is invariant under conjugation by automorphisms;

(iii) Suppose that the following diagram of groups

$$\begin{array}{ccccccccc} 1 & \longrightarrow & G_1 & \longrightarrow & G_2 & \longrightarrow & G_3 & \longrightarrow & 1 \\ & & \downarrow{\scriptstyle f_1} & & \downarrow{\scriptstyle f_2} & & \downarrow{\scriptstyle \mathrm{id}} & & \\ 1 & \longrightarrow & G_1 & \longrightarrow & G_2 & \longrightarrow & G_3 & \longrightarrow & 1 \end{array}$$

commutes, has exact rows and its vertical arrows are automorphisms. Then

$$\rho^{(2)}(f_2) \;=\; \chi(BG_3) \cdot \rho^{(2)}(f_1);$$

(iv) Let $f\colon G \to G$ be an automorphism of a group. Then for all integers $n \geq 1$

$$\rho^{(2)}(f^n) \;=\; n \cdot \rho^{(2)}(f);$$

(v) Suppose that G contains a subgroup G_0 of finite index $[G : G_0]$. Let $f\colon G \to G$ be an automorphism with $f(G_0) = G_0$. Then

$$\rho^{(2)}(f) = \frac{1}{[G : G_0]} \cdot \rho^{(2)}(f|_{G_0});$$

(vi) We have $\rho^{(2)}(f) = 0$ if G satisfies one of the following conditions:

(a) All the L^2-Betti numbers of G vanish;

(b) G contains an amenable infinite normal subgroup.

Example 11.33 (Automorphisms of surfaces)**.** Using Theorem 11.27 one can compute the L^2-torsion of the automorphism $\pi_1(f)$ for an automorphism $f\colon S \to S$ of a compact connected orientable surface, possibly with boundary. Suppose that f is irreducible. Then the following statements are equivalent: i.) f is pseudo-Anosov, ii.) The mapping torus T_f has a hyperbolic structure and iii.) $\rho^{(2)}(\pi_1(f)) < 0$. Moreover, f is periodic if and only if $\rho^{(2)}(\pi_1(f)) = 0$. (see [80, Subsection 7.4.2]).

The L^2-torsion of a Dehn twist is always zero since the associated mapping torus contains no hyperbolic pieces in his Jaco-Shalen-Johannson-Thurston splitting.

Remark 11.34 (Weaker finiteness conditions)**.** The definition of the L^2-torsion of a group automorphism above still makes sends and has still most of the properties above, if one weakens the condition that there is a finite model for BG to the assumption that there is a finite model for the classifying space of proper G-actions $\underline{E}G = E_{\mathcal{FIN}}(G)$. This is explained in [80, Subsection 7.4.4].

12 Novikov-Shubin Invariants

In this section we briefly discuss Novikov-Shubin invariants. They were originally defined in terms of heat kernels. We will focus on their algebraic definition and aspects.

12.1 Definition of Novikov-Shubin Invariants

Let M be a finitely presented $\mathcal{N}(G)$-module. Choose some exact sequence $\mathcal{N}(G)^m \xrightarrow{f} \mathcal{N}(G)^n \to M \to 0$. Let F_f be the spectral density function of f (see Definition 11.1). Recall that F_f is a monotone increasing right continuous function $[0, \infty) \to [0, \infty)$. Define the *Novikov-Shubin invariant* of M by

$$\alpha(M) = \liminf_{\lambda \to 0^+} \frac{\ln(F_f(\lambda) - F_f(0))}{\ln(\lambda)} \in [0, \infty],$$

provided that $F_f(\lambda) > F_f(0)$ holds for all $\lambda > 0$. Otherwise, one puts formally

$$\alpha(M) \; = \; \infty^+.$$

It measures how fast $F_f(\lambda)$ approaches $F_f(0)$ for $\lambda \to 0^+$. For instance, if $F_f(\lambda) = \lambda^\alpha$ for $\lambda > 0$, then $\alpha(M) = \alpha$. The proof that $\alpha(M)$ is independent of the choice of f is analogous to the proof of [80, Theorem 2.55 (1)].

Definition 12.1 (Novikov-Shubin invariants). *Let X be a G-CW-complex of finite type. Define its* p-th Novikov-Shubin invariant *by*

$$\alpha_p(X;\mathcal{N}(G)) \; = \; \alpha\left(H_{p-1}^{(2)}(X;\mathcal{N}(G))\right) \quad \in [0,\infty] \amalg \{\infty^+\}.$$

If the group G is clear from the context, we write $\alpha_p(X) \; = \; \alpha_p(X;\mathcal{N}(G))$.

Notice that $H_{p-1}^{(2)}(X;\mathcal{N}(G))$ is finitely presented since $\mathcal{N}(G)$ is semihereditary (see Theorem 1.6) and $C_k^{(2)}(X)$ is a finitely generated free $\mathcal{N}(G)$-module for all $k \in \mathbb{Z}$ because X is by assumption of finite type.

Remark 12.2 (Analytic definition of Novikov-Shubin invariants). Novikov-Shubin invariants were originally analytically defined by Novikov and Shubin (see [91], [92]). For a cocompact smooth G-manifold M without boundary and with G-invariant Riemannian metric one can assign to its p-th Laplace operator Δ_p a density function $F_{\Delta_p}(\lambda) = \operatorname{tr}_{\mathcal{N}(G)}(E_\lambda)$ for $\{E_\lambda \mid \lambda \in [0,\infty)\}$ the spectral family associated to the essentially selfadjoint operator Δ_p. Define $\alpha_p^\Delta(M;\mathcal{N}(G)) \in [0,\infty] \amalg \{\infty^+\}$ by the same expression as appearing in the definition of $\alpha(M)$ above, only replace F_f by F_{Δ_p}. Then $\alpha_p^\Delta(M)$ agrees with $\frac{1}{2} \cdot \min\{\alpha_p(K), \alpha_{p+1}(K)\}$ for any equivariant triangulation K of M. For a proof of this equality see [31] or [80, Section 2.4]. One can define the analytic Novikov-Shubin invariant $\alpha_p^\Delta(M;\mathcal{N}(G))$ also in terms of heat kernels. It measures how fast the function $\int_{\mathcal{F}} \operatorname{tr}_{\mathbb{C}}(e^{-t\Delta_p}(x,x)) \; dvol_x$ approaches for $t \to \infty$ its limit

$$b_p^{(2)}(M) \; = \; \lim_{t\to\infty} \int_{\mathcal{F}} \operatorname{tr}_{\mathbb{C}}(e^{-t\Delta_p}(x,x)) \; dvol_x.$$

The "thinner" the spectrum of Δ_p is at zero, the larger is $\alpha_p^\Delta(M;\mathcal{N}(G))$.

In view of this original analytic definition the result due to Gromov and Shubin [48] that the Novikov-Shubin invariants are homotopy invariants, is rather surprising.

Remark 12.3 (Analogy to finitely generated $\mathbb{Z}$-modules). Recall Slogan 1.16 that the group von Neumann algebra $\mathcal{N}(G)$ behaves like the ring of integers $\mathbb{Z}$, provided one ignores the properties integral domain and Noetherian. Given a finitely generated abelian group M, the $\mathbb{Z}$-module $M/\operatorname{tors}(M)$ is finitely generated free, there is a $\mathbb{Z}$-isomorphism $M \cong M/\operatorname{tors}(M) \oplus \operatorname{tors}(M)$ and the rank as abelian group of M is $\dim_{\mathbb{Q}}(\mathbb{Q} \otimes_{\mathbb{Z}} M)$ and of $\operatorname{tors}(M)$ is 0. In analogy, given a finitely generated $\mathcal{N}(G)$-module M, then $\mathbf{P}M := M/\mathbf{T}M$ is a finitely generated projective$\mathcal{N}(G)$-module, there is an $\mathcal{N}(G)$-isomorphism $M \cong \mathbf{P}M \oplus \operatorname{tors}(M)$ and we get $\dim_{\mathcal{N}(G)}(M) = \dim_{\mathcal{U}(G)}(\mathcal{U}(G) \otimes_{\mathcal{N}(G)} M)$ and $\dim_{\mathcal{N}(G)}(\mathbf{T}M) = 0$. Define the so called *capacity* $c(M) \in [0, \infty] \cup \{0^-\}$ of a finitely presented $\mathcal{N}(G)$-module M by

$$c(M) = \begin{cases} \frac{1}{\alpha(M)} & \text{if } \alpha(M) \in (0, \infty); \\ \infty & \text{if } \alpha(M) = 0; \\ 0 & \text{if } \alpha(M) = \infty; \\ 0^- & \text{if } \alpha(M) = \infty^+. \end{cases}$$

Then the capacity $c(M)$ contains the same information as $\alpha(M)$ and corresponds under the dictionary between $\mathbb{Z}$ and $\mathcal{N}(G)$ to the order of the finite group $\operatorname{tors}(M)$. Notice for a finitely presented $\mathcal{N}(G)$-module M that $M = 0$ is true if and only if both $\dim_{\mathcal{N}(G)}(M) = 0$ and $c(M) = 0^-$ hold. The capacity is at least subadditive, i.e. for an exact sequence $1 \to M_0 \to M_1 \to M_2 \to 0$ of finitely presented $\mathcal{N}(G)$-modules we have $c(M_1) \leq c(M_0) + c(M_2)$ (with the obvious interpretation of $+$ and $\leq$). In particular we get $c(M) \leq c(N)$ for an inclusion of finitely presented $\mathcal{N}(G)$-modules $M \subseteq N$.

Remark 12.4 (Extension to arbitrary $\mathcal{N}(G)$-modules and G-spaces)**.** The algebraic approach presented above has been independently developed in [33] and [74]. The notion of capacity has been extended by Lück-Reich-Schick [82] to so called cofinal-measurable $\mathcal{N}(G)$-modules, i.e. $\mathcal{N}(G)$-modules such that each finitely generated $\mathcal{N}(G)$-submodule is a quotient of a finitely presented $\mathcal{N}(G)$-module with trivial von Neumann dimension. This allows to define Novikov-Shubin invariants for arbitrary G-spaces and also for arbitrary groups G.

12.2 Basic Properties of Novikov-Shubin Invariants

We briefly list some properties of Novikov-Shubin invariants. The proof of the following theorem can be found in [80, Theorem 2.55] and [80, Lemma 13.45].

Theorem 12.5 (Novikov-Shubin invariants)**.**

(i) Homotopy invariance

Let $f\colon X \to Y$ be a G-map of free G-CW-complexes of finite type. Suppose that the map $H_p(f;\mathbb{C})\colon H_p(X;\mathbb{C}) \to H_p(Y;\mathbb{C})$ induced on homology with complex coefficients is an isomorphism for $p \leq d-1$. Then we get

$$\alpha_p(X;\mathcal{N}(G)) = \alpha_p(Y;\mathcal{N}(G)) \quad \textit{for } p \leq d.$$

In particular we get $\alpha_p(X;\mathcal{N}(G)) = \alpha_p(Y;\mathcal{N}(G))$ for all $p \geq 0$ if f is a weak homotopy equivalence;

(ii) Poincaré duality

Let M be a cocompact free proper G-manifold of dimension n which is orientable. Then $\alpha_p(M;\mathcal{N}(G)) = \alpha_{n+1-p}(M,\partial M;\mathcal{N}(G))$ for $p \geq 1$;

(iii) First Novikov-Shubin invariant

Let X be a connected free G-CW-complex of finite type. Then G is finitely generated and

(a) $\alpha_1(X)$ is finite if and only if G is infinite and virtually nilpotent. In this case $\alpha_1(X)$ is the growth rate of G;

(b) $\alpha_1(X)$ is ∞^+ if and only if G is finite or non-amenable;

(c) $\alpha_1(X)$ is ∞ if and only if G is amenable and not virtually nilpotent;

(iv) Restriction to subgroups of finite index

Let X be a free G-CW-complex of finite type and $H \subseteq G$ a subgroup of finite index. Then $\alpha_p(X;\mathcal{N}(G)) = \alpha_p(\mathrm{res}_G^H X;\mathcal{N}(H))$ holds for $p \geq 0$;

(v) Extensions with finite kernel

Let $1 \to H \to G \to Q \to 1$ be an extension of groups such that H is finite. Let X be a free Q-CW-complex of finite type. Then we get $\alpha_p(p^*X;\mathcal{N}(G)) = \alpha_p(X;\mathcal{N}(Q))$ for all $p \geq 1$;

(vi) Induction

Let H be a subgroup of G and let X be a free H-CW-complex of finite type. Then $\alpha_p(G \times_H X;\mathcal{N}(G)) = \alpha_p(X;\mathcal{N}(H))$ holds for all $p \geq 1$.

A product formula and a formula for connected sums can also be found in [80, Theorem 2.55]. If X is a finite G-CW-complex such that $b_p^{(2)}(X;\mathcal{N}(G)) = 0$ for $p \geq 0$ and $\alpha_p(X;\mathcal{N}(G)) > 0$ for $p \geq 1$, then X is det-L^2-acyclic [80, Theorem 3.93 (7)].

12.3 Computations of Novikov-Shubin Invariants

Example 12.6 (Novikov-Shubin invariants of $\widetilde{T^n}$)**.** The product formula can be used to show $\alpha_p(\widetilde{T^n}) = n$ if $1 \leq p \leq n$, and $\alpha_p(\widetilde{T^n}) = \infty^+$ otherwise (see [80, Example 2.59].)

Example 12.7 (Novikov-Shubin invariants for finite groups)**.** If G is finite, then $\alpha_p(X;\mathcal{N}(G)) = \infty^+$ for each $p \geq 1$ and G-CW-complex X of finite type. This follows from Example 11.2. This shows that the Novikov-Shubin invariants are interesting only for infinite groups G and have no classical analogue in contrast to L^2-Betti numbers and L^2-torsion.

Example 12.8 (Novikov-Shubin invariants for $G = \mathbb{Z}$)**.** Let X be a free $\mathbb{Z}$-CW-complex of finite type. Since $\mathbb{C}[\mathbb{Z}]$ is a principal ideal domain, we get $\mathbb{C}[\mathbb{Z}]$-isomorphisms

$$H_p(X;\mathbb{C}) \cong \mathbb{C}[\mathbb{Z}]^{n_p} \oplus \left(\bigoplus_{i_p=1}^{s_p} \mathbb{C}[\mathbb{Z}]/((z-a_{p,i_p})^{r_{p,i_p}}) \right)$$

for $a_{p,i_p} \in \mathbb{C}$ and $n_p, s_p, r_{p,i_p} \in \mathbb{Z}$ with $n_p, s_p \geq 0$ and $r_{p,i_p} \geq 1$, where z is a fixed generator of $\mathbb{Z}$. Then we get from [80, Lemma 2.58]

$$b_p^{(2)}(X;\mathcal{N}(\mathbb{Z})) = n_p.$$

If $s_p \geq 1$ and $\{i_p = 1, 2 \ldots, s_p, |a_{p,i_p}| = 1\} \neq \emptyset$, then

$$\alpha_{p+1}(X;\mathcal{N}(\mathbb{Z})) = \min\left\{ \frac{1}{r_{p,i_p}} \mid i_p = 1, 2 \ldots, s_p, |a_{p,i_p}| = 1 \right\},$$

and otherwise

$$\alpha_{p+1}(X;\mathcal{N}(\mathbb{Z})) = \infty^+.$$

Remark 12.9 (Novikov-Shubin invariants and S^1-actions)**.** Under the conditions of Theorem 3.8 and of Theorem 3.9 one can show $\alpha_p(\widetilde{X}) \geq 1$ for all $p \geq 1$ (see [80, Theorem 2.61 and Theorem 2.63]).

Remark 12.10 (Novikov-Shubin invariants of symmetric spaces of non–compact type)**.** The Novikov-Shubin invariants of symmetric spaces of non-compact type with cocompact free G-action are computed by Olbricht [93, Theorem 1.1], the result is also stated in [80, Section 5.3].

Remark 12.11 (Novikov-Shubin invariants of universal coverings of 3-manifolds)**.** Partial results about the computation of the Novikov-Shubin invariants of universal coverings of compact orientable 3-manifolds can be found in [69] and [80, Theorem 4.2].

12.4 Open conjectures about Novikov-Shubin invariants

The following conjecture is taken from [69, Conjecture 7.1 on page 56].

Conjecture 12.12. (Positivity and rationality of Novikov-Shubin invariants). *Let G be a group. Then for any free G-CW-complex X of finite type its Novikov-Shubin invariants $\alpha_p(X)$ are positive rational numbers unless they are ∞ or ∞^+.*

This conjecture is equivalent to the statement that for every finitely presented $\mathbb{Z}G$-module M the Novikov-Shubin invariant of $\mathcal{N}(G) \otimes_{\mathbb{Z}G} M$ is a positive rational number, ∞ or ∞^+.

Here is some evidence for Conjecture 12.12. Unfortunately, all the evidence comes from computations, no convincing conceptual reason is known. Conjecture 12.12 is true for $G = \mathbb{Z}$ by the explicit computation appearing in Example 12.8. Conjecture 12.12 is true for virtually abelian G by [66, Proposition 39 on page 494]. Conjecture 12.12 is also true for a free group G. Details of the proof appear in the Ph.D. thesis of Roman Sauer [100] following ideas of Voiculescu. The essential ingredients are non-commutative power series and the question whether they are algebraic or rational. All the computations mentioned above are compatible and give evidence for Conjecture 12.12.

Conjecture 12.13 (Zero-in-the-spectrum Conjecture). *Let G be a group such that BG has a closed manifold as model. Then there is $p \geq 0$ with $H_p^G(EG; \mathcal{N}(G)) \neq 0$.*

Remark 12.14 (The original zero-in-the-spectrum Conjecture). The original zero-in-the-spectrum Conjecture, which appears for the first time in Gromov's article [43, page 120], says the following: Let $\widetilde{M}$ be a complete Riemannian manifold. Suppose that $\widetilde{M}$ is the universal covering of an aspherical closed Riemannian manifold M (with the Riemannian metric coming from M). Then for some $p \geq 0$ zero is in the spectrum of the minimal closure

$$(\Delta_p)_{\min} \colon \operatorname{dom}((\Delta_p)_{\min}) \subseteq L^2\Omega^p(\widetilde{M}) \to L^2\Omega^p(\widetilde{M})$$

of the Laplacian acting on smooth p-forms on $\widetilde{M}$.

It follows from [80, Lemma 12.3] that this formulation is equivalent to the homological algebraic formulation appearing in Conjecture 12.13.

Remark 12.15 (Status of the zero-in-the-spectrum Conjecture). The zero-in-the-spectrum Conjecture is true for G if there is a closed manifold model for BG which is Kähler hyperbolic [45], or whose universal covering is hyper-Euclidean [43] or is uniformly contractible with finite asymptotic dimension [110]. The zero-in-the-spectrum Conjecture is true for G if the strong Novikov Conjecture holds for G [67]. More information about zero-in-the-spectrum Conjecture can be found for instance in [67] and [80, Section 12].

Remark 12.16 (Variations of the zero-in-the-spectrum Conjecture). One may ask whether one can weaken the condition in Conjecture 12.13 that BG has a closed manifold model to the condition that there is a finite CW-complex model for BG. This would rule out Poincaré duality from the picture. Or one could only require that BG is of finite type. Without any finiteness conditions on G Conjecture 12.13 is not true in general. For instance $H_p^G(EG; \mathcal{N}(G)) = 0$ holds for all $p \geq 0$ if G is $\prod_{i=1}^{\infty} \mathbb{Z} * \mathbb{Z}$.

The condition aspherical cannot be dropped. Farber and Weinberger [35] proved the existence of a closed non-aspherical manifold M with fundamental group π a product of three copies of $\mathbb{Z} * \mathbb{Z}$ such that $H_p^\pi(\widetilde{M};\mathcal{N}(\pi))$ vanishes for all $p \geq 0$. Later Higson-Roe-Schick [52] proved that one can find for every finitely presented group π, for which $H_p^\pi(E\pi;\mathcal{N}(\pi)) = 0$ holds for $p = 0, 1, 2$, a closed manifold M with π as fundamental group such that $H_p^\pi(\widetilde{M};\mathcal{N}(\pi))$ vanishes for all $p \geq 0$.

Remark 12.17 (Novikov-Shubin invariants and quasi-isometry)**.** Since one has that $\alpha_1(\mathbb{Z}^n) = n$ for $n \geq 1$, the Novikov-Shubin invariants are not invariant under measure equivalence. It is not known whether they are invariant under quasi-isometry. At least it is known that two quasi-isometric amenable groups G_1 and G_2 which possess finite models for BG_1 and BG_2 have the same Novikov-Shubin invariants [100]. Compare also Theorem 8.2, Remark 8.5 and Conjecture 11.30.

13 A combinatorial approach to L^2-invariants

In this section we want to give a more combinatorial approach to L^2-invariants such as L^2-Betti numbers, Novikov-Shubin invariants and L^2-torsion. The point is that it is in general very hard to compute the spectral density function of an $\mathcal{N}(G)$-map $f\colon \mathcal{N}(G)^m \to \mathcal{N}(G)^n$. However in the geometric situation these morphisms are induced by matrices over the integral group ring $\mathbb{Z}G$. We want to exploit this information to get an algorithm which produces a sequence of rational numbers converging to the L^2-Betti number or the L^2-torsion in question.

Let $A \in M(m, n; \mathbb{C}G)$ be an (m, n)-matrix over $\mathbb{C}G$. It induces by right multiplication an $\mathcal{N}(G)$-homomorphism $r_A : \mathcal{N}(G)^m \to \mathcal{N}(G)^n$. We define an involution of rings on $\mathbb{C}G$ by sending $\sum_{g\in G} \lambda_g \cdot g$ to $\sum_{g\in G} \overline{\lambda_g} \cdot g^{-1}$, where $\overline{\lambda_g}$ is the complex conjugate of λ_g. Denote by A^* the (m, n)-matrix obtained from A by transposing and applying the involution above to each entry. Define the *$\mathbb{C}G$-trace* of an element $u = \sum_{g\in G} \lambda_g \cdot g \in \mathbb{C}G$ by the complex number $\operatorname{tr}_{\mathbb{C}G}(u) := \lambda_e$ for e the unit element in G. This extends to a trace of square (n, n)-matrices A over $\mathbb{C}G$ by

$$\operatorname{tr}_{\mathbb{C}G}(A) \quad := \quad \sum_{i=1}^{n} \operatorname{tr}_{\mathbb{C}G}(a_{i,i}) \quad \in \mathbb{C}. \tag{13.1}$$

We get directly from the definitions that the $\mathbb{C}G$-trace $\operatorname{tr}_{\mathbb{C}G}(u)$ for $u \in \mathbb{C}G$ agrees with the von Neumann trace $\operatorname{tr}_{\mathcal{N}(G)}(u)$ introduced in Definition 1.8.

Let $A \in M(m, n; \mathbb{C}G)$ be an (m, n)-matrix over $\mathbb{C}G$. In the sequel let K be any positive real number satisfying $K \geq ||r_A^{(2)}||$, where $||r_A^{(2)}||$ is the operator norm of the bounded G-equivariant operator $r_A^{(2)} : l^2(G)^m \to l^2(G)^n$ induced by right multiplication with A. For $u = \sum_{g \in G} \lambda_g \cdot g \in \mathbb{C}G$ define $||u||_1$ by $\sum_{g \in G} |\lambda_g|$. Then a possible choice for K is

$$K = \sqrt{(2n-1)m} \cdot \max\left\{||a_{i,j}||_1 \mid 1 \leq i \leq n, 1 \leq j \leq m\right\}.$$

Definition 13.2. *The* characteristic sequence *of a matrix $A \in M(m,n;\mathbb{C}G)$ and a non-negative real number K satisfying $K \geq ||r_A^{(2)}||$ is the sequence of real numbers given by*

$$c(A,K)_p := \operatorname{tr}_{\mathbb{C}G}\left(\left(1 - K^{-2} \cdot AA^*\right)^p\right).$$

We have defined $\dim_{\mathcal{N}(G)}(\ker(r_A))$ in Definition 1.12 and $\det_{\mathcal{N}(G)}(r_A)$ in Definition 11.4. The proof of the following result can be found in [73] or [80, Theorem 3.172].

Theorem 13.3. (Combinatorial computation of L^2-invariants). *Let $A \in M(m,n;\mathbb{C}G)$ be an (m,n)-matrix over $\mathbb{C}G$. Let K be a positive real number satisfying $K \geq ||r_A^{(2)}||$. Then:*

(i) Monotony

The characteristic sequence $(c(A,K)_p)_{p \geq 1}$ is a monotone decreasing sequence of non-negative real numbers;

(ii) Dimension of the kernel

We have

$$\dim_{\mathcal{N}(G)}(\ker(r_A)) = \lim_{p \to \infty} c(A,K)_p;$$

(iii) Novikov-Shubin invariants of the cokernel

Define $\beta(A) \in [0,\infty]$ to be equal to

$$\sup\left\{\beta \in [0,\infty) \;\middle|\; \lim_{p \to \infty} p^\beta \cdot \left(c(A,K)_p - \dim_{\mathcal{N}(G)}(\ker(r_A))\right) = 0\right\}.$$

If $\alpha(\operatorname{coker}(r_A)) < \infty$, then $\alpha(\operatorname{coker}(r_A)) \leq \beta(A)$ and if $\alpha(\operatorname{coker}(r_A)) \in \{\infty, \infty^+\}$, then $\beta(A) = \infty$;

(iv) Fuglede-Kadison determinant

The sum of positive real numbers

$$\sum_{p=1}^{\infty} \frac{1}{p} \cdot \left(c(A,K)_p - \dim_{\mathcal{N}(G)}(\ker(r_A))\right)$$

converges if and only if r_A is of determinant class and in this case

$$\begin{aligned}\ln(\det(r_A)) &= (n - \dim_{\mathcal{N}(G)}(\ker(r_A))) \cdot \ln(K) \\ &\quad - \frac{1}{2} \cdot \sum_{p=1}^{\infty} \frac{1}{p} \cdot \left(c(A,K)_p - \dim_{\mathcal{N}(G)}(\ker(r_A))\right);\end{aligned}$$

(v) Speed of convergence

Suppose $\alpha(\operatorname{coker}(r_A)) > 0$. *Then* r_A *is of determinant class. Given a real number* α *satisfying* $0 < \alpha < \alpha(\operatorname{coker}(r_A))$, *there is a real number* C *such that we have for all* $L \geq 1$

$$0 \leq c(A,K)_L - \dim_{\mathcal{N}(G)}(\ker(r_A)) \leq \frac{C}{L^\alpha}$$

and

$$\begin{aligned} 0 \quad \leq \quad & -\ln(\det(r_A)) + (n - \dim_{\mathcal{N}(G)}(\ker(r_A))) \cdot \ln(K) \\ & -\frac{1}{2} \cdot \sum_{p=1}^{L} \frac{1}{p} \cdot \left(c(A,K)_p - \dim_{\mathcal{N}(G)}(\ker(r_A))\right) \quad \leq \quad \frac{C}{L^\alpha}. \end{aligned}$$

Remark 13.4 (Vanishing of L^2-Betti numbers and the Atiyah Conjecture). Suppose that the Atiyah Conjecture 4.1 is satisfied for $(G, d, \mathbb{C})$. If we want to show the vanishing of $\dim_{\mathcal{N}(G)}(\ker(r_A))$, it suffices to show that for some $p \geq 0$ we have $c(A,K)_p < \frac{1}{d}$. It is possible that a computer program spits out such a value after a reasonable amount of calculation time.

14 Miscellaneous

The analytic aspects of L^2-invariants are also very interesting. We have already mentioned that L^2-Betti numbers were originally defined by Atiyah [2] in context with his L^2-index theorem. Other L^2-invariants are the L^2-Eta-invariant and the L^2-Rho-invariant (see Cheeger-Gromov [13], [14]). The L^2-Eta-invariant appears in the L^2-index theorem for manifolds with boundary due to Ramachandran [97]. These index theorems have generalizations to a C^*-setting due to Miščenko-Fomenko [89]. There is also an L^2-version of the signature. It plays an important role in the work of Cochran, Orr and Teichner [16] who show that there are non-slice knots in 3-space whose Casson-Gordon invariants are all trivial. Chang and Weinberger [11] show using L^2-invariants that for a closed oriented smooth manifold M of dimension $4k+3$ for $k \geq 1$ whose fundamental group has torsion there are infinitely many smooth manifolds which are homotopy equivalent to M (and even simply and tangentially homotopy equivalent to M) but not homeomorphic to M. The L^2-cohomology has also been investigated for complete non-necessarily compact Riemannian manifolds without a group action. For instance algebraic and arithmetic varieties have been studied. In particular, the Cheeger-Goresky-MacPherson Conjecture [12] and the Zucker Conjecture [111] have created a lot of activity. They link the L^2-cohomology of the regular part with the intersection homology of an algebraic variety.

Finally we mention other survey articles which deal with L^2-invariants: [30], [39], [46, Section 8], [67], [75], [78], [79], [87] and [96].

References

[1] P. Ara and D. Goldstein. A solution of the matrix problem for Rickart C^*-algebras. *Math. Nachr.*, 164:259–270, 1993.

[2] M. F. Atiyah. Elliptic operators, discrete groups and von Neumann algebras. *Astérisque*, 32-33:43–72, 1976.

[3] W. Ballmann and J. Brüning. On the spectral theory of manifolds with cusps. *J. Math. Pures Appl. (9)*, 80(6):593–625, 2001.

[4] M. E. B. Bekka and A. Valette. Group cohomology, harmonic functions and the first L^2-Betti number. *Potential Anal.*, 6(4):313–326, 1997.

[5] S. K. Berberian. *Baer *-rings*. Springer-Verlag, New York, 1972. Die Grundlehren der mathematischen Wissenschaften, Band 195.

[6] A. Borel. The L^2-cohomology of negatively curved Riemannian symmetric spaces. *Ann. Acad. Sci. Fenn. Ser. A I Math.*, 10:95–105, 1985.

[7] M. G. Brin and C. C. Squier. Groups of piecewise linear homeomorphisms of the real line. *Invent. Math.*, 79(3):485–498, 1985.

[8] K. S. Brown and R. Geoghegan. An infinite-dimensional torsion-free fp_∞ group. *Invent. Math.*, 77(2):367–381, 1984.

[9] D. Burghelea, L. Friedlander, T. Kappeler, and P. McDonald. Analytic and Reidemeister torsion for representations in finite type Hilbert modules. *Geom. Funct. Anal.*, 6(5):751–859, 1996.

[10] J. W. Cannon, W. J. Floyd, and W. R. Parry. Introductory notes on Richard Thompson's groups. *Enseign. Math. (2)*, 42(3-4):215–256, 1996.

[11] S. Chang and S. Weinberger. On invariants of Hirzebruch and Cheeger. Preprint, 2003.

[12] J. Cheeger, M. Goresky, and R. MacPherson. L^2-cohomology and intersection homology of singular algebraic varieties. In *Seminar on Differential Geometry*, pages 303–340. Princeton Univ. Press, Princeton, N.J., 1982.

[13] J. Cheeger and M. Gromov. Bounds on the von Neumann dimension of L^2-cohomology and the Gauss-Bonnet theorem for open manifolds. *J. Differential Geom.*, 21(1):1–34, 1985.

[14] J. Cheeger and M. Gromov. On the characteristic numbers of complete manifolds of bounded curvature and finite volume. In *Differential geometry and complex analysis*, pages 115–154. Springer-Verlag, Berlin, 1985.

[15] J. Cheeger and M. Gromov. L_2-cohomology and group cohomology. *Topology*, 25(2):189–215, 1986.

[16] T. Cochran, K. Orr, and P. Teichner. Knot concordance, Whitney towers and L^2-signatures. Preprint, to appear in *Ann. of Math.*, 1999.

[17] M. M. Cohen. *A course in simple-homotopy theory.* Springer-Verlag, New York, 1973. Graduate Texts in Mathematics, Vol. 10.

[18] A. Connes and D. Shlyakhtenko. L^2-homology for von Neumann algebras. Preprint, 2003.

[19] M. Davis and I. Leary. The L^2-cohomology of Artin groups. Preprint, to appear in LMS, 2001.

[20] M. W. Davis and B. Okun. Vanishing theorems and conjectures for the ℓ^2-homology of right-angled Coxeter groups. *Geom. Topol.*, 5:7–74 (electronic), 2001.

[21] P. de la Harpe. *Topics in geometric group theory.* University of Chicago Press, Chicago, IL, 2000.

[22] W. Dicks and T. Schick. The spectral measure of certain elements of the complex group ring of a wreath product. *Geom. Dedicata*, 93:121–137, 2002.

[23] J. Dixmier. *Von Neumann algebras.* North-Holland Publishing Co., Amsterdam, 1981. With a preface by E. C. Lance, translated from the second French edition by F. Jellett.

[24] J. Dodziuk. De Rham-Hodge theory for L^2-cohomology of infinite coverings. *Topology*, 16(2):157–165, 1977.

[25] J. Dodziuk. L^2harmonic forms on rotationally symmetric Riemannian manifolds. *Proc. Amer. Math. Soc.*, 77(3):395–400, 1979.

[26] H. Donnelly and F. Xavier. On the differential form spectrum of negatively curved Riemannian manifolds. *Amer. J. Math.*, 106(1):169–185, 1984.

[27] B. Eckmann. Lattices and l_2-Betti numbers. Preprint, Zürich, 2002.

[28] B. Eckmann. Lattices with vanishing first L^2-Betti number and deficiency one. Preprint, Zürich, 2002.

[29] B. Eckmann. 4-manifolds, group invariants, and l_2-Betti numbers. *Enseign. Math. (2)*, 43(3-4):271–279, 1997.

[30] B. Eckmann. Introduction to l_2-methods in topology: reduced l_2-homology, harmonic chains, l_2-Betti numbers. *Israel J. Math.*, 117:183–219, 2000. Notes prepared by G. Mislin.

[31] A. Efremov. Combinatorial and analytic Novikov Shubin invariants. Preprint, 1991.

[32] G. Elek. The rank of finitely generated modules over group rings. Preprint, 2003.

[33] M. Farber. Homological algebra of Novikov-Shubin invariants and Morse inequalities. *Geom. Funct. Anal.*, 6(4):628–665, 1996.

[34] M. Farber. Geometry of growth: approximation theorems for L^2-invariants. *Math. Ann.*, 311(2):335–375, 1998.

[35] M. Farber and S. Weinberger. On the zero-in-the-spectrum conjecture. *Ann. of Math. (2)*, 154(1):139–154, 2001.

[36] A. Furman. Gromov's measure equivalence and rigidity of higher rank lattices. *Ann. of Math. (2)*, 150(3):1059–1081, 1999.

[37] A. Furman. Orbit equivalence rigidity. *Ann. of Math. (2)*, 150(3):1083–1108, 1999.

[38] D. Gaboriau. Invariants l^2 de relation d'équivalence et de groupes. Preprint, Lyon, 2001.

[39] D. Gaboriau. On orbit equivalence of measure preserving actions. Preprint, Lyon, 2001.

[40] R. I. Grigorchuk, P. A. Linnell, T. Schick, and A. Żuk. On a question of Atiyah. *C. R. Acad. Sci. Paris Sér. I Math.*, 331(9):663–668, 2000.

[41] R. I. Grigorchuk and A. Żuk. The Lamplighter group as a group generated by a 2-state automaton, and its spectrum. *Geom. Dedicata*, 87(1-3):209–244, 2001.

[42] M. Gromov. Volume and bounded cohomology. *Inst. Hautes Études Sci. Publ. Math.*, (56):5–99 (1983), 1982.

[43] M. Gromov. Large Riemannian manifolds. In *Curvature and topology of Riemannian manifolds (Katata, 1985)*, pages 108–121. Springer-Verlag, Berlin, 1986.

[44] M. Gromov. Hyperbolic groups. In *Essays in group theory*, pages 75–263. Springer-Verlag, New York, 1987.

[45] M. Gromov. Kähler hyperbolicity and L_2-Hodge theory. *J. Differential Geom.*, 33(1):263–292, 1991.

[46] M. Gromov. Asymptotic invariants of infinite groups. In *Geometric group theory, Vol. 2 (Sussex, 1991)*, pages 1–295. Cambridge Univ. Press, Cambridge, 1993.

[47] M. Gromov. *Metric structures for Riemannian and non-Riemannian spaces.* Birkhäuser Boston Inc., Boston, MA, 1999. Based on the 1981 French original [MR 85e:53051], with appendices by M. Katz, P. Pansu and S. Semmes, Translated from the French by S. M. Bates.

[48] M. Gromov and M. A. Shubin. Von Neumann spectra near zero. *Geom. Funct. Anal.*, 1(4):375–404, 1991.

[49] J.-C. Hausmann and S. Weinberger. Caractéristiques d'Euler et groupes fondamentaux des variétés de dimension 4. *Comment. Math. Helv.*, 60(1):139–144, 1985.

[50] J. Hempel. 3-*Manifolds.* Princeton University Press, Princeton, N. J., 1976. Ann. of Math. Studies, No. 86.

[51] E. Hess and T. Schick. L^2-torsion of hyperbolic manifolds. *Manuscripta Math.*, 97(3):329–334, 1998.

[52] N. Higson, J. Roe, and T. Schick. Spaces with vanishing l^2-homology and their fundamental groups (after Farber and Weinberger). *Geom. Dedicata*, 87(1-3):335–343, 2001.

[53] J. A. Hillman. Deficiencies of lattices in connected Lie groups. Preprint, 1999.

[54] N. Hitchin. Compact four-dimensional Einstein manifolds. *J. Differential Geometry*, 9:435–441, 1974.

[55] C. Hog, M. Lustig, and W. Metzler. Presentation classes, 3-manifolds and free products. In *Geometry and topology (College Park, Md., 1983/84)*, pages 154–167. Springer-Verlag, Berlin, 1985.

[56] J. Howie. On locally indicable groups. *Math. Z.*, 180(4):445–461, 1982.

[57] N. Ivanov. Foundations of the theory of bounded cohomology. *J. Soviet Math.*, 37:1090–1114, 1987.

[58] F. E. A. Johnson and D. Kotschick. On the signature and Euler characteristic of certain four-manifolds. *Math. Proc. Cambridge Philos. Soc.*, 114(3):431–437, 1993.

[59] J. Jost and Y. L. Xin. Vanishing theorems for L^2-cohomology groups. *J. Reine Angew. Math.*, 525:95–112, 2000.

[60] R. V. Kadison and J. R. Ringrose. *Fundamentals of the theory of operator algebras. Vol. II.* Academic Press Inc., Orlando, FL, 1986. Advanced theory.

[61] E. Kaniuth. Der Typ der regulären Darstellung diskreter Gruppen. *Math. Ann.*, 182:334–339, 1969.

[62] A. W. Knapp. *Representation theory of semisimple groups.* Princeton University Press, Princeton, NJ, 1986. An overview based on examples.

[63] D. Kotschick. Four-manifold invariants of finitely presentable groups. In *Topology, geometry and field theory*, pages 89–99. World Sci. Publishing, River Edge, NJ, 1994.

[64] P. A. Linnell. Zero divisors and $L^2(G)$. *C. R. Acad. Sci. Paris Sér. I Math.*, 315(1):49–53, 1992.

[65] P. A. Linnell. Division rings and group von Neumann algebras. *Forum Math.*, 5(6):561–576, 1993.

[66] J. Lott. Heat kernels on covering spaces and topological invariants. *J. Differential Geom.*, 35(2):471–510, 1992.

[67] J. Lott. The zero-in-the-spectrum question. *Enseign. Math. (2)*, 42(3-4):341–376, 1996.

[68] J. Lott. Deficiencies of lattice subgroups of Lie groups. *Bull. London Math. Soc.*, 31(2):191–195, 1999.

[69] J. Lott and W. Lück. L^2-topological invariants of 3-manifolds. *Invent. Math.*, 120(1):15–60, 1995.

[70] A. Lubotzky. Group presentation, p-adic analytic groups and lattices in $SL_2(\mathbb{C})$. *Ann. of Math. (2)*, 118(1):115–130, 1983.

[71] W. Lück. *Transformation groups and algebraic K-theory.* Springer-Verlag, Berlin, 1989. Mathematica Gottingensis.

[72] W. Lück. Approximating L^2-invariants by their finite-dimensional analogues. *Geom. Funct. Anal.*, 4(4):455–481, 1994.

[73] W. Lück. L^2-torsion and 3-manifolds. In *Low-dimensional topology (Knoxville, TN, 1992)*, pages 75–107. Internat. Press, Cambridge, MA, 1994.

[74] W. Lück. Hilbert modules and modules over finite von Neumann algebras and applications to L^2-invariants. *Math. Ann.*, 309(2):247–285, 1997.

[75] W. Lück. L^2-Invarianten von Mannigfaltigkeiten und Gruppen. *Jahresber. Deutsch. Math.-Verein.*, 99(3):101–109, 1997.

[76] W. Lück. Dimension theory of arbitrary modules over finite von Neumann algebras and L^2-Betti numbers. I. Foundations. *J. Reine Angew. Math.*, 495:135–162, 1998.

[77] W. Lück. Dimension theory of arbitrary modules over finite von Neumann algebras and L^2-Betti numbers. II. Applications to Grothendieck groups, L^2-Euler characteristics and Burnside groups. *J. Reine Angew. Math.*, 496:213–236, 1998.

[78] W. Lück. L^2-invariants and their applications to geometry, group theory and spectral theory. In *Mathematics Unlimited—2001 and Beyond*, pages 859–871. Springer-Verlag, Berlin, 2001.

[79] W. Lück. L^2-invariants of regular coverings of compact manifolds and CW-complexes. In *Handbook of geometric topology*, pages 735–817. North-Holland, Amsterdam, 2002.

[80] W. Lück. *L^2-invariants: theory and applications to geometry and K-theory*, volume 44 of *Ergebnisse der Mathematik und ihrer Grenzgebiete. 3. Folge*. Springer-Verlag, Berlin, 2002.

[81] W. Lück and H. Reich. The Baum-Connes and the Farrell-Jones conjectures in K- and L-theory. Preprintreihe SFB 478 — Geometrische Strukturen in der Mathematik, Heft 324, Münster, arXiv:math.GT/0402405, accepted for the handbook of K-theory, Springer, 2003.

[82] W. Lück, H. Reich, and T. Schick. Novikov-Shubin invariants for arbitrary group actions and their positivity. In *Tel Aviv Topology Conference: Rothenberg Festschrift (1998)*, pages 159–176. Amer. Math. Soc., Providence, RI, 1999.

[83] W. Lück and M. Rørdam. Algebraic K-theory of von Neumann algebras. *K-Theory*, 7(6):517–536, 1993.

[84] W. Lück and T. Schick. L^2-torsion of hyperbolic manifolds of finite volume. *Geometric and Functional Analysis*, 9:518–567, 1999.

[85] R. C. Lyndon and P. E. Schupp. *Combinatorial group theory*. Springer-Verlag, Berlin, 1977. Ergebnisse der Mathematik und ihrer Grenzgebiete, Band 89.

[86] V. Mathai. L^2-analytic torsion. *J. Funct. Anal.*, 107(2):369–386, 1992.

[87] V. Mathai. L^2-invariants of covering spaces. In *Geometric analysis and Lie theory in mathematics and physics*, pages 209–242. Cambridge Univ. Press, Cambridge, 1998.

[88] F. I. Mautner. The structure of the regular representation of certain discrete groups. *Duke Math. J.*, 17:437–441, 1950.

[89] A. S. Miščenko and A. T. Fomenko. The index of elliptic operators over C^*-algebras. *Izv. Akad. Nauk SSSR Ser. Mat.*, 43(4):831–859, 967, 1979. English translation in *Math. USSR-Izv.* 15 (1980), no. 1, 87–112.

[90] J. W. Morgan. On Thurston's uniformization theorem for three-dimensional manifolds. In *The Smith conjecture (New York, 1979)*, pages 37–125. Academic Press, Orlando, FL, 1984.

[91] S. P. Novikov and M. A. Shubin. Morse inequalities and von Neumann II_1-factors. *Dokl. Akad. Nauk SSSR*, 289(2):289–292, 1986.

[92] S. P. Novikov and M. A. Shubin. Morse inequalities and von Neumann invariants of non-simply connected manifolds. *Uspekhi. Matem. Nauk*, 41(5):222–223, 1986. in Russian.

[93] M. Olbrich. L^2-invariants of locally symmetric spaces. Preprint, Göttingen, 2000.

[94] D. S. Ornstein and B. Weiss. Ergodic theory of amenable group actions. I. The Rohlin lemma. *Bull. Amer. Math. Soc. (N.S.)*, 2(1):161–164, 1980.

[95] P. Pansu. Cohomologie L^p: Invariance sous quasiisometries. Preprint, Orsay, 1995.

[96] P. Pansu. Introduction to L^2-Betti numbers. In *Riemannian geometry (Waterloo, ON, 1993)*, pages 53–86. Amer. Math. Soc., Providence, RI, 1996.

[97] M. Ramachandran. Von Neumann index theorems for manifolds with boundary. *J. Differential Geom.*, 38(2):315–349, 1993.

[98] H. Reich. *Group von Neumann algebras and related algebras.* PhD thesis, Universität Göttingen, 1999. http://www.math.uni-muenster.de/u/lueck/publ/diplome/reich.dvi.

[99] H. Reich. On the K- and L-theory of the algebra of operators affiliated to a finite von Neumann algebra. *K-theory*, 24:303–326, 2001.

[100] R. Sauer. *L^2-Invariants of groups and discrete measured groupoids.* PhD thesis, Universität Münster, 2002.

[101] T. Schick. Integrality of L^2-Betti numbers. *Math. Ann.*, 317(4):727–750, 2000.

[102] T. Schick. L^2-determinant class and approximation of L^2-Betti numbers. *Trans. Amer. Math. Soc.*, 353(8):3247–3265 (electronic), 2001.

[103] E. Thoma. Über unitäre Darstellungen abzählbarer, diskreter Gruppen. *Math. Ann.*, 153:111–138, 1964.

[104] T. tom Dieck. *Transformation groups.* Walter de Gruyter & Co., Berlin, 1987.

[105] C. T. C. Wall. Rational Euler characteristics. *Proc. Cambridge Philos. Soc.*, 57:182–184, 1961.

[106] C. Wegner. *L^2-invariants of finite aspherical CW-complexes with fundamental group containing a non-trivial elementary amenable normal subgroup.* PhD thesis, Westfälische Wilhelms-Universität Münster, 2000.

[107] C. Wegner. L^2-invariants of finite aspherical CW-complexes. Preprintreihe SFB 478 — Geometrische Strukturen in der Mathematik, Heft 152, Münster, 2001.

[108] C. A. Weibel. *An introduction to homological algebra.* Cambridge University Press, Cambridge, 1994.

[109] K. Whyte. Amenability, bi-Lipschitz equivalence, and the von Neumann conjecture. *Duke Math. J.*, 99(1):93–112, 1999.

[110] G. Yu. The Novikov conjecture for groups with finite asymptotic dimension. *Ann. of Math. (2)*, 147(2):325–355, 1998.

[111] S. Zucker. L_2-cohomology and intersection homology of locally symmetric varieties. In *Singularities, Part 2 (Arcata, Calif., 1981)*, pages 675–680. Amer. Math. Soc., Providence, RI, 1983.

Notation

Index

Constructing non-positively curved spaces and groups

Jon McCammond*

Abstract

The theory of non-positively curved spaces and groups is tremendously powerful and has enormous consequences for the groups and spaces involved. Nevertheless, our ability to construct examples to which the theory can be applied has been severely limited by an inability to test – in real time – whether a random finite piecewise Euclidean complex is non-positively curved. In this article I focus on the question of how to construct examples of non-positively curved spaces and groups, highlighting in particular the boundary between what is currently do-able and what is not yet feasible. Since this is intended primarily as a survey, the key ideas are merely sketched with references pointing the interested reader to the original articles.

Over the past decade or so, the consequences of non-positive curvature for geometric group theorists have been thoroughly investigated, most prominently in the book by Bridson and Haefliger [26]. See also the recent review article by Kleiner in the Bulletin of the AMS [59] and the related books by Ballmann [4], Ballmann-Gromov-Schroeder [5] and the original long article by Gromov [48]. In this article I focus not on the consequences of the theory, but rather on the question of how to construct examples to which it applies. The structure of the article roughly follows the structure of the lectures I gave during the Durham symposium with the four parts corresponding to the four talks. Part 1 introduces the key problems and presents some basic decidability results, Part 2 focuses on practical algorithms and low-dimensional complexes, Part 3 presents case studies involving special classes of groups such as Artin groups, small cancellation groups and ample-twisted face pairing 3-manifolds. In Part 4 I explore the weaker notion of conformal non-positive curvature and introduce the notion of an angled n-complex. As will become clear, the topics covered have a definite bias towards research in which I have played some role. These are naturally the results with which I am most familiar and I hope the reader will pardon this lack of a more impartial perspective.

*Partially supported by NSF grant no. DMS-0101506

1 Negative curvature

Although the definitions of δ-hyperbolic and CAT(0) spaces / groups are well-known, it is perhaps under-appreciated that there are at least seven potentially distinct classes of groups which all have some claim to the description "negatively-curved." Sections 1.1 and 1.2 briefly review the definitions needed in order to describe these classes, the known relationships between them, and the current status of the (rather optimisitic) conjecture that all seven of these classes are identical. Since the emphasis in this article is on the construction of examples, finite piecewise Euclidean complexes play a starring role. Section 1.3 describes how, in theory, one can decide whether such a complex is non-positively curved and Section 1.4 concludes with a few comments about the length spectrum of such a complex.

1.1 Varieties of negative curvature

In this section I define the basic objects of study and highlight the distinctions between classes of groups which are often treated interchangably by newcomers to the field. Because the main definitions and results in this section are well-known, the presentation is brief and impressionistic. A more detailed overview covering similar territory can be found in the expository article by Martin Bridson [23] or the one by Ruth Charney [35]. The most basic objects are δ-hyperbolic spaces and CAT(κ) spaces. Recall that a *geodesic* is a length minimizing curve and that a *geodesic metric space* is a metric space in which every pair of points is connected by a geodesic.

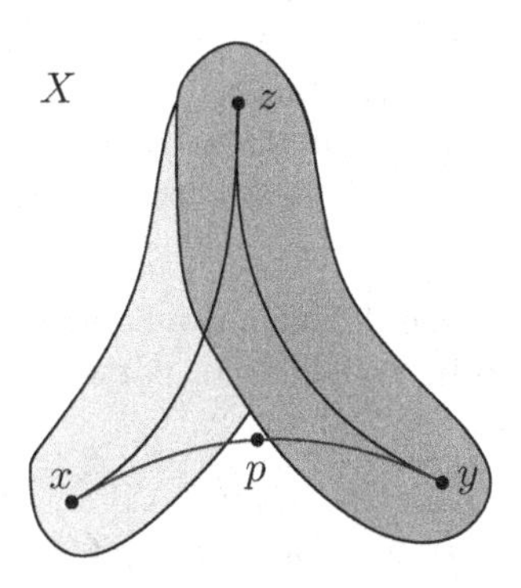

Figure 1: A triangle which is not quite δ-thin.

Definition 1.1 (δ-hyperbolic) *A geodesic metric space X is* δ-hyperbolic *if there is a fixed $\delta \geq 0$ such that for all points $x, y, z \in X$ and for all geodesics connecting x, y, and z and for all points p on the chosen geodesic connecting*

x to y, the distance from p to the union of the other two geodesics is at most δ.

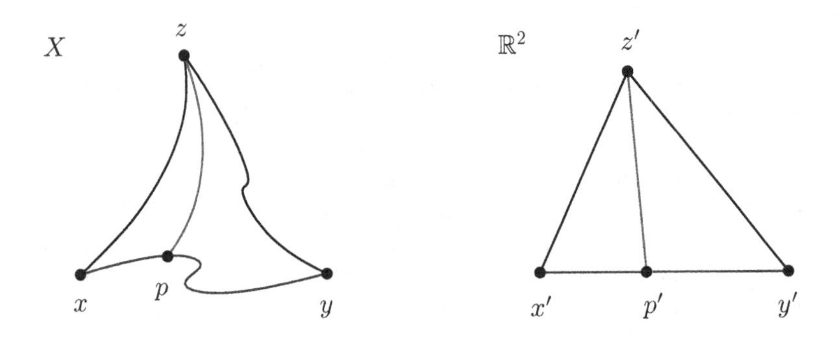

Figure 2: A geodesic triangle and the corresponding triangle in $\mathbb{R}^2$.

Definition 1.2 (CAT(κ)) *A geodesic metric space X is said to be (globally)* CAT(0) *if for all points $x, y, z \in X$, for all geodesics connecting x, y, and z and for all points p on the chosen geodesic connecting x to y, $d(p, z) \leq d(p', z')$ where the second distance is calculated in the corresponding configuration in the Euclidean plane. A* corresponding configuration *is the unique configuation (up to isometry) of labeled points in $\mathbb{R}^2$ where the distance in X between labeled points matches the Euclidean distance between the corresponding labeled points in $\mathbb{R}^2$. See Figure 2 for an illustration. The conditions* CAT(1) *and* CAT(−1) *are defined similarly using $\mathbb{S}^2$ and $\mathbb{H}^2$, respectively - with restrictions on x, y, and z in the spherical case, since not all spherical comparison triangles are constructible. More generally,* CAT(κ) *is defined using the complete simply-connected 2-dimensional space with constant curvature κ. See [26] for further details.*

As should be clear from the definition, increasing the value of κ weakens the conditions that need to be satisfied. In particular, the following result is immediate.

Theorem 1.3 (Increasing κ) *If X is a* CAT(κ) *space, then for all $k' \geq k$, X is also a* CAT(κ') *space. In particular, every* CAT(−1) *space is a* CAT(0) *space.*

Since hyperbolic n-space, $\mathbb{H}^n$, is a common inspiration for both of these theories of negative curvature, it is not too surprising that it is both δ-hyperbolic and a CAT(−1) space. Notice that the definition of δ-hyperbolic only implies that the large scale curvature is negative: we get no information about local structure. On the other hand, CAT(0) and CAT(−1) spaces have good local curvature properties which can be formalized as follows.

Definition 1.4 (Locally CAT(κ)) *A complete geodesic metric space is* locally CAT(κ) *if every point has a neighborhood which satisfies the* CAT(κ) *condition in the induced local metric.*

Note that the completeness of the metric is asssumed primarily to rule out such pathologies as the lateral surface of a right circular cone minus its apex. This is a perfectly good locally CAT(0) space (except for completeness), but its universal cover is not even a geodesic metric space. Assuming completeness excludes such oddities. The connection between this local CAT(κ) definition and the previous global definition is now straightforward: it is known as the Cartan-Hadamard Theorem. See [26, p. 193].

Theorem 1.5 (Local-to-global) *If X is a complete geodesic metric space and $\kappa \leq 0$ is a fixed constant, then the space X is* CAT(κ) *if and only if X is locally* CAT(κ) *and simply-connected.*

Notice that one consequence of this lemma is that a local examination of a compact complex is sufficient to establish that its universal cover satisfies the global version. In keeping with recent practice, the phrases "locally CAT(0)" and "non-positively curved" are used interchangeably. Similarly, the phrase "locally CAT(-1)" is often equated with the phrase "negatively curved", but since I am trying to highlight the various classes of spaces and groups with a legitimate claim to this title, the phrase "negatively curved" is never used in this sense in this article.

Shifting attention from spaces to groups, consider those groups which act properly and cocompactly by isometries on these types of spaces. Following Cannon [33] I call such an action *geometric.* Of the seven classes of groups to be delineated, one acts geometrically on δ-hyperbolic spaces and the other six on CAT(κ) spaces with various restrictions. A group G is *hyperbolic* (or word-hyperbolic or Gromov-hyperbolic) if for some δ it acts geometrically on some δ-hyperbolic space. Similarly, a group G is a CAT(κ) *group* if it acts geometrically on some CAT(κ) space. Deciding whether a group is word-hyperbolic is made easier by the following result.

Lemma 1.6 (Testing hyperbolicity) *If A is a finite generating set for a group G, then G is word-hyperbolic if and only if there exists a δ such that the Cayley graph of G with respect to A is δ-hyperbolic.*

Thus hyperbolicity can be detected from a close examination of the metric space associated with any finite generating set. By way of contrast, showing that a group is CAT(κ) first requires the construction of an appropriate CAT(κ) space and there is no natural candidate with which to begin. This lack of a natural space on which the group should act is one of the main sources of difficulty for proving (and especially for disproving) that a group is a CAT(κ) group.

The first type of restriction on CAT(κ) groups I wish to consider involves the existence of flats in the spaces upon which they act. A *flat* is an isometric embedding of a Euclidean space $\mathbb{R}^n$ with $n > 1$. Because an isometrically embedded flat plane would violate the thin triangle condtion, every CAT(-1) space is also a CAT(0) space with no flats. Also, every CAT(-1) space is δ-hyperbolic. The strongest result along these lines is the flat plane theorem. See [26, Theorem 3.1 on p. 459] or [22].

Theorem 1.7 (Flat Plane Theorem) *If X is a* CAT(0)*-space and G is a group which acts geometrically on X, then X is δ-hyperbolic (and G is hyperbolic) if and only if X does not contain an isometrically embedded copy of the Euclidean plane.*

Notice that the quotient of X by its G-action is a non-positively-curved space with fundamental group G and universal cover X. Conversely, if we start with a compact non-positively curved space with fundamental group G, then G acts geometrically on its universal cover. The rough algebraic equivalent to the existence of an isometrically embedded flat plane is a rank 2 free abelian subgroup. The following is a small portion of the result known as the flat torus theorem ([26, Theorem 7.1 on p. 244]).

Theorem 1.8 (Flat Torus Theorem) *Let X be a* CAT(0)*-space and let G be a group which acts geometrically on X. If G contains a $\mathbb{Z} \times \mathbb{Z}$ subgroup then X contains an isometrically embedded flat plane.*

An obvious question at this point is whether the converse also holds. Unfortunately this question is wide open.

Problem 1.9 (Flat vs. $\mathbb{Z} \times \mathbb{Z}$) *If X is a* CAT(0)*-space and a group G acts geometrically on X, is it true that there is an isometrically embedded flat plane in X if and only if there is a $\mathbb{Z} \times \mathbb{Z}$ subgroup in G?*

A good test case is a compact non-positively curved 2-complex constructed entirely out of unit Euclidean squares. Problem 1.9 remains open even in this highly restricted setting and it is not even clear that the answer should be yes. One phenomenon which frustrates many naive approaches to this problem is the following construction due to Dani Wise [86].

Theorem 1.10 (Isolated aperiodic flats) *There exists a compact piecewise Euclidean non-positively curved space constructed entirely out of unit Euclidean squares which contains an aperiodic flat that is not the limit of periodic flats.*

A *periodic flat* is one on which a $\mathbb{Z} \times \mathbb{Z}$ subgroup of G acts cocompactly. Alternatively, a flat is *aperiodic* if no such subgroup exists. The final assertion of the theorem essentially means that there does not exist a sequence of periodic flats whose intersections with the aperiodic flat increase so as to exhaust the aperiodic flat in the limit. The specific construction can be found in Section 7 of [85] or in [86]. Wise's example shows that attempts to prove that the existence of a flat in X implies the existence of a $\mathbb{Z} \times \mathbb{Z}$ subgroup in G cannot start by closely examining a randomly chosen flat in X with the hope of showing that it is periodic – or even that there is a periodic flat nearby. The route from the flat to the subgroup, if indeed it exists, is thus likely to start from a flat which has been carefully selected to satisfy additional conditions.

The final type of restriction I wish to consider severely constrains the type of spaces on which the group G can act. In particular, suppose we restrict our attention to the spaces constructed out of polyhedral pieces taken from one of the three standard constant curvature models: the n-sphere $\mathbb{S}^n$, Euclidean n-space $\mathbb{R}^n$, or hyperbolic n-space $\mathbb{H}^n$. A rough definition of these M_κ-complexes is given below. See [26] for a more precise definition.

Definition 1.11 (M_κ-complexes) *A* piecewise spherical / euclidean / hyperbolic complex X *is a polyhedral complex in which each polytope is given a metric with constant curvature* 1 / 0 / -1 *and the induced metrics agree on overlaps. In the spherical case, the cells must be convex polyhedral cells in $\mathbb{S}^n$ in the sense that they must be embeddable in an open hemisphere of $\mathbb{S}^n$. The generic term is M_κ-complex, where κ is the curvature common to each of its cells.*

It is an early foundational result of Bridson that compact M_κ-complexes are indeed geodesic metric spaces [21, 24]. In fact, Bridson showed that non-compact M_κ-complexes are also geodesic metric spaces so long as one assumes that there are only finitely many isometry types of cells. For ease of exposition, we restrict our attention to compact complexes. A second key result is that one can check whether an M_κ-complex ($\kappa \leq 0$) is non-positively curved by checking for the existence of short geodesic loops in the links of cells.

Definition 1.12 (Links) *Let X be an M_κ-complex. The* link of a point x *in X is the set of unit tangent vectors at x. Notice that the link of x comes equipped with a natural piecewise spherical structure. The* link of a cell *is the set of unit tangent vectors at one of its points, but restricted to those which are orthogonal to all of the tangent vectors lying in the cell. For example, the link of a point in the interior of an edge in a tetrahedron is a spherical lune, but the link of the edge which contains it is a metric circular arc whose length is equal to the dihedral angle along this edge.*

Definition 1.13 (Geodesics) *A* piecewise geodesic γ *in X is a path $\gamma : [a,b] \to X$ where $[a,b]$ can be subdivided into a finite number of subintervals so that the restriction of γ to each closed subinterval is a path lying entirely in some closed cell σ of X. There is a further restriction that this portion of the path is the unique geodesic connecting its endpoints in the metric of*

σ. *A* local geodesic *is a local isometric embedding of an interval into X and a* closed geodesic loop *is a local isometric embedding of a metric circle into X. In an M_κ-complex, the structure of a local geodesic (or a closed geodesic loop) is always that of a piecewise geodesic, and to test whether a piecewise geodesic is a local geodesic it is sufficient to check whether at each of the transition points, the "angles are large", meaning that the distance between the in-coming tangent vector and the out-going tangent vector is at least π in the link of this point.*

Using this language the precise statement is the following.

Theorem 1.14 (Gromov's link condition) *For $\kappa \leq 0$, an M_κ-complex is locally* CAT(κ) $\Leftrightarrow$ *the link of each vertex is globally* CAT(1) $\Leftrightarrow$ *the link of each cell is an piecewise spherical complex which contains no closed geodesic loop of length less than 2π.*

Thus, showing that piecewise Euclidean complexes are non-positively curved hinges on showing that piecewise spherical complexes have no short closed geodesic loops. I return to this theme in Section 1.3. Finally, notice that a compact M_κ-complex can be described using only a finite amount of data. This observation is important since it makes M_κ-complexes suitable for computer investigation. More specifically, it is sufficient to give the curvature κ, the simplicial structure of (a simplicial subdivision of) the complex together with the exact length of each of its edges. This is sufficient because the shape of each simplex (of curvature κ) is completely determined by its edge length data. This section concludes with an exercise.

Exercise 1.15 (Feasible edge lengths) *What restrictions on a set of edge lengths ℓ_{ij}, $1 \leq i, j \leq n$, are necessary in order for a nondegenerate piecewise Euclidean n-simplex to be constructible with these lengths? The answer is classical, but the necessary and sufficient conditions are not necessarily easily reconstructed from first principles. Historically, the full answer consists of a set of equations (due to Cayley) and some inequalities (due to Menger) [69], although an alternative description in terms of quadratic forms was found by Schoenberg [79].*

1.2 Curvature conjecture

The primary goal of this section is to state and discuss the curvature conjecture. In order to state the conjecture seven potentially distinct classes of groups need to be defined.

Definition 1.16 (7 classes of groups) *Consider the set of all groups G which act geometrically on*

1. *δ-hyperbolic spaces,*

2. CAT(-1) *spaces,*
3. CAT(0) *spaces with no isometrically embedded flat planes,*
4. CAT(0) *spaces with no* $\mathbb{Z} \times \mathbb{Z}$ *subgroups in* G,
5. *piecewise hyperbolic* CAT(-1) *spaces,*
6. *piecewise Eulcidean* CAT(0) *spaces with no isometrically embedded flat planes, or*
7. *piecewise Euclidean* CAT(0) *spaces with no* $\mathbb{Z} \times \mathbb{Z}$ *subgroups in* G.

Notice that first two classes define word-hyperbolic groups and CAT(-1) groups, respectively. The known relationships between these classes are summarized in Figure 3. In the figure, PH and PE stand for piecewise hyperbolic and piecewise Euclidean, respectively.

$$\begin{array}{ccccc}
\text{PH CAT}(-1) & \Rightarrow & \text{CAT}(-1) & & \\
(?) & & \Downarrow & & \\
\text{PE CAT}(0) \text{ no flats} & \Rightarrow & \text{CAT}(0) \text{ no flats} & \Rightarrow & \text{word-hyperbolic} \\
\Downarrow & & \Downarrow & & \\
\text{PE CAT}(0) \text{ no } \mathbb{Z} \times \mathbb{Z} & \Rightarrow & \text{CAT}(0) \text{ no } \mathbb{Z} \times \mathbb{Z} & &
\end{array}$$

Figure 3: The known relationships between various classes of groups

Although the following is formulated as a conjecture, I should note that many experts in the area firmly believe the conjecture is false.

Conjecture 1.17 (Curvature conjecture) *All seven classes of groups described above are identical.*

Conjecture 1.17 is a natural analogue of Thurston's hyperbolization conjecture. In particular, if the hyperbolization conjecture is true then this conjecture is true when restricted to 3-manifolds and their fundamental groups. As we noted above, many experts expect the general conjecture to be false. A resolution either way would be a major advance for geometric group theory and its quest to understand the nature of negative curvature.

Perhaps the most likely source of a counterexample to Conjecture 1.17 is a quotient of a cocompact lattice in a high-dimensional quaternionic hyperbolic space by a long relation, the idea being that the non-existence of a CAT(0) space on which this group acts geometrically can be established by a suitable extension of super-rigidity. Several groups of researchers have been working towards resolving this conjecture either positively or negatively (Mineyev-Yu

[70], Burger-Monod [27, 28], etc.). Nevertheless, even should this approach yield a counterexample, the more general question of *which* word-hyperbolic groups can act geometrically on CAT(0) complexes will remain wide open. It is at least conceivable that essentially the only counterexamples are ones constructible from quaternionic hyperbolic spaces. As (admittedly rather weak) evidence for this, there is the well-known propensity of the quarternions and octonions to be involved whenever there are only a few exceptions to an otherwise cleanly stated classification theorem [2, 41].

One aspect of Figure 3 which might be surprising at first is the lack of a relationship between piecewise hyperbolic CAT(-1) groups and piecewise Euclidean CAT(0) groups. A naive approach would be to simply replace each hyperbolic simplex with the Euclidean simplex which has the same edge lengths. This procedure does tend to lengthen piecewise geodesics in vertex links (where the break points are described using barycentric coordinates), but the problem is that the status of such a path as a local geodesic is not stable under this inflation. In particular, there might exist a short piecewise geodesic loop in the hyperbolic version which is not a local geodesic, but which becomes a slightly longer (although still less than 2π) local geodesic in the Euclidean version. Thus, the path that shows the Euclidean version is not CAT(0) may not even be considered in the hyperbolic version since it is not a local geodesic. Another more serious problem is that the effect of this replacement on the length of piecewise geoedesics in links of edges and of other higher dimensional faces is completely unclear: some may lengthen, others may shrink, depending on the intricate details of the shapes of the original hyperbolic simplices. A partial positive result in this direction has been shown by Charney, Davis and Moussong [34].

Theorem 1.18 (hyperbolic vs. PE CAT(0)) *If M is a compact manifold which supports a hyperbolic metric (i.e. a Riemannian metric with constant sectional curvature -1), then M also supports a metric which is piecewise Euclidean and locally* CAT(0).

Their proof looks at the orbit of a point in the hyperboloid model of hyperbolic space under the action of the fundamental group of M and then considers the piecewise Euclidean boundary of the convex hull of these points. They then quotient this piecewise Euclidean complex by the piecewise linear action of the fundamental group. Notice, however, that this proof scheme is highly dependent on the fact that M is a constant curvature manifold. The corresponding result for compact n-manifolds with *variable* negative Riemannian curvature or a locally CAT(-1) piecewise hyperbolic structure is still open. In the other direction, there is also a partial result, at least in low-dimensions.

Proposition 1.19 (No tight loops) *If M is a 2-dimensional piecewise Euclidean* CAT(0) *complex, and there does not exist a closed geodesic loop in* *the link of a vertex in M which has length exactly 2π, then M can also be given a piecewise hyperbolic metric which is* CAT(-1).

The argument is easy and goes roughly as follows. Give each simplex the metric of a simplex with identical edge lengths but with constant curvature ϵ where ϵ is negative but close to 0. By chosing ϵ sufficiently close to 0, the length of the closed geodesics in the vertex links (which are metric graphs in this case) do not shrink below 2π. Thus by Gromov's criterion (Theorem 1.14), the result is CAT(ϵ). Rescaling now produces a metric of curvature -1. The restriction that M be 2-dimensional is crucial. As far as I can tell from the literature, the following stronger conjecture has neither been established nor disproved.

Conjecture 1.20 (No tight loops) *Suppose M is a piecewise Euclidean* CAT(0) *complex, and there does not exist a closed geodesic loop of length exactly 2π in the link of any simplex in M, is it true that this particular cell structure for M can also be given a piecewise hyperbolic metric which is* CAT(-1)*?*

As mentioned above, the potential problems lie in the links of faces other than vertices since a simple substitution of slightly hyperbolic simplices for the Euclidean ones has unpredictable effects on the metrics in these links. Note also that the no-tight-loops restriction in Proposition 1.19 is absolutely necessary for this result to hold in dimension 2. Specifically, Noel Brady and John Crisp [12] and Misha Kapovich [58] have produced several examples with the following properties.

Theorem 1.21 (PH CAT(-1) vs. PE CAT(0)) *There is a group which acts geometrically on a 3-dimensional piecewise hyperbolic* CAT(-1) *space, and on a 2-dimensional piecewise Euclidean* CAT(0) *space, but not on any 2-dimensional piecewise hyperbolic* CAT(-1) *space.*

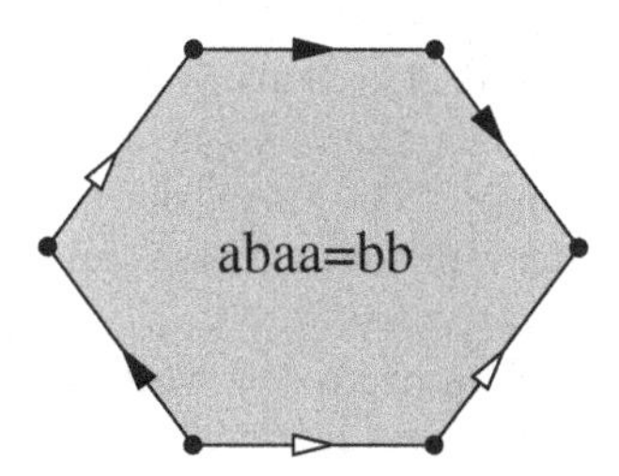

Figure 4: A one-relator group satisfying Theorem 1.21.

The simplest example of such a group is the one-relator group shown in Figure 4. The moral here is that although it is conceivable that the class of piecewise hyperbolic CAT(-1) groups and the class of piecewise Euclidean CAT(0) groups without flat planes are identical, a change in cell structure – and dimension! – is sometimes necessary in order to pass between them. In other words, higher dimensions are sometimes necessary in order to smooth things out. In particular, the space which proves a group is CAT(0) might have a higher dimension than its geometric dimension [12, 25] and, if CAT(-1) is desired, another space of even higher dimension might be required. Despite the exceptions under special circumstances noted above (Theorem 1.18 and Proposition 1.19) the behavior noted by Brady-Crisp and Kapovich is likely to be fairly common. For an explicit illustration where higher dimensions are useful see Section 3.2. On the other hand, the passage to higher-dimensions is not without perils of its own, as we shall see.

If our goal is to create complexes with good local curvature properties on which a particular word-hyperbolic group can act, there is at least one obvious candidate: the Rips complex, or some variant on it. Let $P_d(G, A)$ denote the flag complex on the graph whose vertices are labeled by G and which has an edge connecting g and h iff gh^{-1} is represented by a word of length at most d in the generators A. Recall that a *flag complex* is a simplicial complex which contains a simplex if and only if it contains the 1-skeleton of that simplex. This complex has extremely nice properties for large values of d [1].

Theorem 1.22 (Rips complex) *If G is word-hyperbolic and d is large relative to the hyperbolicity constant δ, then the Rips complex $P_d(G, A)$ is contractible, finite dimensional, and G acts discretely and compactly on $P_d(G, A)$.*

As defined above, the Rips complex is a simplicial complex with no natural metric. One approach to the curvature conjecture would be to try and add a metric to the Rips complex and then to show that the result is a CAT(0) space. Unfortunately, this can be quite difficult even in the simplest of cases. Let G be a word-hyperbolic group and suppose we carefully pick a generating set A, pick a d very large and declare each simplex in $P_d(G, A)$ to be a regular Euclidean simplex with every edge length 1. Is the result a CAT(0) space? Is this true when G is free and A is a basis? Although it is embarrassing to admit, I believe the answer is that no one knows, even in the case of a free group generated by a basis. The moral here is that our ability to test whether a compact constant curvature metric space is locally CAT(0) or locally CAT(-1) is *very* primitive.

Finally, there is another class of groups which deserves to be highlighted in connection with the seven classes listed above, even though they are not negatively curved. These are the groups which act geometrically on CAT(0) spaces with isolated flats.

Remark 1.23 (Spaces with isolated flats) *A* CAT(0) *space X has the* isolated flats property *if there is a collection $\mathcal{F}$ of isometrically embedded Euclidean spaces (each of dimension at least* 2*) such that this collection is* *maximal and isolated. It should be maximal in the sense that every flat in X lies in a k-neighborhood of a flat in $\mathcal{F}$, and isolated in that k-neighborhoods of distinct flats in $\mathcal{F}$ intersect in bounded sets whose diameter is a function of k alone. This definition was introduced and extensively studied by Chris Hruska in his dissertation [52], although it was implicit in earlier works by other authors [57, 85]. There are a number of results for hyperbolic groups which readily extend to groups acting geometrically on* CAT(0) *spaces with isolated flats, but which do not extend to* CAT(0) *groups in general. For example, if G is such a group (and it also has the closely related relative fellow traveler property), then there is a well-defined notion of a quasiconvex subgroup in G which is independent of the* CAT(0) *space with isolated flats on which G acts and there is a well-defined boundary up to homeomorphism. Both of these results were shown by Gromov for hyperbolic groups but fail for general* CAT(0) *groups.*

1.3 Decidability

When constructing examples of non-positively curved spaces, attention has naturally focused on the class of constant curvature complexes. This attention is partly due, no doubt, to the reduction allowed by Gromov's link condition (Theorem 1.14) and to the fact that these complexes can be described using only a finite amount of data. But this leads to a fundamental question. Let X be a finite M_κ-complex, say piecewise Euclidean. Is there an algorithm to determine whether this metric space is non-positively curved? The answer, fortunately, is yes, but the proof is not as straightforward as one might hope. In [45] Murray Elder and I proved the following result.

Theorem 1.24 (CAT(κ) is decidable) *Given a compact M_κ-complex X, there is an explicit algorithm which decides whether X is locally* CAT(κ)*. In particular, it is possible to determine if a finite piecewise Euclidean complex is non-positively curved.*

The word "algorithm" in the theorem should really be placed in quotation marks since the algorithm we described in the course of the proof is not one which can be used in practice except in the most trivial of situations where the answer is already obvious. The difficulty is that the proof relies, ultimately, on a result of Tarski's about the decidability of the first order theory of the reals. In outline, the proof can be described as follows. First, reduce the problem to the existence of certain types of paths in special piecewise spherical complexes that we call circular galleries. Next, encode the existence of these paths into a real semi-algebraic set. Recall that a *real semi-algebraic set* is a subset of Euclidean space which can be described using Boolean combination of polynomial equations and inequalities. The name is derived from the fact that these subsets are also boolean combinations of (projections of) real algebraic varieties. Finally, Tarski's theorem about the decidability of first order sentences over the reals implies that there is an algorithm which decides whether a real semi-algebraic set is empty or not. The rest of this section describes the proof in more detail. The key concept is that of a gallery.

Definition 1.25 (Galleries) *Let X be a piecewise constant curvature complex and let γ be a local geodesic in X. Consider the sequence of open cells through which γ passes. This sequence is essentially what we call a* linear gallery. *If γ is a closed geodesic, then this sequence is given a cyclic ordering and the sequence is called a* circular gallery. *Instead of a precise definition, consider the example shown in Figure 5. The figure in the upper left shows a very simple complex (which is essentially the boundaries of two Euclidean tetrahedra identified along one edge) along with a local geodesic path from x to y. On the upper right is the corresponding linear gallery. The lower left is result of concatenating the closed cells through which the local geodesic passes rather than the open cells, and the difference between these two is the* boundary *of the gallery. The circular version constructs a similar complex homotopy equivalent to a circle.*

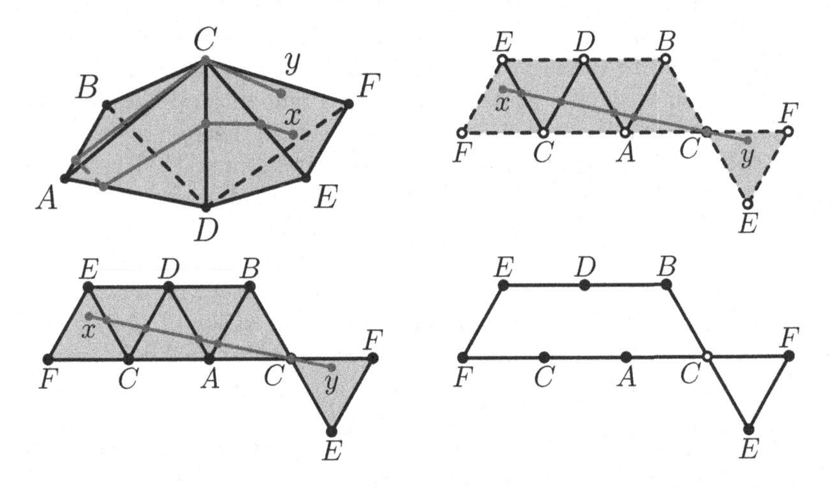

Figure 5: A 2-complex, a linear gallery, its closure and its boundary.

The reasoning now goes as follows. If there is a cell in X whose link is a piecewise spherical complex containing a closed geodesic loop of length less than 2π, then this geodesic determines a circular gallery. Moreover, since X is finite, this circular gallery involves only finitely many shapes. There is a result of Bridson [21] which asserts that in this context the length of the geodesic is quasi-isometric to the number of simplices in the circular gallery and that the quasi-isometry constants only depends on the shapes of the cells involved. Thus, if we are able to (1) check whether any particular piecewise spherical circular gallery contains a short closed geodesic loop with winding number 1 and (2) determine whether any of the closed geodesic loops found remain locally geodesic when the circular gallery is immersed back into the cell link from which it came, then we are done. Carrying out this procedure for each of the finite number of circular galleries containing a bounded number of simplices completes the test. To highlight the dependence of the argument on the quasi-isometry constant derived from the shapes, consider the situation where the metric is not given in advance.

Remark 1.26 (Testing for the existence of a CAT(0) metric) *If X is a finite cell complex endowed with a piecewise Euclidean metric then the argument outlined above can determine whether or not this particular metric is non-positively curved. It remains an open question, however, (in dimensions greater than 2) whether there is an algorithm to determine whether a finite simplicial complex (with no pre-assigned metric) supports a metric of non-positive curvature. The problem is that without knowing in advance what the metric is, it is not clear what the quasi-isometry constants are, and thus it is not clear how large the circular galleries are which need to be considered. In particular, there is no* a priori *bound on the combinatorial length of the galleries we need to check. Dimension 2 is special in this regard because its vertex links are metric graphs. Thus, the only closed circular galleries which need to be considered are the finite number of simple closed loops in these graphs – regardless of the metric. Therefore, a (highly impractical) algorithm does exist in dimension 2.*

The remainder of the argument is a complicated conversion process which encodes the existence of a closed geodesic loops in a circular gallery into a Boolean combination of polynomial equations and inequalities. The full details can be found in [45], but as an indication of the issues involved, here are two brief remarks. First, even determining whether a piecewise geodesic in a piecewise spherical complex is longer than π is a non-trivial task using only polynomials. Analytically, one would simply think of the transition points as unit vectors and add up the arccosines of the appropriate dot products. Since arccosines are prohibited, we create instead a 2-dimensional "model space" such as the one pictured in Figure 6 and then test whether the second coordinate of the final point plotted is positive.

The second remark is that to conclude that a particular piecewise geodesic is a local geodesic necessarily involves an induction on dimension: it is a local geodesic if and only if at each transition point the distance between in the in-coming tangent vector and the out-going tangent vector are distance at least π apart in the link of this point. To show this one would need to show that for every possible linear gallery (up to a certain combinatorial length) connecting these two points in this lower dimensional piecewise spherical complex and for every possible local geodesic in each of these galleries, the distance (calculated

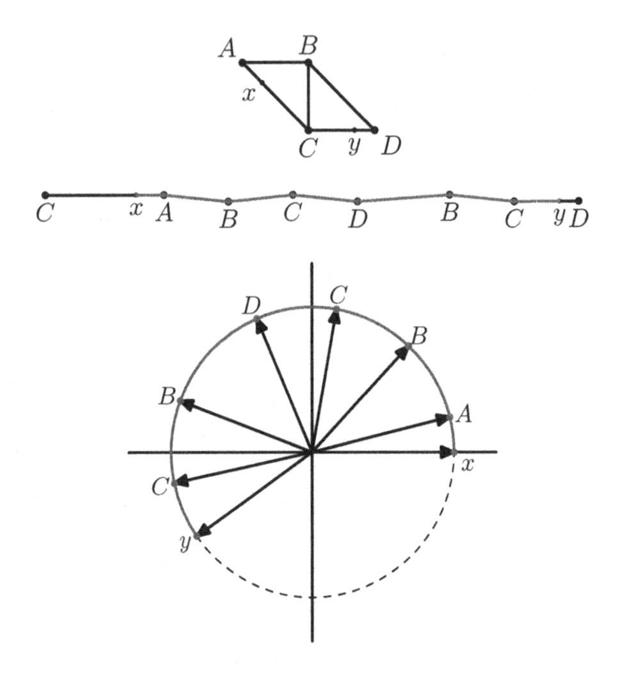

Figure 6: A 1-complex, a linear gallery and its model space.

using a 2-dimensional model space) is at least π. By an induction on dimension, one can assume that the appropriate logical combination of equations and inequalities exists to test each of these conditions, completing the proof. The result is that by inducting through dimensions it is possible to construct a real semi-algebraic set whose points are in one-to-one correspondence with the closed geodesic loops of length less than 2π in the specific circular gallery under consideration. Tarski's theorem can then be used to test whether this set is empty or not.

Remark 1.27 (Why is this so hard?) *At this point it would be natural to wonder whether it is really necessary to use real semi-algebraic sets in order to decide this type of problem. Perhaps there is an elementary solution waiting to be discovered. Perhaps. On the other hand, the problem of determining whether a complex is* CAT(0) *involves finding* 1*-dimensional paths in high dimensional piecewise spherical complexes, and problems involving high codimension (where "high" typically means codimension* 2 *or more) can often be surprisingly hard. As an example, consider the question of finding the unit volume Euclidean* 3*-polytope with the smallest* 1*-skeleton (where the size of the* 1*-skeleton is measured by adding up all of the edge lengths). The answer, in this case, is not known, although the solution is probably the regular triangular* *prism shown in Figure 7 (all edges have equal length, all faces are regular, and rescaled to have unit volume). This is known as the Melzak problem and it is a classical problem in geometric analysis which has stubbornly resisted all attempts to resolve it. For a summary of recent partial progress see [76].*

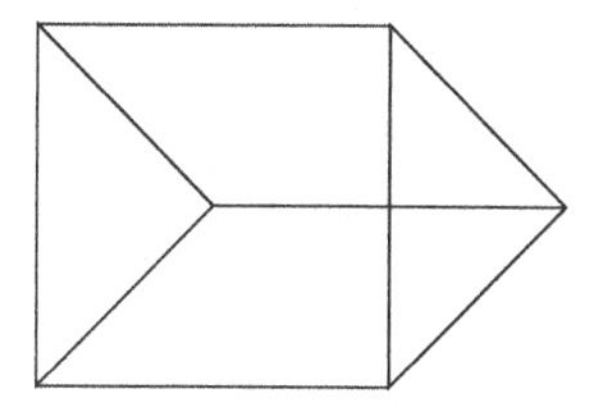

Figure 7: A regular triangular prism.

1.4 Length spectrum

The ability (in theory) to test whether a finite constant curvature complex is non-positively curved is not the only result which can be shown using the methodology described in the previous section. A similar approach allows one to prove that length spectrum of such a complex must be discrete.

Definition 1.28 (Length spectrum) *Let X be a compact M_κ-complex and let x and y be points in X. The lengths of all the local geodesics from x to y is called the length spectrum from x to y. The lengths of all the closed geodesic loops in X is simply its* length spectrum.

In [26], Bridson and Haefliger use a relatively easy compactness argument to establish the discreteness of the length spectrum between specific points. Noel Brady and I were recently able to show that the same restriction holds for lengths of all closed geodesic loops as well [14]. The proof, on the other hand, requires a type of argument which is reminenscent of the proof of Theorem 1.24.

Theorem 1.29 (Length spectrum) *The length spectrum of any compact M_κ-complex is discrete.*

As one might expect, the compactness hypothesis can be weakened to finitely many isometry types of cells without altering the proof. A sketch of the proof goes as follows. If the length spectrum is not discrete then we can find a sequence of closed geodesic loops in X whose lengths are distinct but bounded. By passing to a subsequence we can assume that all of these loops determine the same circular gallery (this reduction uses the quasi-isometry constant alluded to above). By establishing that the function which calculates the length of certain piecewise geodesic paths is analytic, one can use the Morse-Sard theorem to show that the length spectrum for this particular circular gallery has measure zero in $\mathbb{R}$. Moreover, by encoding these piecewise geodesics as a real semi-algebraic set, one can show that the topology on the set of closed geodesic loops has only finitely many connected components. Combining these two ingredients shows that the length spectrum of this particular circular gallery is finite, providing the necessary contradiction.

One reason for proving a result of this type has to do with locally geodesic surfaces. Call an isometric immersion $f : D \to X$ of a metric polyhedral surface *locally geodesic* if for all points $d \in D$, the link of d is sent to a local geodesic in the link of $f(d)$. In a 2-dimensional piecewise Euclidean complex, every null-homotopic curve bounds a locally geodesic surface, but this fails in dimensions 3 and higher. This is also one of the key reasons why some theorems in dimension 2 fail to generalize easily to higher dimensions. Chris Hruska, for example, proves in [52] that in dimension 2 the isolated flats property is equivalent to the relative fellow travel property and to the relatively thin triangles property. He conjectures, moreover, that the same result should hold in all dimensions. In his proof, he heavily uses the fact that the curvature of points in locally geodesic surfaces mapped into finite 2-dimensional complexes is bounded away from 0. By the *curvature of a point* in the interior of D we mean 2π minus the length of its link. See also Section 4.2 where a more general definition of curvature is discussed. An immediate corollary of Theorem 1.29 is that this portion of his proof immediately generalizes to arbitrary dimenisions.

Corollary 1.30 (Quantizing curvature) *If $D \to X$ is a locally geodesic surface in a compact piecewise Euclidean complex X, then those points in the interior of D with nonzero curvature have curvatures uniformly bounded away from 0. Moreover, this bound depends only on X and not on D. As a consequence, if the total amount of negative (or positive) curvature is known, then there is a bound on the number of points in D whose curvature is less than (or more than) zero.*

2 Algorithms

In this part the focus shifts from theorectical results to practical algorithms which can actually be implemented on computers to obtain results. In Section 2.1 a practical algorithm in dimension 3 is discussed and in Section 2.2 this algorithm is used to solve a conjecture about 3-manifolds. Finally, in Section 2.3 the inherent difficulties in higher dimensions are considered.

2.1 Algorithm in dimension 3

As pointed out above, the use of Tarski's theorem in the course of the proof makes the implementation of the "algorithm" described in Theorem 1.24 impractical if one is really interested in actually carrying out a test on a particular concrete example. Thus, it makes sense to continue to search for alternative approaches in those special cases where real algebraic geometry can be avoided. In [44] Murray Elder and I were able to find a rather elementary geometric argument when the complexes under consideration are restricted to dimension at most 3.

Theorem 2.1 (Practical algorithm) *There exists an elementary practical algorithm which decides whether or not a 3-dimensional piecewise Euclidean complex is non-positively curved.*

As with the word "algorithm" in Theorem 1.24, the word "practical" in Theorem 2.1 should really be in quotation marks since the procedure is reasonably fast for each circular gallery, but the number of circular galleries which need to be tested in even a modest size example can quickly grow to unmanageable levels. The results described in Section 2.2 illustrate this phenomenon. The heart of the argument uses only elementary 3-dimensional geometry and has been implemented in both GAP (Groups, Algorithms, Programming) and PARI (a number theory package). The packages are called `cat.g` and `cat.gp`, and are available to download [43].

The key to the argument is the restricted nature of circular galleries in low dimensions. If X is a 3-dimensional piecewise Euclidean complex, the only links of cells which need to be examined are the links of vertices. The piecewise spherical complexes which result are 2-dimensional and the circular galleries to which they lead can be classified as either annular galleries, Möbius galleries or necklace galleries.

Definition 2.2 (Types of gallery) *If a hypothetical geodesic does not pass through a vertex of the link, then the circular gallery it produces will be homeomorphic to either an annulus or a Möbius strip, whereas those which pass through a vertex look like a beaded necklace. We call these* annular galleries, Möbius galleries, *and* necklace galleries, *respectively. The portion of a necklace gallery from one vertex to the next is called a* bead.

Annular galleries and Möbius galleries can be "cut open" and then developed onto a 2-sphere in an essentially unique way up to isometry. Once this is done, an easy construction using dot products and cross products enables one to test whether or not the original circular gallery contained a short closed geodesic loop. Similarly, each bead in a necklace gallery can be developed onto a 2-sphere in an essentially unique way and there is an easy procedure to check whether there is a local geodesic in this bead which connects its endpoints. These processes are illustrated in Figures 8, 9 and 10. See [44] for further details.

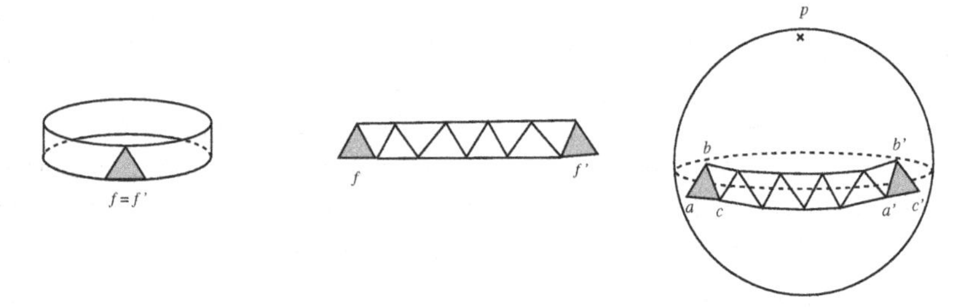

Figure 8: An annular gallery, cut open and developed.

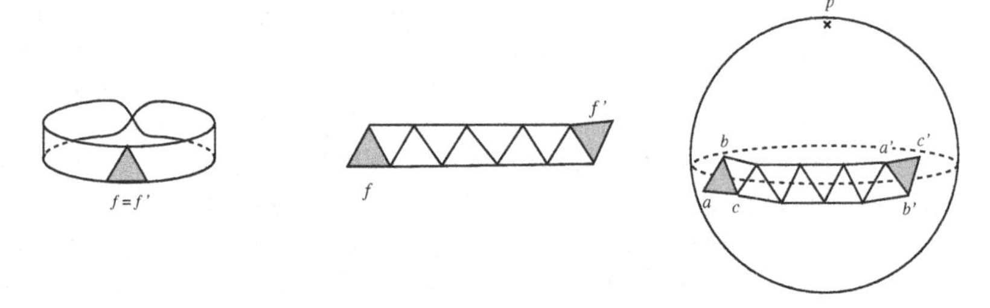

Figure 9: A Möbius gallery, cut open and developed.

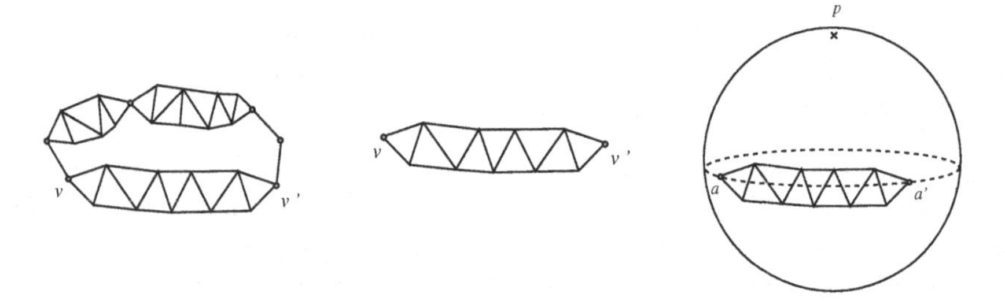

Figure 10: A necklace gallery with one of its beads developed.

As mentioned above, the main reason why the algorithm described is not completely practical is because of the sheer number of circular galleries which need to be examined in any reasonable example. In practice, this search can be restricted to unshrinkable geodesics. A geodesic is *unshrinkable* if there does not exist a homotopy through rectifiable curves of non-increasing length to a curve of strictly shorter length. One corollary of a result established by Brian Bowditch in [11] is the following.

Theorem 2.3 (Unshrinkable geodesics) *If an M_κ-complex (with $k \leq 0$) contains a cell whose link contains a closed geodesic loop of length less than 2π, then it also contains a cell whose link contains an* unshrinkable *closed geodesic loop of length less than 2π.*

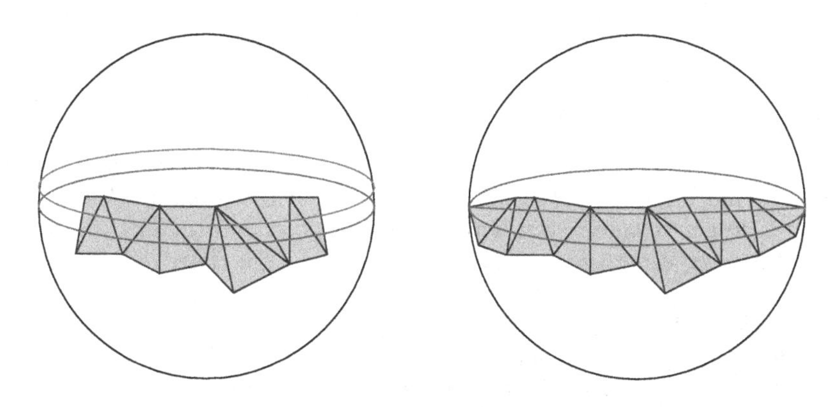

Figure 11: Shrinking annular galleries and long beads.

Using a slightly more delicate argument, we can then show that when testing whether an M_κ-complex X is non-positively curved it is sufficient to test whether there are piecewise spherical circular galleries arising from the links of cells in X which contain closed geodesic loops that can neither be shrunk nor homotoped until it meets the boundary of its gallery without increasing length. All annular galleries, Möbius galleries of length at least π, and beads of length at least π are shrinkable in this expanded sense. Figure 11 illustrates the main ideas. The left hand side shows how a geodesic in a cut-open and developed annular gallery can be homotoped through latitude lines to a strictly shorter curve, while the right hand side shows how a geodesic in a bead of length at least π can be homotoped in a non-length-increasing way to a piecewise geodesic which intersects the boundary of the bead. This reduction simplifies the search for short geodesics immensely.

This section concludes with two examples showing the type of results produced by the software implementing this algorithm.

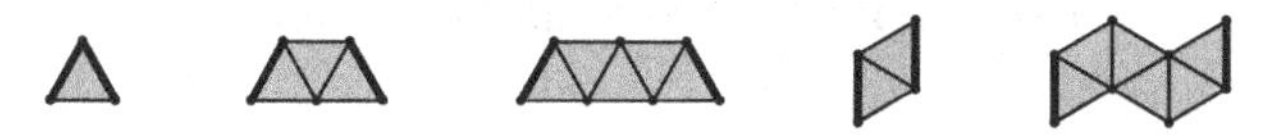

Figure 12: 5 Möbius galleries.

Example 2.4 (Regular tetrahedra) *Let X be a piecewise Euclidean complex built entirely out of regular tetrahedra with each edge length equal to* 1.

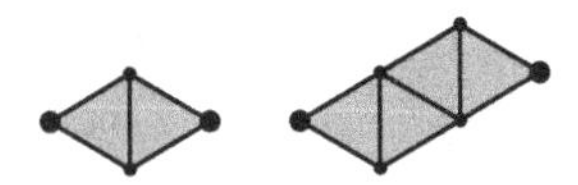

Figure 13: Two non-trivial beads.

Each of the vertex links in X is piecewise spherical 2-complexes constructed out of equilateral spherical triangles with side length $\pi/3$. The software produces a list of circular galleries constructed out of triangles of this type which can carry short closed geodesic loops. The task of determining whether any of these circular galleries can actually be immersed into any of the vertex links of X and, more importantly, whether any of the closed geodesic loops they carry remain local geodesics under these immersions has not been implemented. Currently, the researcher must check these conditions by hand.

In the situation under consideration, the circular galleries carrying unshrinkable short closed geodesics consist of 22 necklace galleries and the 5 Möbius galleries shown schematically in Figure 12. In order to produce the Möbius strips the thick edges on the left and the right should be identified with a half-twist. The 22 necklaces are those which can be strung together using the one edge type (which is a spherical arc of length $\pi/3$) and the two non-trivial beads shown in Figure 13. If we call these beads types A, B, and C, respectively, then the 22 necklaces which are can strung together to contain a closed geodesic loop of length less than 2π are described by the sequences A, A^2, A^3, A^4, A^5, B, BA, BA^2, BA^3, BA^4, B^2, B^2A, B^2A^2, $BABA$, B^3, C, CA, CA^2, CA^3, CB, CBA, and C^2. The final result is that X is non-positively curved if and only if it does not contain a vertex whose link contains an isometric immersion of one of these 5 Möbius galleries or one of these 22 necklaces galleries where the short closed geodesic loop it contains survives as a local geodesic under this immersion.

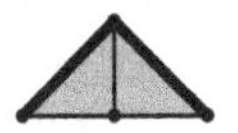

Figure 14: A Möbius gallery to avoid.

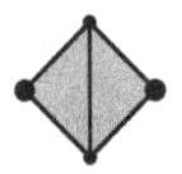

Figure 15: A non-trivial bead to avoid.

Example 2.5 (Coxeter shapes) *Let X be a piecewise Euclidean 3-complex built entirely out of metric tetrahedra with 4 edges of length $\sqrt{3}$ and two non-adjacent edges of length 2. Tetrahedra of this type arise in the theory of Euclidean Coxeter groups. In particular, there is a regular tiling of $\mathbb{R}^3$ using these tetrahedra. In the standard notation it is an $\widetilde{A}_3$ shape. The vertex links in X will be piecewise spherical 2-complexes built out of isometric isosceles spherical triangles. This is because all four vertices of our standard tetrahedron have isometric links. The software in this case produces a single Möbius gallery and a list of 19 necklace galleries. The Möbius gallery is shown in Figure 14.*

The 19 necklaces are again strung together using 3 beads. In this case, there are two types of edges (of different lengths) in the vertex links and a single non-trivial bead shown in Figure 15. Labeling these A, B, and C as before, the 19 necklace galleries which can be strung together to contain a closed geodesic loop of length less than 2π are those described by the sequences A^2, A^4, A^6, A^2B, A^2B^2, A^2B^3, $ABAC$, A^2C, A^2C^2, A^4B, $CA^4\ B$, B^2, B^3, B^4, B^5, C, C^2, and C^3. The final result is that X is non-positively curved if and only if it does not contain a vertex whose link contains an isometric immersion of this Möbius gallery or one of these 19 necklaces galleries where the short closed geodesic loop it contains survives as a local geodesic under this immersion.

2.2 3-manifold results

In this section we discuss how the software described in the previous section can be used in conjunction with a sort of spherical small cancellation theory to prove a result about 3-manifolds originally conjectured by Bill Thurston. The theorem is a result is about $5/6^*$ manifolds.

Definition 2.6 ($5/6^*$-triangulations) *Let M be a closed triangulated 3-manifold. The triangulation is a* 5/6-triangulation *if every edge has degree* 5 *or* 6 *(where the degree of an edge equals the combinatorial length of its link). The triangulation is a* $5/6^*$-triangulation *if, in addition, each 2-cell contains at most one edge of degree* 5.

The conjecture that Thurston makes is that every closed 3-manifold with a $5/6^*$-triangulation has a word-hyperbolic fundamental group. In [46] Murray Elder, John Meier and I proved a slightly stronger result.

Theorem 2.7 *Every $5/6^*$-triangulation of a closed 3-manifold M admits a piecewise Euclidean metric of non-positive curvature with no isometrically embedded flat planes in its universal cover. Thus $\pi_1 M$ is word-hyperbolic.*

The proof involves a mixture of traditional combinatorial group theory and computations carried out by the software `cat.g`. The combinatorial group theory portion is described first. The link of a vertex in M is a triangulated 2-sphere in which every vertex has valence 5 or 6 and no two vertices of valence 5 are connected by an edge. Thus, the dual is a tiling of the 2-sphere by pentagons and hexagons in which every vertex has valence 3 and no two pentagons have an edge in common. Since the smallest tiling with these properties is that of a soccer ball, we call these *soccer tilings* and a simply-connected subdiagram of a soccer tiling homeomorphic to a disc is called a *soccer diagram.* A typical example is shown in Figure 16.

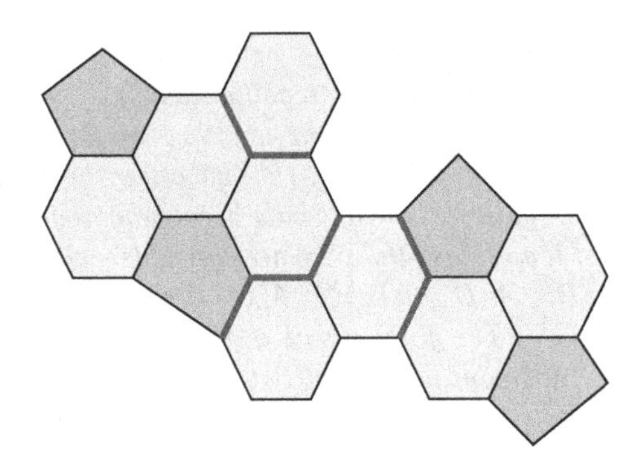

Figure 16: A soccer diagram.

Curvature considerations imply that every soccer tiling has exactly 12 pentagons. If γ is a simple closed curve in the 1-skeleton of a soccer tiling, cutting along this curve splits the 2-sphere into two soccer diagrams, at least one of which has at most 6 pentagons. Since every vertex in a soccer tiling has valence 3, the vertices on the boundary of a soccer diagram have valence either two or three. The boundary vertices of valence 2 are called *right turns* and those of valenced 3 are called *left turns.* The terminology arises from imagining that every corner of each polygon has been assigned an angle of $2\pi/3$ and we are traversing the boundary cycle clockwise. An analysis of soccer diagrams reminiscent of small cancellation theory yields the following key result.

Lemma 2.8 (Special diagrams) *The only soccer diagrams D with $\partial D \leq 14$, at most six pentagons, and no three consecutive right turns are the two diagrams shown in Figure 17.*

This combinatorial result reappears at the end of the argument. Returning to the triangulated 3-manifold M, we assign a piecewise Euclidean metric to M as follows. Each edge of degree 5 is assigned a length of 2 and each edge of degree 6 is assigned a length of $\sqrt{3}$. The corresponding metrics on the 2-cells and 3-cells are the unique Euclidean metrics with these specific edge lengths.

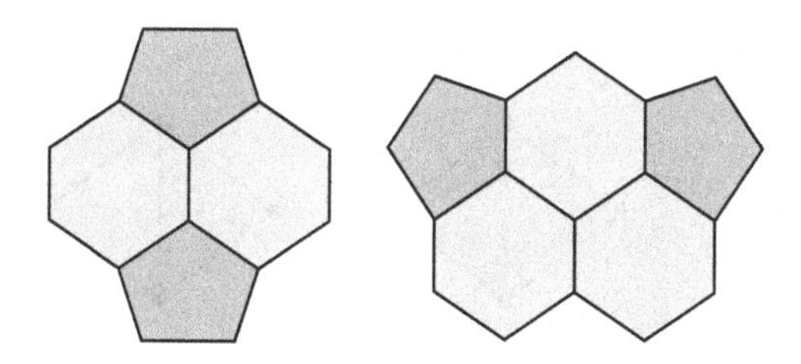

Figure 17: The only two soccer diagrams satisfying the conditions of Lemma 2.8

Since no tetrahedron in M contains edges of degree 5 which are adjacent, there are exactly three types of metric tetrahedra which can be found in M. These are shown in Figure 18. The thicker edges are the ones of length 2. Notice that the tetrahedron on the left is regular and the one on the right is the Coxeter shape considered in Example 2.5.

Figure 18: The 3 metric tetraheda which arise in a $5/6^*$ 3-manifold.

It is easy to show that the links of edges are metric circles with length at least 2π by calculating the sizes of the various dihedral angles. Thus, the only question which remains is whether the links of vertices contain closed geodesic loops of length less than 2π. Since annular galleries are shrinkable and Möbius galleries cannot immerse into 2-spheres, only necklaces galleries need to be examined. Using the software `cat.g`, we calculated the list of beads which can be used to string together a necklace gallery containing a short closed geodesic loop. Using the simplifed search algorithm that only looks for beads whose geodesic has length less than π, it takes less than an hour to produce the list of 75 beads. In the process, the program examines over 116,000 circular galleries. Before restricting the search to unshrinkable geodesics, the program needed to search through more the 3 million circular galleries and it found more than a thousand beads which could be used. Moreover, performing this search required over 2 months of computation time, and this in a problem which only involves three metric tetrahedra! The problem under consideration is thus only barely feasible using current technology. Slight modifications are likely to quickly lead to intractable complications. In this instance, the output

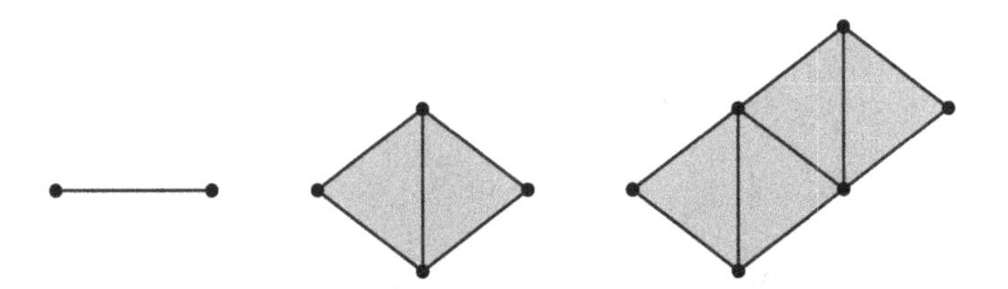

Figure 19: The three combinatorial types of beads in the output.

of 75 beads is small enough to be successfully analyzed. In fact, all 75 beads in the output belong to one of the three combinatorial types shown in Figure 19. More specifically, the output consisted of 4 edges of varying length, 26 beads consisting of exactly 2 triangles as shown in the middle figure, and 45 beads with 4 triangles as shown in the figure on the right. The different beads of each type were combinatorially identical but metrically distinct.

Labeling the beads of each type as type A, B, and C, it was possible to calculate the length of the shortest geodesic in each case. The results are given in Table 1. Using this explicit output, we can string together a rough list of all possible necklace types.

Type	**Minimum length**
A	$.302\pi$
B	$.5\pi$
C	$.833\pi$

Table 1: The minimum geodesic length for each type.

If M is not non-positively curved, then there is a vertex link in M which contains a short closed geodesic loop. This loop determines one of the combinatorial types of necklaces galleries in our list. Moreover, it is possible to perturb this loop slightly so that it misses all of the vertice in the link. The circular gallery determined by this new path will be an annular gallery, and it is straightforward to argue using the list of combinatorial types that the annular gallery which results contains at most 14 triangles. As a result, the dual of this annular gallery is a simple closed curve of length is at most 14 in the 1-skeleton of a soccer tiling of $\mathbb{S}^2$, and an analysis of dihedral angles shows that no three consecutive left or right turns can occur as this path is traversed.

Using Lemma 2.8 we can now conclude that if a short geodesic exists, this process leads to an approximation of this geodesic which bounds one of the two diagrams shown in Figure 17 in the dual soccer tiling. A careful inspection of the possible left/right sequences coming from our three explicit (combinatorial) beads shows this is impossible. This contradiction proves that vertex links cannot contain short closed geodesic loops and that M is locally CAT(0). Finally, the argument that the universal cover of M contains no isometrically embedded flat planes is relatively easy, but omitted. See [46] for details.

Remark 2.9 (The role of computers) *One particularly appealing aspect of this proof scheme is the way in which computer computations and traditional proofs are intimately intertwined. After a metric is assigned, the computer performs a series of calculations, the output provides leads to an investigation of a particular type of disc diagram, and finally, following this analysis, the consequences must be compared back with the computer output in order to reach the final contradiction. It does not seem too far-fetched to predict that in the near future, this type of human-computer interaction will be increasingly common.*

The ubiquity of $5/6*$ triangulations is suggested by the following result obtained in collaboration with Noel Brady and John Meier.

Theorem 2.10 (Bounding edge degrees) *Each closed orientable* 3*-manifold has a triangulation in which each edge has degree* 4*,* 5*, or* 6*.*

The proof essentially uses the universality of the figure eight knot complement and a careful triangulation of various 3-manifolds with boundary that we use as building blocks. See [15] for details.

Remark 2.11 (Foams) *As a final remark, we note that the structures that originally prompted Thurston to consider these types of* 3*-manifolds are* positively curved 3*-manifolds called foams. A* foam *is a* $5/6$*-triangulation in which no* 2*-cell contain more than one edge of degree* 6*. The name and the conditions are derived from the structure of the bubbles in chemical foams. See [81] for details and further references.*

2.3 Higher dimensions

Although the focus of this part has been on low-dimensional complexes, there are, of course, some classes of higher dimensional complexes where easy algorithms are known which check non-positive curvature. Most of these follow from either Gromov's lemma or its generalization known as Moussong's lemma [71]. Recall that a simplicial complex is *flag* if every 1-skeleton of a simplex is actually filled with a simplex, and a metric complex is *metric flag* if every 1-skeleton of a metrically feasible simplex is filled with a simplex.

Lemma 2.12 (Gromov's lemma) *If every edge in a piecewise spherical complex K has length $\pi/2$, then K is globally* CAT(1) *if and only if it is* *a flag complex. In particular, a piecewise Euclidean complex built out of unit cubes (of various dimensions) is non-positively curved if and only if its vertex links are flag complexes.*

Lemma 2.13 (Moussong's lemma) *Suppose that every edge in a piecewise spherical complex K has length at least $\pi/2$. Then K is globally* CAT(1) *if and only if it is a metric flag complex.*

As a consequence of his lemma, Moussong was able to establish that all Coxeter groups are CAT(0) groups. By way of contrast, no such easily checked combinatorial conditions have been found for any class of high dimensional piecewise Euclidean *simplicial* complexes. One might initially harbor the hope that every CAT(−1) group could be made to act geometrically on a non-positivley piecewise Euclidean cube complex. Graham Niblo and Lawrence Reeves [74] have shown that this is not the case.

Theorem 2.14 (Cube complexes are not sufficient) *There exist locally* CAT(−1) *Riemannian manifolds which are not homotopy equivalent to any finite dimensional, locally* CAT(0) *cube complex.*

In other words, the class of groups which act geometrically on CAT(0) cube complexes is strictly smaller then the class of CAT(0) groups. As an aside, we note that the manifolds alluded to in the theorem are constructed using cocompact lattices in quaternionic hyperbolic space and that they only exist in (real) dimension 8 and higher. As a consequence of Theorem 2.14, working with general simplicial complexes is probably inevitable when attacking questions such as the curvature conjecture.

Finally, Section 1.3 concluded with a remark trying to explain why testing for short closed geodesics in high dimensional complexes is difficult. After seeing the 3-dimensional algorithm (Section 2.1), one might think that searching only for *unshrinkable* closed geodesics might be easier. Unfortunately, even this restricted task is likely to remain intractable.

Remark 2.15 (Why is testing shrinkability difficult?) *Consider a sequence of n all-right spherical tetrahedra (i.e. all edges have length $\frac{\pi}{2}$), and let $\mathcal{G}$ be the circular gallery formed by identifying an edge in one tetrahedron to an edge in the next in such a way that the two edges used in each tetrahedron are non-adjacent. In this example, every piecewise geodesic determined by selecting one point on each of the shared edges and connecting them has the same length, namely $\frac{n\pi}{2}$. A closer examination would also show that all of these loops are closed geodesics. It is also instructive to consider a slight modification of this example where every edge which is not shared between two tetrahedra has length $\frac{\pi}{2}+\epsilon$. When ϵ is small and positive $\mathcal{G}$ contains no closed geodesic loops, and when ϵ is small and negative $\mathcal{G}$ contains exactly one. This extreme sensitivity to shape is one difficulty inherent in problems of this type.*

More generally, consider the same combinatorial configuration but with a more random metric. Even if one were able to find a closed geodesic loop in this complex with winding number 1*, it would not necessarily be easy to determine whether it was a shrinkable geodesic. Since each transition point has essentially one degree of freedom, the question of shrinkability can be reframed as a question about whether the Hessian of the multivariable function which calculates length from the positions of the transition points is positive definite or not. A moment's reflection indicates that the Hessian in this case is a tri-diagonal real symmetric matrix since the local distance calculation only depends on the neighboring transition points. Next, one can determine whether the matrix is positive definite by calculating the determinants of its principal minors. One more recoding transforms this sequence of determinants into a sequence of inequalities involving the initial portions of a finite continued fraction. Since continued fractions are notoriously sensitive to small perturbations and rather chaotic in their dependence on the precise numerators and denominators, we can conclude that whether or not a closed geodesic loop in this complex is shrinkable can be quite sensitive to the precise lengths of the edges of these simplices. Under these circumstances, it is highly unlikely that an elementary extension of the algorithm described in Section 2.1 exists, and even if it does, the restriction of the search to only unshrinkable geodesics is no longer likely to be feasible in practice.*

3 Special classes of groups

Since high-dimensional complexes are necessary in order to fully address questions such as the curvature conjecture, and since arbitrary metric simplical complexes in high dimensions are hard to work with, one option is to restrict our attention to those special classes of groups where special types of complexes can be used. In this part, I discuss recent progress along these lines for three such classes of groups. Section 3.1 considers the relationship between non-positive curvature and Artin groups, Section 3.2 does the same for small cancellation groups, and Section 3.3 considers non-positive curvature and the fundamental groups of ample twisted face-pairing 3-manifolds.

3.1 Coxeter groups and Artin groups

Ever since Moussong's dissertation it has been known that all Coxeter groups are CAT(0) groups. The situation for Artin groups, however, remains far from clear.

Definition 3.1 (Artin and Coxeter groups) *Let* Γ *be a finite graph with edges labeled by integers greater than* 1*, and let* $\langle a,b\rangle^n$ *denote the length* n *prefix of* $(ab)^n$*. The* Artin group A_Γ *is the group generated by a set in one-to-one correspondence with the vertices of* Γ *with a relation of the form* $\langle a,b\rangle^n =$

$\langle b, a\rangle^n$ whenever a and b correspond to vertices joined by an edge labeled n. The Coxeter group *W_Γ is the Artin group A_Γ modulo the additional relations $a^2 = 1$ for each generator a. There is also an alternative convention for associating diagrams with Coxeter groups and Artin groups which is derived from a consideration of the finite Coxeter groups. In that case, the graph Γ (using the above convention) is always a complete graph with most of the edges labeled 2 or 3. The alternate convention simplifies the diagram by removing all edges labeled 2 and leaving the label implicit for the edges labeled 3.*

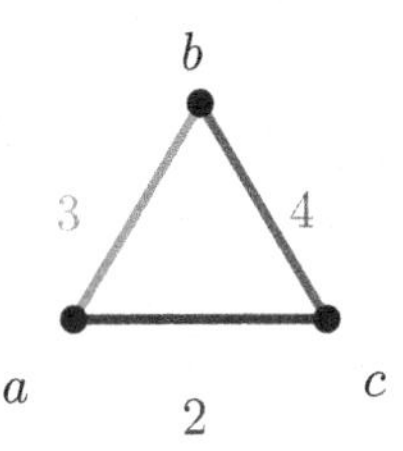

Figure 20: A labeled graph used to define a Coxeter group and an Artin group.

Example 3.2 *Let Γ denote the labeled graph shown in Figure 20. The presentation of the Artin group A_Γ is $\langle a, b, c |\ aba = bab, ac = ca, bcbc = cbcb\rangle$ and the presentation $\langle a, b, c | aba = bab, ac = ca, bcbc = cbcb, a^2 = b^2 = c^2 = 1\rangle$ defines the Coxeter group W_Γ.*

Coxeter groups first arose in the classification of finite groups acting on Euclidean space which are generated by reflections (i.e. isometries fixing a codimension 1 hyperplane). In fact, the collection of finite Coxeter groups is the same as the collection of such finite reflection groups. The well-known classification of irreducible Coxeter groups divides them into type A_n $(n \geq 1)$, B_n $(n \geq 2)$, D_n $(n \geq 4)$, E_6, E_7, E_8, F_4, H_3, H_4 and $I_2(m)$ $(m \geq 2)$. The corresponding diagrams (using the alternative convention) are shown in Figure 21. For more details see some of the standard references for Coxeter groups such as Bourbaki [10], Humphreys [54] or Kane [56].

An Artin group defined by the same labeled graph as a finite Coxeter group is called an *Artin group of finite-type*. There are a number of reasons for believing that all finite-type Artin groups might be CAT(0) groups. Most importantly, they act on several different spaces in ways which almost qualify. First, every finite-type Artin group is, essentially, the fundamental group of a locally Euclidean space derived from a complexified hyperplane arrangement. Being locally Euclidean, it is also locally CAT(0) but this space is neither compact nor complete. There is another space called the Deligne

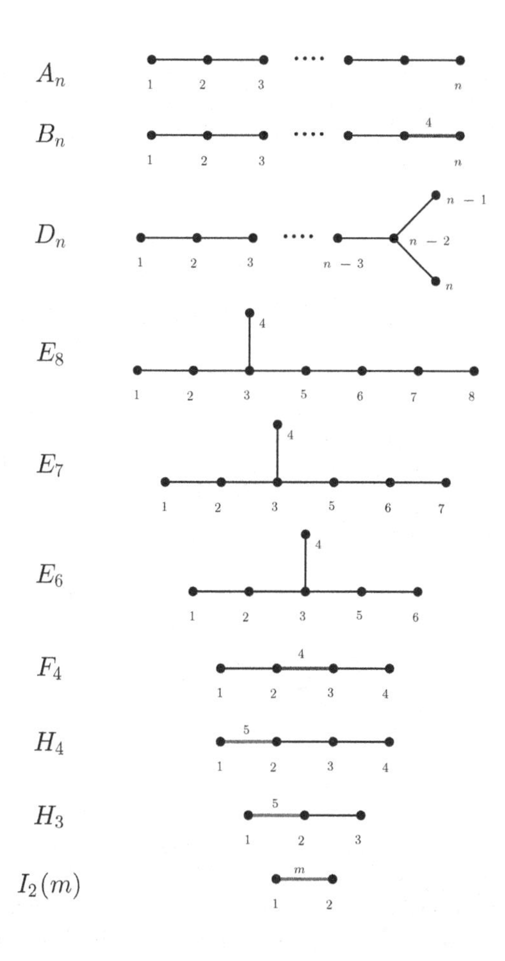

Figure 21: Diagrams for the irreducible finite Coxeter groups.

complex which is complete and homotopy equivalent to the one derived from the hyperplane complement, but the action of the Artin group on the universal cover of this new complex is not proper; there are infinite stabilizers. See [35] and the references therein for details. Moreover, Mladen Bestvina has used a weak version of non-positive curvature to show that finite-type Artin groups have essentially all of the expected group-theoretic consequences of non-positive curvature [8]. Finally, there is a finite Eilenberg-Maclane space for the braid groups constructed independently by Tom Brady [17] and Daan Krammer [60, 61] and generalized to arbitrary Artin groups of finite-type by David Bessis [7] and by Tom Brady and Colum Watt [20]. If a piecewise Euclidean metric is assigned to one of these *Brady-Krammer complexes* then the corresponding Artin group will act geometrically on its universal cover. The only question is whether a metric can be assigned to these complexes which is non-positively curved. Thus, although Artin groups of finite-type do not yet qualify as CAT(0) groups, given these different spaces and actions, they are extremely good candidates.

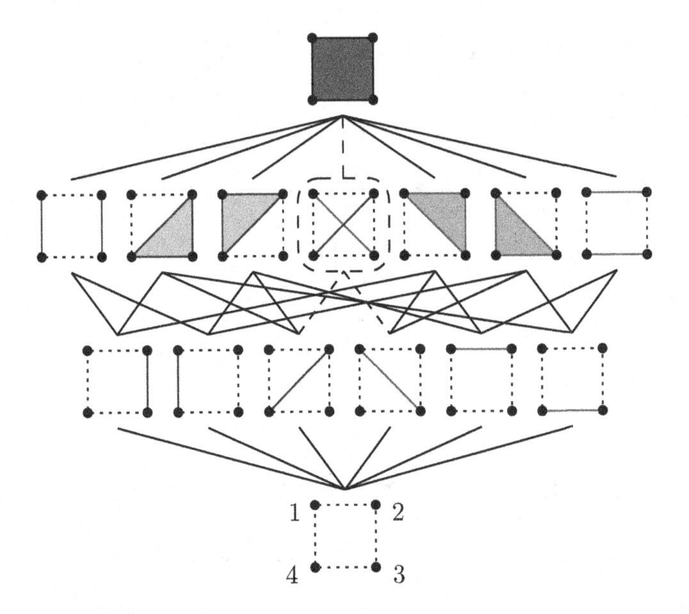

Figure 22: The (noncrossing) partition lattice for $n = 4$.

The general construction of the Brady-Krammer complexes is too complicated to review here, but it is possible to give a rough idea of their structure. In the case of the braid groups, the complex is closely connected with a well-known combinatorial object called the noncrossing partition lattice. For a survey of the connection between braid groups and noncrossing partitions see [65].

Definition 3.3 (Noncrossing partition) *A* noncrossing partition *is a partition of the vertices of a regular n-gon so that the convex hulls of the partitions are disjoint. One noncrossing partition σ is contained in another τ if each block of σ is contained in a block of τ. The noncrossing partition lattice for $n = 4$ is shown in Figure 22. The only "crossing partition" is the one inside the dashed line; the partition* $\{\{1,3\},\{2,4\}\}$.

More generally, there is a version of the noncrossing partition lattice for each (irreducible) finite-type Artin group. As an illustration, the partially ordered set for type F_4 is shown in Figure 23. The connection between the poset and the corresponding Brady-Krammer complex is that the geometric

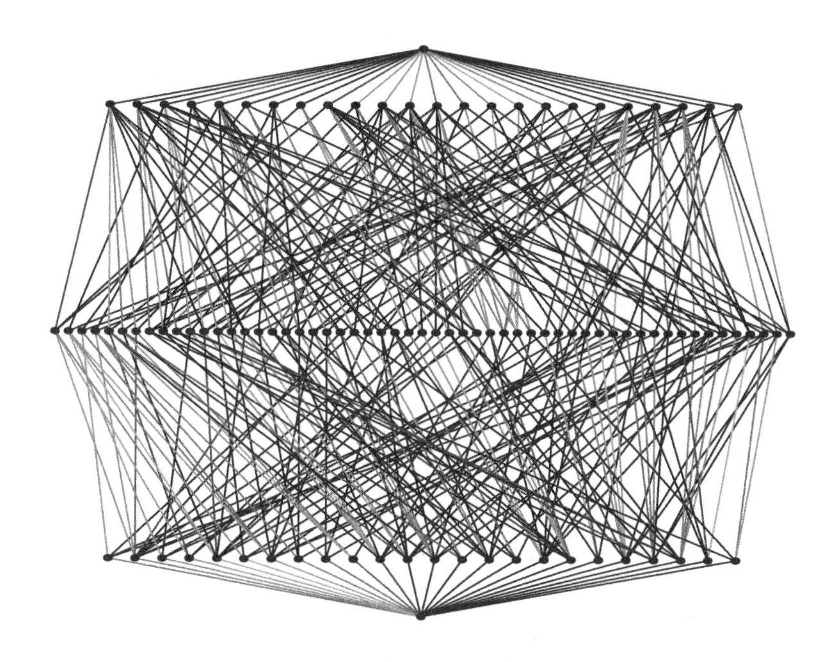

Figure 23: The poset used to construct the F_4 Brady-Krammer complex.

realization of this poset is a fundamental domain for the universal cover of the Brady-Krammer complex. In particular, every increasing path from the lowest vertex to the highest vertex in Figure 23 creates a distinct 4-simplex in the fundamental domain for the F_4 Brady-Krammer complex.

Although the Brady-Krammer complexes do not come with a metric pre-assigned, after examining their structure in low-dimensions, one is quickly led to consider a metric that Tom Brady and I have called the "natural metric." The metric views the edges in a maximal chain as mutually orthogonal steps in a Euclidean space. This metric is natural in the sense that the metric it assigns to the geometric realization of a Boolean lattice is that of a Euclidean cube. See Figure 24. In other words, one of the nicest possible lattices is converted into one of the nicest possible shapes. Also notice that the link of the long diagonal in this n-cube is the Coxeter complex for a symmetric group.

In several instances Tom Brady and I were able to show that the Brady-Krammer complexes with the natural metric were non-positively curved [16, 19, 18].

Theorem 3.4 (Low-dimensions) *The finite-type Artin groups with at most 3 generators are* CAT(0)*-groups and the Artin groups A_4 and B_4 are* CAT(0) *groups.*

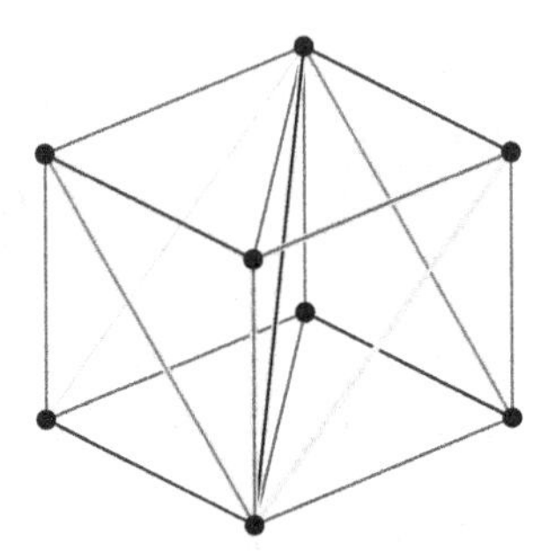

Figure 24: The natural metric on the geometric realization of a Boolean lattice.

When the Artin group has at most 3 generators, the complexes involved are very low dimensional and their links are easy to examine by hand. The A_4 and B_4 cases are slightly more complicated to carry out by hand, but the software `coxeter.g` (described below) is now able to quickly verify that there are no short closed geodesic loops in the various links in this case. Given this evidence, it made sense to conjecture that the Brady-Krammer complexes using the natural metric is non-positively curved for all Artin groups of finite type. Recently, however, a graduate student of mine, Woonjung Choi, disproved this conjecture. In fact, she was able to prove the following stronger result [40].

Theorem 3.5 (D_4 and F_4) *The Brady-Krammer complexes for D_4 and F_4 do not support any piecewise Euclidean metrics which respect the symmetries of their defining diagrams and at the same time are non-positively curved.*

By "respecting the symmetries" she means that all of the symmetries of the defining diagrams should be reflected in the metrics. Since the argument is the same in each case, both complexes are discussed simultaneously. The main idea behind the proof is to first use the cyclic center of these groups, together with the splitting theorem [26, Theorem 6.21 on p. 239], to force there to be a piecewise Euclidean structure on a well-defined 3-dimensional "cross-section" complex. Next, she determined which Euclidean metrics on the tetrahedra in the cross-section complex result in dihedral angles which ensure that the edge links (which are finite graphs) contain no closed geodesic loops of length less than 2π. Because of the sheer size of these computations, this portion of the proof was carried out by GAP using routines that she wrote specifically for this purpose. The symmetries of the defining diagrams were used to cut down on the complexity of the computation. This process drastically reduced the number of metrics which needed to be considered. Finally, for each feasible metric, the one 2-dimensional piecewise spherical vertex link was examined and found to always contains a short closed geodesic loop. Thus, every piecewise Euclidean metric on the cross section complex either produces a short closed geodesic loop in one of the edge links or in the vertex link. As a result, the cross section complex does not support any non-positively curved piecewise Euclidean metric, and this completes the proof.

Remark 3.6 (The software) *The file* `coxeter.g` *is the name of a set of GAP routines used to examine Brady-Kramer complexes and they are available to download [39]. I initially developed some of these routines to test the curvature of the low-dimensional Brady-Krammer complexes exclusively using the natural metric. These early routines were extensively modified and extended by Woonjung Choi so that they are now able to first find the simplicial structure of the* 3*-dimensional cross-section complex, find representive vertex and edge links (up to automorphism), find the graphs representing the various edge links, find the simple cycles in these graphs, find the linear system of inequalities which need to be satisfied by the dihedral angles of the tetrahedra in order for these simple cycles to have length at least* 2π*, and finally, to simplify this large system of inequalities by removing redundancies.*

When the software is run for the D_4 *Artin group, the result is an initial* 4*-dimensional complex with* 162 4*-simplices, the cross section consists of* 15 *tetrahedra, and in the end the software produces a list of* 13 *simplified inequalities in* 9 *variables which needs to be analyzed. When the* F_4 *Artin group is considered, the software finds* 432 4*-simplices in the original complex,* 18 *tetrahedra in the cross section, and produces a system of* 27 *simplified inequalities in* 13 *variables which needs to be analyzed.*

A key result that Choi uses in her study of piecewise Euclidean metrics on Brady-Krammer complexes is the following fact about dihedral angle rigidity. It is almost assuredly classic, but I cannot find a reference to it in the literature.

Theorem 3.7 (Dihedral angle rigidity) *Let* σ *and* τ *be Euclidean simplices of dimension* n *and let* f *be a bijection between their vertices. If the dihedral angle at each codimension* 2 *face of* σ *is at least as big as the dihedral angle at the corresponding codimension* 2 *face of* τ*, then* σ *and* τ *are isometric up to a scale factor.*

The proof is remarkably easy. Let $\vec{u}_i$ and $\vec{v}_i$, $i = 1, \ldots, n$ be the external unit normal vectors for the facets of σ and τ, numbered so that corresponding facets receive corresponding subscripts. It is a result due to Minkowski that there are positive numbers $a_i > 0$ such that $\sum_i a_i \vec{u}_i = \vec{0}$. In fact, the numbers a_i can be chosen to be the volume of the ith facet. Noel Brady pointed out to me, that this can also be thought of as a consequence of the divergence theorem applied to this simplex. The proof consists of the following sequence of equations and inequalities.

$$0 = ||\sum_i a_i \vec{u}_i||^2 = \sum_i \sum_j a_i a_j (\vec{u}_i \cdot \vec{u}_j) \geq \sum_i \sum_j a_i a_j (\vec{v}_i \cdot \vec{v}_j) = ||\sum_i a_i \vec{v}_i||^2 \geq 0$$

The first inequality follows from the assumption on dihedral angles and the positivity of the a_i. Comparing the ends of the sequence we conclude that the inequality must be an equality and that each of the corresponding dihedral angles must be exactly equal in size. Finally, it is easy to show that the collection of dihedral angles determine the isometry type of the simplex up to rescaling.

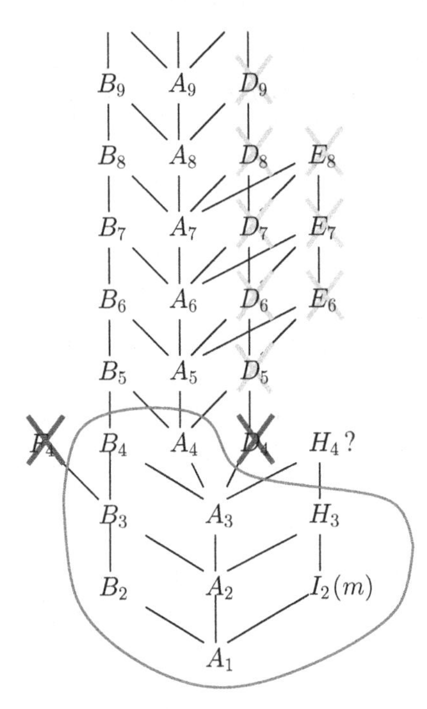

Figure 25: CAT(0) and Brady-Krammer complexes.

Remark 3.8 (CAT(0) and Brady-Krammer complexes) *A quick summary of the results to date is given in Figure 25. The circled cases are the ones which are* CAT(0) *groups (and all of these use the natural metric). The two which are boldly crossed out are the cases which do not support non-positively curved piecewise Euclidean in any reasonable sense. The higher rank examples which are lightly crossed out are also not going to support non-positively curved piecewise Euclidean metrics in any reasonable sense precisely because they contain the D_4 complex as part of their structure. And finally, there is a* *question mark next to the Artin group of type H_4 since this case is half-way analyzed, but the analysis is not yet complete.*

Remark 3.9 (The H_4 complex) *The case of H_4 is more difficult to analyze than the D_4 or F_4 cases because its defining diagram has no symmetries. This greatly increases the number of equations and variables involved in the computations. In particular, the software applied to the H_4 Artin group produces a list of* 1350 *4-simplices in the original complex,* 23 *tetrahedra in the cross section complex, and it produces a list of* 638 *simplified inequalities in* 96 *variables which needs to be analyzed. Although it is in principle feasible to plug these inequalities into linear programming software, neither Woonjung nor I have yet done so. The D_4 and F_4 cases produced systems which were small enough to analyze by hand; this system for H_4 is not.*

3.2 Small cancellation groups

The most successful reduction of curvature to combinatorial conditions is, of course, the case of complexes built entirely out of Euclidean cubes of various dimensions. Partly due to the ease with which curvature can be checked (Lemma 2.12), these *cube complexes* have become a favorite of geometric group theorists [9, 37, 72, 73, 77, 78]. In this section and the next I discuss how high dimensional cube complexes can be used to prove that various groups are CAT(0) groups. After a brief digression about cube complexes for Coxeter groups, the primary focus in this section is on cube complexes for small cancellation groups.

One source of additional interest in cube complexes is caused by the fact that groups acting on non-positively curved cube complexes have properties which do not hold for more general CAT(0) groups. For example, groups acting geometrically on non-positively curved cube complexes cannot have Kazhdan's property T (which is the property used by Niblo and Reeves [74] to establish Theorem 2.14). Thus, even for classes of groups such as Coxeter groups where they are already known to be CAT(0), it is still of interest to investigate which Coxeter groups are capable of acting geometrically on CAT(0) cube complexes. This question has been (essentially) completely answered by recent work of Graham Niblo and Lawrence Reeves [73] and by Ben Willams [82]. In particular, Niblo and Reeves use the Sageev cube construction ([77, 78]) to establish the following.

Theorem 3.10 (Coxeter groups and cube complexes) *Let W denote a finitely generated Coxeter group. Then W acts properly discontinuously by isometries on some locally finite, finite dimensional* CAT(0) *cube complex.*

As stated the action is not necessarily geometric since it need not be cocompact. On the hand, in his dissertation [82], Ben Williams was able to characterize exactly when this action is cocompact in terms of the subgroup structure.

Theorem 3.11 (Characterising cocompactness) *The action in the preceeding theorem is cocompact if and only if for any triple p, q, r of positive integers, the Coxeter group W contains only finitely many conjugacy classes of subgroups isomorphic to the p, q, r triangle group*

$$\langle a, b, c | a^2 = b^2 = c^2 = (ab)^p = (bc)^q = (ac)^r = 1 \rangle .$$

Despite this progress, it is would still be of interest to determine more explicitly which Coxeter groups satisfy the conditions isolated by Wiliams.

The situation for small cancellation groups is much less satisfactory. Small cancellation groups are a relatively well-understood class of cohomological dimension 2 (mostly) word-hyperbolic groups whose definition and key properties have been known for nearly thirty years [63]. In fact, they formed one of the key models for the development of word-hyperbolic groups. Nonetheless, the relationship between small cancellation groups and the more local notions of curvature such as CAT(0) has remained completely mysterious until recently. The basic definitions of small cancellation theory can be found in [63] or [67], but a rough description is included for the sake of completeness.

Definition 3.12 (Small cancellation groups) *Let X be a finite combinatorial 2-complex. A* piece *in X is a path in the 1-skeleton which can be ϵ-pushed off the 1-skeleton in at least two distinct ways. In other words, this path must be liftable through the attaching maps of the 2-cells in more than one way. The complex X is called a $C(p)$-complex if for each 2-cell in X its boundary cycle cannot be covered with fewer than p pieces. It is called a $T(q)$-complex if there does not exist an immersed path in any vertex link with combinatorial length strictly between 2 and q. In other words, every 2-complex satisfies the condition $T(3)$ but the condition $T(4)$ represents an actual restriction.*

There is also a metric version of the $C(p)$ condition known as $C'(\alpha)$. The complex is a $C'(\alpha)$-complex if for each piece P in the boundary of a 2-cell R of X, the combinatorial length of P is less than α of the combinatorial length of the boundary of R. It is a basic result of small cancellation theory that $C'(1/4)$-$T(4)$-complexes and $C'(1/6)$-complexes are word-hyperbolic. These two classes are called metric small cancellation groups. *Sergei Ivanov and Paul Schupp have also characterized exactly when the non-metric small cancellation groups, that is the fundamental groups of $C(p)$-$T(q)$ complexes where (p, q) is either $(3, 6)$, $(4, 4)$ or $(6, 3)$, are word-hyperbolic [55].*

Noel Brady and I recently established a theorem concerning small cancellation groups and CAT(0) structures [13], which has also be established by

Dani Wise independently using different techniques [83]. The basic philosophy is described first since it is also used to prove the results in the next section.

Remark 3.13 (Philosophy) *Let X be any finite combinatorial cell complex, let $\mathcal{C}$ be the collection of maximal closed cells in its universal cover $\widetilde{X}$, and let P be the partially ordered set of intersections of elements in $\mathcal{C}$. The poset P is called the* nerve *of $\widetilde{X}$. See Figure 26 for a simple illustration.*

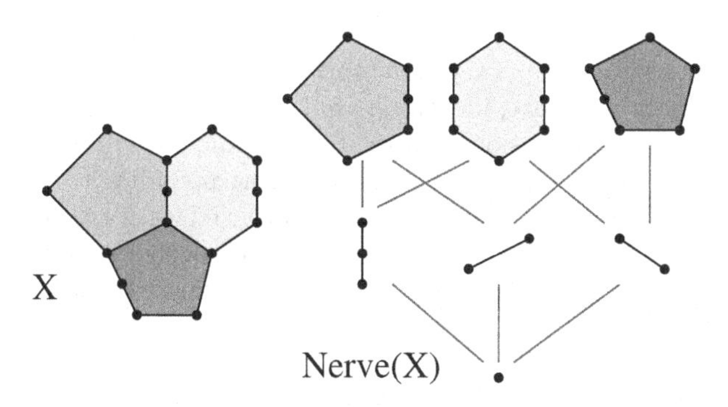

Figure 26: A complex and its nerve.

The main idea is to replace each maximal cell in X with a high-dimensional cell so that they glue together nicely and the nerve of the resulting complex is identical. In this new situation a piece *is defined as a subcomplex of $\widetilde{X}$ which corresponds to an element of the nerve. In other words, a piece P is subcomplex which is an intersection of maximal cells in the universal cover. Notice that for small cancellation complexes this is a slightly modified notion of piece since arbitrary connected subcomplexes of pieces are no longer pieces. We need the following observation. If the pieces of X are contractible and the maximal cells embed in $\widetilde{X}$, then X is homotopy equivalent to the geometric realization of the nerve quotiented by the group action. As a consequence, if we find two complexes with these properties and there is a correspondence between their fundamental groups and their nerves, then these complexes are homotopy equivalent to each other. Various small-cancellation-like conditions on X will guarantee both of these properties. For example, these properties will hold if the overlaps between maximal cells are "small" subcomplexes of its boundary (in some suitable sense) and the links are "large."*

In the case of small cancellation complexes, the idea is to replace each 2-cell and each piece with a high dimensional Euclidean cube, preserving the nerve and all of the other necessary properties described above. The following is a sample theorem along these lines. This result demonstrates how high-dimensional cube complexes really can be useful, even when studying objects whose cohomological dimension is quite small.

Theorem 3.14 **($C'(1/4)$-$T(4)$ groups)** *Each $C'(1/4)$-$T(4)$ group is the fundamental group of a compact high-dimensional non-positively curved cube complex.*

The key idea behind the proof is to subdivide the 1-skeleton of the standard 2-complex for G and then embed it into the 1-skeleton of a high-dimensional cube so that the pattern of intersections between the maximal cells in its universal cover, i.e. its nerve, remains the same. The proof proceeds roughly as follows. First subdivide every edge once so that every 2-cell has an even length boundary cycle. Next, identify each 2-cell R whose boundary cycle has length $2n$ with an n-dimensional cube. In addition, we select a path in the 1-skeleton of this n-cube which we think of as corresponding to the boundary of R. The choice is made so that any n consecutive edges of the path all travel in distinct directions and the antipodal edges are parallel. See Figure 27 for the path in a 3-cube which corresponds to an original hexagon.

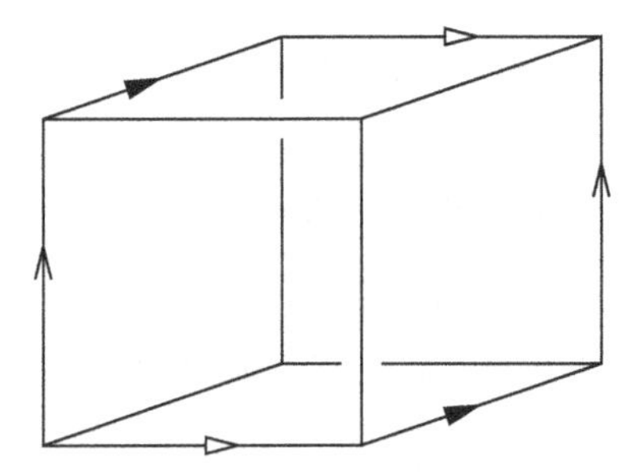

Figure 27: The cube corresponding to a hexagon.

Finally, we glue these cubes together, exactly as the original 2-cells were glued, but with one crucial difference. If R_1 and R_2 shared a piece P of length m, then the cubes corresponding to R_1 and R_2 will be identified along the m-cubes spanned by P in R_1 and R_2. For example, if P has length 2, then a square in one cube will be glued to a square in the other. It is now relatively easy to check that the result is a non-positively curved cube complex satisfying all the necessary conditions.

I should note that the hypothesis of the theorem can be weakened slightly without altering the proof: it is actually sufficient for the total length of any two consecutive pieces in R to be at most half the length of ∂R. Also, Dani Wise has extended this result to (a similar generalization of) $C'(1/6)$ groups. This case is more complicated and requires a more delicate construction. The next step would be, of course, to try and extend these results to all small cancellation groups, including the non-metric ones.

Conjecture 3.15 (Small cancellation groups and cubes) *Every small cancellation group is the fundamental group of a compact high-dimensional non-positively curved cube complex.*

The key difficulty here is to deal with the fact that 2-cells can now overlap a significant amount (even along more than half of their boundary cycles), so long as this large overlap is compensated by the existence of only very small overlaps along the remainder of the boundary path. A solution to this difficulty is likely to involve a more wholesale modification of the original 2-complex (perhaps along the lines described in Section 4.3 for one-relator groups) as a preparation for the transformation into a high-dimensional cube complex.

3.3 Ample twisted face-pairing groups

The techniques and the philosophy described in the previous section are, of course, applicable more generally. As an illustration, Noel Brady and I have also shown that one of the ample twisted fair-pairing 3-manifolds used as an illustration in [29] can be expanded in this way to 6-dimensional non-positively curved cube complex [13]. This brief tour of special cases concludes with a short account of this construction and its consequences.

Twisted face-pairing 3-manifolds were introduced by Jim Cannon, Bill Floyd and Walter Parry in [30], and their properties were further developed in [31] and [32]. Roughly speaking, one starts with a faceted 3-ball and a pairing of its faces. Then, one subdivides the 1-skeleton of this 3-ball according to well-defined formulae and finally, one slightly alters the face-pairing maps by adding a small twist to each. Surprisingly, the result is always a 3-manifold. Moreover, in [29] they established that if the original faceted 3-ball satisfies certain elementary conditions that they call *ample*, then the 3-manifold constructed always has a word-hyperbolic fundamental group.

One of the primary motivations behind their construction was to construct explicit examples of word-hyperbolic 3-manifolds which were not already known to be hyperbolic (Kleinian) as a testing ground for various approaches to Thurston's geometrization conjecture. Having established that these 3-manifolds have word-hyperbolic fundamental groups, they have asked whether any of these constructions can be given a metric of non-positive curvature. The high-dimensional cube complex Noel Brady and I construct answers this question for at least one example. The natural conjecture to make is the following.

Conjecture 3.16 (Ample twisted 3-manifolds and cubes) *If M is an ample twisted face-pairing 3-manifold, then $\pi_1 M$ is not only a word-hyperbolic group, but also the fundamental group of a compact non-positively curved cube complex of high dimension.*

Given the flexibility of our construction, it is already clear that Conjecture 3.16 does holds for an infinite number of ample twisted face-pairing 3-manifolds, and possibly for all of them. If one were even bolder, one could conjecture that for every compact, orientable hyperbolic 3-manifold M, there is a finite high-dimensional cube complex with fundamental group $\pi_1 M$. Rubinstein and Sageev's results on the k-plane property [77] are totally compatible with this conjecture, and, in fact, make it seem a bit more plausible. Moreover, this process of constructing high-dimensional non-positively curved cube complexes for word-hyperbolic 3-manifold groups might be interesting in another respect. Our cube complex construction forces the original 3-manifold to contain a π_1-injective immersed surface. Thus, constructions of this type might be used as a way of approaching (at the very least) special cases of the virtual fibering conjecture.

4 Combinatorial notions of curvature

In the last two parts I examined the possibility of restricting the complexes under consideration to low dimensions where good implementable algorithms exist, and of restricting the classes of groups under consideration so that special complexes such as cube complexes could be used. In this final part I consider a third possibility: weakening the notion of curvature itself. Rather than assigning piecewise Euclidean metrics to each of the polytopes in a complex, this more combinatorial version simply assigns values representing the "size" of each internal and external angle. Section 4.1 briefly reviews the various identities satisfied by these angles in a Euclidean polytope since this is the situation the more "conformal" version is modeled on. Next, Section 4.2 discusses the 2-dimensional combinatorial Gauss-Bonnet theorem and proposes a number of ways in which this theorem could be extended to higher dimensions. Finally, Section 4.3 discusses the notion of conformally CAT(0) 2-complexes, sectional curvature, and the potential relationship between these concepts, special polyhedra and the coherence of one-relator groups.

4.1 Angles in Polytopes

Every mathematician learns early on that the sum of the internal angles in a triangle equals π and that sum of the external angles of a convex polygon is 2π, but far fewer people learn the corresponding relations in higher dimensional Euclidean polytopes. For example, how many students learn that the sum of the dihedral angles in a tetrahedron can vary depending on its shape? This section is a short digression into the various identities among the internal and external angles of a Euclidean polytope particularly in high dimensions. All of this material is well-known and classical [49, 68, 80], but perhaps less well-known in our own subfield than it should be. The main goal of this section is to review the background material necessary to state and prove a fairly general high-dimensional combinatorial Gauss-Bonnet type theorem in the next section. The first concept to review is the precise definition of normalized internal and external angles in a Euclidean polytope.

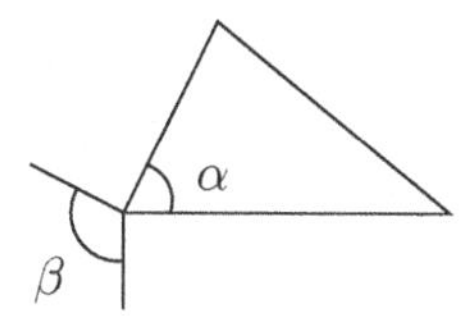

Figure 28: Internal and external angles.

Definition 4.1 (Internal and external angles) *Let P be any Euclidean polytope and let F be a face of P. The* internal angle, *$\alpha(F, P)$, is the proportion of unit vectors perpendicular to F which point into P (i.e. the measure of this set of vectors divided by the measure of the sphere of the appropriate dimension). The* external angle, *$\beta(F, P)$, is the proportion of unit vectors perpendicular to F so that there is a hyperplane with this unit normal which contains F and the remainder of P is on the other side. Also, by convention $\alpha(P, P) = \beta(P, P) = 1$. As an example, consider the vertex in the triangle shown in Figure 28. The internal angle (marked by α and the external angle (marked by β) are the radian measures of these angles divided by 2π (which is the normalization). The angles considered here are always normalized in this sense, since in higher dimensions it is the normalized versions which satisfy the most elegant relations.*

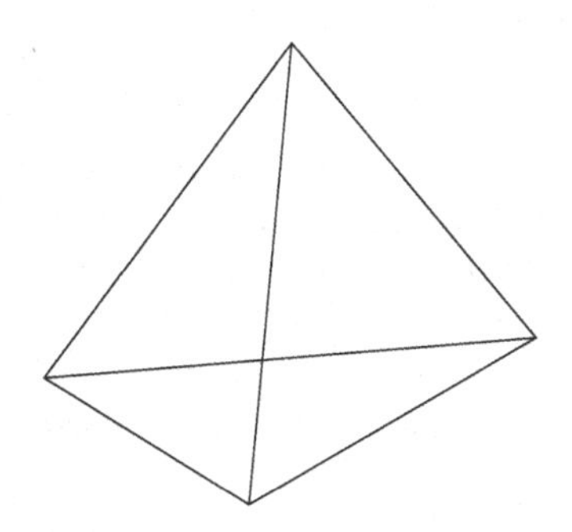

Figure 29: A regular tetrahedron.

Example 4.2 (Angles in a regular tetrahedron) *Consider the tetrahedron shown in Figure 29. Let t denote the tetrahedron itself containing a* *face f which contains an edge e which contains a vertex v. Both $\alpha(v,e)=\frac{1}{2}$ and $\beta(v,e)=\frac{1}{2}$ since each contains a single unit vector and the 0-sphere consists of two unit vectors. Similarly, $\alpha(e,f)=\beta(e,f)=\alpha(f,t)=\beta(f,t)=\frac{1}{2}$. The value of $\alpha(v,f)=\frac{1}{6}$ since $\frac{\pi}{3}$ is one-sixth of a circle. Also, since the dihedral angles along the edges are $\cos^{-1}\left(\frac{1}{3}\right)$ and the area of the vertex link is $3\cos^{-1}\left(\frac{1}{3}\right)-\pi$, the values of $\alpha(e,t)$ and $\alpha(v,t)$ are $\frac{1}{2\pi}\cos^{-1}\left(\frac{1}{3}\right)$ and $\frac{3}{4\pi}\cos^{-1}\left(\frac{1}{3}\right)-\frac{1}{4}$ respectively. The normalized external angles are $\beta(e,t)=\frac{1}{2}-\frac{1}{2\pi}\cos^{-1}\left(\frac{1}{3}\right)$, $\beta(v,f)=\frac{1}{3}$ and $\beta(v,t)=\frac{1}{4}$.*

From the definition of external angles and the fact that a hyperplane approaching a polytope from a generic direction first encounters the polytope at a vertex, the following result is immediate.

Theorem 4.3 (Summing external vertex angles) *If P is any Euclidean convex polytope then the result of summing the normalized external angles $\beta(v,P)$ over all the vertices $v\in P$ is 1.*

This result is one explanation for the relative simplicity of the external angles of the regular tetrahedron as compared with its internal angles. In order to easily describe the other identities which hold between the internal and external angles in a Euclidean polytope a short digression into combinatorics is necessary.

Definition 4.4 (Face lattice of a polytope) *Associated to any Euclidean polytope P is a partially ordered set called its face lattice. The* face lattice *of a polytope is its poset of faces under inclusion. Conventionally, the empty set and the whole polytope are included as faces. When these elements are included the poset is a lattice in the combinatorialist's sense (i.e. every pair of elements has a unique least upper bound and a unique greatest lower bound). Notice that the set of all normalized internal angles of a polytope can be viewed as a single real-valued function on those pairs of elements in its face lattice which are comparable in the poset ordering. I call this function α. The normalized external angles can be viewed as a similar function which I call β. There is a small issue about extending the values of α and β to pairs of faces where the smaller face is the empty set since these values were not defined above. The conventional assignments, derived from thinking of a Euclidean polytope as a rescaled version of a very small spherical polytope, are as follows.*

$$\alpha(\emptyset,F)=\begin{cases}1 & \text{if} \quad F=\emptyset\\ \frac{1}{2} & \text{if} \quad \dim F=0\\ 0 & \text{if} \quad \dim F>0\end{cases} \qquad \beta(\emptyset,F)=\begin{cases}1 & \text{if} \quad F=\emptyset\\ \frac{1}{2} & \text{if} \quad \dim F\geq 0\end{cases}$$

The justifications are that if F is a very small spherical polytope of dimension n, then as seen from the origin $\alpha(\emptyset,F)$ is one-half of a 0-sphere when F is a vertex and nearly zero when F is higher dimensional. Similarly, the *external angle is one-half of a* 0*-sphere when F is a vertex and nearly one-half when F is higher dimensional.*

The collection of all functions of this type is known as the incidence algebra of the poset and the algebraic structure of this object is quite useful for expressing identities involving the internal and external angle functions α and β.

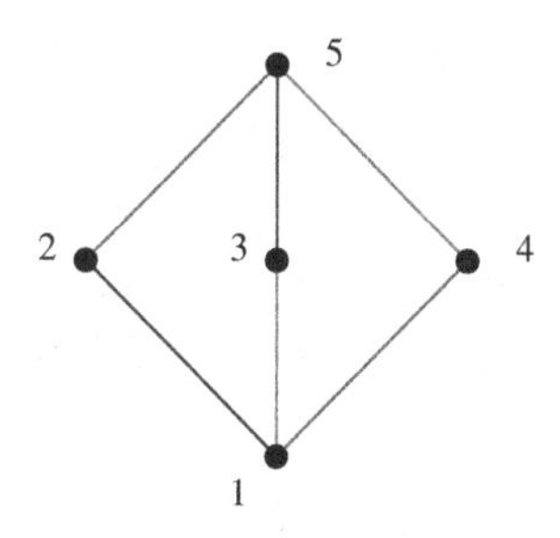

Figure 30: A sample poset.

Definition 4.5 (Incidence algebra of a poset) *Let P be a finite partially ordered set whose elements are labeled by the set $\{1, 2, \ldots, n\}$. The set of $n \times n$ matrices over the reals with $a_{ij} \neq 0$ only when $i \leq j$ in P is called the* incidence algebra of P, *or $I(P)$. Notice that for any finite poset P the ordering of the vertices can be chosen so it is consistent with its partial order. For example, the numbering of the vertices in Figure 30 is consistent. Using such an ordering, the incidence algebra is a set of upper triangular matrices. Alternatively, and more abstractly, the elements of $I(P)$ can also be thought of as functions $f : P \times P \to \mathbb{R}$ where $f(x, y) = 0$ whenever $x \not\leq y$ in P. Because of this possibility, elements of $I(P)$ are often referred to as functions rather than matrices. Finally, matrix multiplication defines a corresponding multiplication of functions, called their* convolution.

Inside the incidence algebra, there are three elements that are particularly crucial: the delta function, the zeta function and the Möbius function.

$$\zeta = \begin{bmatrix} 1 & 1 & 1 & 1 & 1 \\ 0 & 1 & 0 & 0 & 1 \\ 0 & 0 & 1 & 0 & 1 \\ 0 & 0 & 0 & 1 & 1 \\ 0 & 0 & 0 & 0 & 1 \end{bmatrix} \qquad \mu = \begin{bmatrix} 1 & -1 & -1 & -1 & 2 \\ 0 & 1 & 0 & 0 & -1 \\ 0 & 0 & 1 & 0 & -1 \\ 0 & 0 & 0 & 1 & -1 \\ 0 & 0 & 0 & 0 & 1 \end{bmatrix}$$

Figure 31: The functions ζ and μ for the sample poset.

Definition 4.6 (δ, ζ, and μ) *The identity matrix is represented by the* delta function *where* $\delta(x,y) = 1$ *iff* $x = y$. *The* zeta function *is the function* $\zeta(x,y) = 1$ *if* $x \leq y$ *in* P *and* 0 *otherwise (i.e.* 1*'s wherever possible). The* Möbius function *is function represented by the matrix inverse of the matrix representing* ζ. *Note that* $\mu \cdot \zeta = \zeta \cdot \mu = \delta$. *The matrices that represent* ζ *and* μ *for the sample poset are given in Figure 31.*

Remark 4.7 (Euler characteristics) *There is a close connection between the values of the Möbius function and various Euler characteristics. In particular, the value of* $\mu(x,y)$ *is the reduced Euler characteristic of the geometric realization of the portion of the poset that lies strictly between* x *and* y. *For example, the value of the Möbius function on the interval of the sample poset from vertex* 1 *to vertex* 5 *is* 2 *(i.e. the value in the upper left corner of the matrix in Figure 31). The portion of the poset between the first and fifth elements is the three incomparable elements labeled* 2, 3, *and* 4, *the geometric realization is three points with the discrete topology, and the reduced Euler characteristic of this topological space is also* 2.

The connection between Möbius functions and Euler characteristics enables a quick calculation of the Möbius function for the face lattice of a polytope.

Lemma 4.8 (Möbius functions for polytopes) *If* F *and* G *are faces of a polytope* P *with* $F \leq G$, *then the value of the Möbius function of the face lattice evaluated on the interval* (F,G) *is* $\mu(F,G) = (-1)^{\dim G - \dim F}$.

The idea is that the geometric realization of the portion of the face lattice between F and G is topologically an n-sphere, where n is determined by the difference in dimension between F and G. Thus, $\mu(F,G)$ is the reduced Euler characteristic on this n-sphere. This connection between Möbius functions and Euler characteristics was also recently used by myself and John Meier to calculate the ℓ^2 Betti numbers of the pure symmetric automorphism groups [66].

Remark 4.9 (ℓ^2 Betti numbers) *Elsewhere in this book, Wolfgang Lück describes various theorems about* ℓ^2 *Betti numbers which make them easier to calculate [62]. For the groups that John Meier and I studied, we first used a spectral sequence argument to show that all of the* ℓ^2 *Betti numbers were trivial - except for the one in top dimension. As a consequence, this final Betti number must be the Euler characteristic of the group. Next, we show this value must equal the reduced Euler characteristic of the fundamental domain of the space which is in turn a geometric realization of the poset of hypertrees. The hypertree poset for* $n = 4$ *is shown in Figure 32. The size of the hypertree posets grows quite rapidly. In fact, the poset for* $n = 5$ *already has* 311 *elements. Using the connection between reduced Euler characteristics*

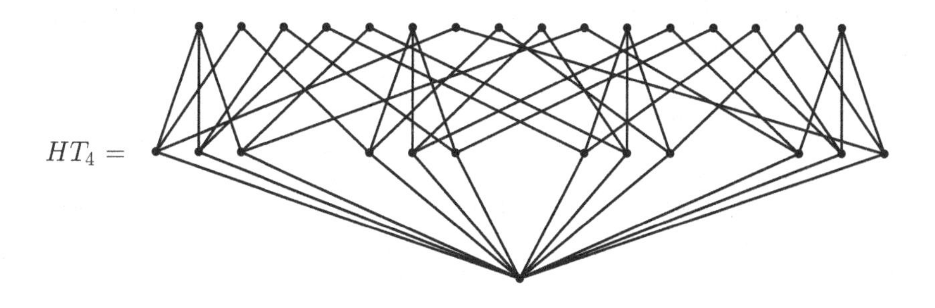

Figure 32: The poset of hypertrees on a set of size 4.

and Möbius functions, the final calculation could be completed using standard generating function techniques from enumerative combinatorics. The moral here is that combinatorial ideas such as Möbius functions of posets can be relevant, even to seemingly distant areas such as the calculation of ℓ^2 Betti numbers.

The final identities I wish to highlight (Sommerville's identities and McMullen's identity) involve the convolution product of elements already defined, although for Sommerville's identities, it is convenient to introduce a signed version of the internal and external angles. Let $\bar{\alpha}(F,G)$ denote a *signed normalized internal angle* where the sign is given by $\mu(F,G)$. In other words $\bar{\alpha}(F,G) = \mu(F,G)\alpha(F,G)$. There is also a *signed normalized external angle*, $\bar{\beta}$ defined as $\bar{\beta}(F,G) = \mu(F,G)\beta(F,G)$. Signed normalized external angles resurface in the discussion of vertex curvature in Section 4.2.

Theorem 4.10 (Sommerville's identity) *If P is a Euclidean polytope and α, $\bar{\alpha}$, β, $\bar{\beta}$ and μ are defined as above, then $\mu \cdot \alpha = \bar{\alpha}$ and $\beta \cdot \mu = \bar{\beta}$. In other words, for all faces $F \leq H$ in P, the sum of $\mu(F,G)\alpha(G,H)$ over all G with $F \leq G \leq H$ is equal to $\mu(F,H)\alpha(F,H)$. Similarly, for all faces $F \leq H$ in P, the sum of $\beta(F,G)\mu(G,H)$ over all G with $F \leq G \leq H$ is equal to $\beta(F,H)\mu(F,H)$.*

Since the values of μ are simply signs, Sommerville's identity asserts that summing up the internal angles $\alpha(G,H)$ over all G between F and H with the appropriate signs equals the value of $\alpha(F,H)$ with the appropriate sign. The proof of this result is an easy application of inclusion-exclusion. The second identity listed ($\beta \cdot \mu = \bar{\beta}$) is derived from the first by applying Sommerville's identity to the dual polytopal cone.

Example 4.11 *To illustrate the first identity, consider a regular tetrahedron, let F be one of vertices and H be the tetrahedron itself. Using v, e, f and t*

to denote arbitrary vertices, edges, faces and the tetrahedron itself, the sum under consideration is

$$\mu(v,v)\cdot\alpha(v,t)+3\mu(v,e)\cdot\alpha(e,t)+3\mu(v,f)\cdot\alpha(f,t)+\mu(v,t)\cdot\alpha(t,t)$$

where the non-trivial coefficients account for the fact that there are three edges and three faces which contain the vertex and are contained in the tetrahedron. Numerically this works out to be

$$1\cdot\left(\frac{3}{4\pi}\cos^{-1}\left(\frac{1}{3}\right)-\frac{1}{4}\right)-3\cdot\left(\frac{1}{2\pi}\right)+3\cdot\frac{1}{2}-1\cdot 1,$$

which simplifies to $-\frac{3}{4\pi}\cos^{-1}\left(\frac{1}{3}\right)+\frac{1}{4}=-\alpha(v,t)$ *as claimed.*

Finally, the most interesting angle identity is the nonlinear identity relating internal and external angles discovered by Peter McMullen [68].

Theorem 4.12 (McMullen's identity) *If P is a Euclidean polytope and α, β and ζ are defined as above, then $\alpha\cdot\beta=\zeta$. In other words, for all faces $F\leq H$ in P, the sum of $\alpha(F,G)\beta(G,H)$ over all G with $F\leq G\leq H$ is equal to 1.*

The proof of this remarkable identity is a surprisingly short half page argument. First reduce to the case where F is a vertex and view the neighborhood of F in H as a polytopal cone. The idea is to look at a product of this particular polytopal cone with its dual in $\mathbb{R}^{2n}$ (where n is the dimension of H) and then to integrate the function $f(\vec{x})=\exp(-||\vec{x}||^2)$ over this space in two different ways. As an immediate corollary the identities $\mu\cdot\alpha\cdot\beta=\bar{\alpha}\cdot\beta=\delta$ and $\alpha\cdot\beta\cdot\mu=\alpha\cdot\bar{\beta}=\delta$ must also hold. In other words $\bar{\alpha}$ and β are inverses of each other and α and $\bar{\beta}$ are inverses. Once again, the regular tetrahedron provides a good illustration.

Example 4.13 *Let F be one of the vertices of a regular tetrahedron H. Using the notational abbreviations defined above, the sum under consideration is*

$$\alpha(v,v)\cdot\beta(v,t)+3\alpha(v,e)\cdot\beta(e,t)+3\alpha(v,f)\cdot\beta(f,t)+\alpha(v,t)\cdot\beta(t,t)$$

where the coefficients account for the three edges and three faces between F and H as before. Numerically this works out to be

$$1\cdot\frac{1}{4}+3\cdot\frac{1}{2}\cdot\left(\frac{1}{2}-\frac{1}{2\pi}\cos^{-1}\left(\frac{1}{3}\right)\right)+3\cdot\frac{1}{6}\cdot\frac{1}{2}+\left(\frac{3}{4\pi}\cos^{-1}\left(\frac{1}{3}\right)-\frac{1}{4}\right)\cdot 1$$

which simplifies to 1 as claimed.

4.2 Combinatorial Gauss-Bonnet

In this section I discuss two seemingly different well-known combinatorial versions of the Gauss-Bonnet formula and a very flexible common generalization which is being introduced here for the first time. The first version is a formula due to Cheeger, Müller and Schrader which starts with a piecewise Euclidean complex and assigns a curvature to each vertex (based on certain external angles of the Euclidean polytopes involved) so that the sum of these vertex curvatures is the Euler characteristic of the complex. The second version (rediscovered by a number of different researchers over the years) starts with a polygonal 2-complex, randomly assigns real numbers as external "angles," and then calculates vertex and face curvatures so that the sum of all the vertex and face curvatures is the Euler characteristic. Finally, using the identities from the previous section, I show how both versions are special cases of a very general procedure for assigning random "angles" to high-dimensional polyhedral cell complexes and then calculating combinatorial curvatures which add up to the Euler characteristic of the complex. The first step is to introduce the two versions which are already well-known.

The version of the combinatorial Gauss-Bonnet theorem due to Cheeger, Müller and Schrader is essentially a quick calculation [38]. See also the paper by Charney and Davis [36] where they review this calculation and then use the resulting formula to find a combinatorial analogue of the Hopf conjecture in the context of non-positively curved piecewise Euclidean n-manifolds. Let X be a piecewise Euclidean complex. In the following calculation P denotes an arbitrary (nonempty) cell of X and v an arbitrary vertex. By convention, when both variables are mentioned below a summation sign, it is the first listed which is varying over the course of the summation. Using Theorem 4.3 and an inversion of the order of summation, the Euler characteristic of X can be rewritten as follows.

$$\begin{aligned}
\chi(X) &= \sum_P (-1)^{\dim P} \\
&= \sum_P \sum_{v\in P} (-1)^{\dim P} \beta(v,P) \\
&= \sum_v \sum_{P\ni v} (-1)^{\dim P} \beta(v,P) \\
&= \sum_v \sum_{P\ni v} \bar{\beta}(v,P) = \sum_v \kappa(v)
\end{aligned}$$

where the *vertex curvature* $\kappa(v)$, as might be guessed from the final equality, is defined by summing the signed normalized external angles $\bar{\beta}(v,P)$ over all polytopes P containing v. Notice that if the portion of X containing the vertex v has a unique maximum element P, then by Sommerville's identity (Theorem 4.10) the absolute value of the vertex curvature $\kappa(v)$ will be the normalized external angle $\beta(v, P)$.

The second combinatorial version of the Gauss-Bonnet formula is specifically 2-dimensional, but where the angles are now randomly assigned real numbers rather than numbers derived from a piecewise Euclidean metric.

Definition 4.14 (Angled 2-complex) *An* angled 2-complex X *is one in which arbitrary real numbers are assigned as values for the normalized external angles* $\beta(v, f)$ *for each vertex* v *in a face* f*, but the smaller intervals retain their conventional values. Thus* $\beta(v, v) = \beta(e, e) = \beta(f, f) = 1$ *and* $\beta(v, e) = \beta(e, f) = \frac{1}{2}$ *for each vertex* v *contained in an edge* e *contained in a face* f*. We can then define values for the internal angle function* α *based on the goal of preserving the relationship* $\alpha \cdot \beta = \zeta$*. This forces* $\alpha(v, v) = \alpha(e, e) = \alpha(f, f) = 1$*,* $\alpha(v, e) = \alpha(e, f) = \frac{1}{2}$ *and* $\alpha(v, f) = \frac{1}{2} - \beta(v, f)$ *as one would expect from the Euclidean situation. In any angled 2-complex, one can define a vertex curvature* $\kappa(v)$ *as Cheeger, Müller and Schrader did above. In addition define a* face curvature $\kappa(f)$*, designed as a correction term, which measures how far the sum of the external vertex angles is from* 1*. Concretely, this means that* $\kappa(v)$ *and* $\kappa(f)$ *are defined by the following formulas.*

$$\kappa(v) = \beta(v, v) - \sum_{e \ni v} \beta(v, e) + \sum_{f \ni v} \beta(v, f) = 1 - \frac{1}{2} \mathit{degree}(v) + \sum_{f \ni v} \beta(v, f)$$

$$\kappa(f) = 1 - \sum_{v \in f} \beta(v, f)$$

Summing $\kappa(v)$ over all vertices of X and considering the result term by term, it is easy to see that this sum counts the number of vertices minus the number of edges plus the sum of all the assigned external angles. Similarly, summing $\kappa(f)$ over all faces of X counts the number of faces minus the sum of all the assigned external angles. Combining these two counts yields the following theorem, versions of which have been proven by Lyndon [63], Gersten [47], Pride [75], Ballmann and Buyalo [3], and myself and Wise [67], among others. See Section 4 of [67] for a more detailed discussion of its history.

Theorem 4.15 (2-dimensional Gauss-Bonnet theorem) *If* X *is an angled 2-complex and vertex and face curvatures are defined as above, then*

$$\sum_v \kappa(v) + \sum_f \kappa(f) = \chi(X)$$

Typically, the sum in these references is actually equal to $2\pi\chi(X)$ instead of $\chi(X)$ since in 2-dimensions angles are not normally normalized. As seen in the previous section normalization is crucial when extending these types of results to higher dimensions.

The final goal of this section is to place both of these results in a broader context. The main idea is the following. Let X be a finite polyhedral cell complex and consider the zeta function of its face lattice (including the empty simplex). If α and β are any two elements of its incidence algebra such that $\alpha \cdot \beta = \zeta$ then the reduced Euler characteristic of X can be rewritten as follows. The variables P and Q here represent (possibly empty) cells of X and since they are being identified with particular elements of the face poset, inequalities are used to denote inclusions.

$$\begin{aligned}
\widetilde{\chi}(X) &= \sum_{P \geq \emptyset} (-1)^{\dim P} = \sum_{P \geq \emptyset} (-1)^{\dim P} \ \zeta(\emptyset, P) \\
&= \sum_{P \geq \emptyset} (-1)^{\dim P} \left(\sum_{Q \in [\emptyset, P]} \alpha(\emptyset, Q)\beta(Q, P) \right) \\
&= \sum_{Q \geq \emptyset} (-1)^{\dim Q} \ \alpha(\emptyset, Q) \left(\sum_{P \geq Q} (-1)^{\dim P - \dim Q} \ \beta(Q, P) \right) \\
&= \sum_{Q \geq \emptyset} (-1)^{\dim Q} \ \alpha(\emptyset, Q) \ \kappa^{\uparrow}(Q)
\end{aligned}$$

where $\kappa^{\uparrow}(Q)$ is defined as the obvious signed sum implicit in the final equality. Call this reformulation the *general combinatorial Gauss-Bonnet formula.* For later reference, here is a formal statement of what this calculation shows.

Theorem 4.16 (General combinatorial Gauss-Bonnet) *If X is a finite polyhedral cell complex and α and β are any two elements of the incidence algebra of the face poset of X such that their product $\alpha \cdot \beta$ is the zeta function ζ of this poset, then*

$$\widetilde{\chi}(X) = \sum_{P \geq \emptyset} (-1)^{\dim P} \ \alpha(\emptyset, P) \ \kappa^{\uparrow}(P)$$

where $\kappa^{\uparrow}(P)$ is found by summing $(-1)^{\dim Q - \dim P} \ \beta(P, Q)$ over all cells Q containing P.

Because of the similarities with angled 2-complexes, a polyhedral cell complex together with a particular factorization of its zeta function into $\alpha \cdot \beta$ is called an *angled n-complex.* The resemblance between $\alpha \cdot \beta = \zeta$ and McMullen's identity is, of course, not accidental. I claim that both of the results given above can be reinterpreted in this manner. For the Cheeger-Müller-Schrader result, the values of α and β are the standard ones, most of the values of $\alpha(\emptyset, F)$ are zero, and most of the values of $\beta(\emptyset, F)$ are $\frac{1}{2}$. In particular, using the values of $\alpha(\emptyset, F)$ and $\beta(\emptyset, F)$ established above, the general formula reduces to the following.

$$\begin{aligned}\widetilde{\chi}(X) = -1 + \chi(X) &= \sum_{Q \geq \emptyset} (-1)^{\dim Q}\ \alpha(\emptyset, Q)\ \kappa^{\uparrow}(Q) \\ &= (-1)\kappa^{\uparrow}(\emptyset) + \frac{1}{2}\sum_{v} \kappa^{\uparrow}(v) \\ &= \sum_{Q \geq \emptyset} (-1)^{\dim Q}\ \beta(\emptyset, Q) + \frac{1}{2}\sum_{v} \kappa^{\uparrow}(v) \\ &= -1 + \frac{1}{2}\sum_{Q > \emptyset} (-1)^{\dim Q} + \frac{1}{2}\sum_{v} \kappa^{\uparrow}(v) \\ &= -1 + \frac{1}{2}\chi(X) + \frac{1}{2}\sum_{v} \kappa^{\uparrow}(v)\end{aligned}$$

Cancelling the negative ones, collecting both Euler characteristic terms on one side and factoring out the one-half yields the previous formula. Alternatively, a slight change in the conventions for the values of $\alpha(\emptyset, F)$ and $\beta(\emptyset, F)$ makes the derivation even quicker and more immediate. Let $\alpha(P, Q)$ and $\beta(P, Q)$ denote the normalized internal and external angles as originally defined when P and Q are not empty, but when $P = \emptyset$ set

$$\alpha(\emptyset, F) = \begin{cases} 1 & \text{if} \quad F = \emptyset \\ 1 & \text{if} \quad \dim F = 0 \\ 0 & \text{if} \quad \dim F > 0 \end{cases} \qquad \beta(\emptyset, F) = \begin{cases} 1 & \text{if} \quad F = \emptyset \\ 0 & \text{if} \quad \dim F \geq 0 \end{cases}$$

It is an easy exercise to show that this new extension of α and β to the empty simplex preserves the identity $\alpha \cdot \beta = \zeta$. Thus, the general combinatorial Gauss-Bonnet formula is applicable and the result is the calculation of Cheeger, Müller and Schrader with no rearrangement necessary.

$$\begin{aligned}\widetilde{\chi}(X) = -1 + \chi(X) &= \sum_{P \geq \emptyset} (-1)^{\dim P}\ \alpha(\emptyset, P)\ \kappa^{\uparrow}(P) \\ &= (-1)\kappa^{\uparrow}(\emptyset) + \sum_{v} \kappa^{\uparrow}(v) \\ &= \sum_{P \geq \emptyset} (-1)^{\dim P}\ \beta(\emptyset, P) + \sum_{v} \kappa^{\uparrow}(v) \\ &= -1 + \sum_{v} \kappa^{\uparrow}(v)\end{aligned}$$

The derivation of the 2-dimensional version is even easier. If X is an angled 2-complex, then the values of $\beta(v, f)$ have been assigned at random, and remaining values of $\beta(P, Q)$ and $\alpha(P, Q)$ with Q and P nonempty were either chosen to be the standard ones or derived from the conditions embedded in the equation $\alpha \cdot \beta = \zeta$. For the extension to the empty simplex there is again a degree of choice, but the extension which forces the more recently defined choices for $\alpha(\emptyset, P)$ (i.e. equal to 1 when P is the empty set or a vertex and 0 otherwise) is one feasible possibility. Once made, this choice forces the values of $\beta(\emptyset, P)$ to be 1 when P is empty, 0 when P is a vertex or an edge, and equal to 1 minus the sum of its external angles $\beta(v, f)$ when P is a face f. Notice in particular, that $\beta(\emptyset, f)$ equals the face curvature $\kappa(f)$ and $\kappa^{\uparrow}(v)$ equals the vertex curvature $\kappa(v)$ as previously defined. Applying Theorem 4.16 to this factorization of the zeta function yields the result recorded in Theorem 4.15:

$$
\begin{aligned}
\widetilde{\chi}(X) = -1 + \chi(X) &= \sum_{P \geq \emptyset} (-1)^{\dim P}\ \alpha(\emptyset, P)\ \kappa^{\uparrow}(P) \\
&= (-1)\kappa^{\uparrow}(\emptyset) + \sum_{v} \kappa^{\uparrow}(v) \\
&= \sum_{P \geq \emptyset} (-1)^{\dim P}\ \beta(\emptyset, P) + \sum_{v} \kappa^{\uparrow}(v) \\
&= -1 + \sum_{f} \beta(\emptyset, f) + \sum_{v} \kappa^{\uparrow}(v) \\
&= -1 + \sum_{f} \kappa(f) + \sum_{v} \kappa(v)
\end{aligned}
$$

As these two examples illustrate, the technique of rewriting the reduced Euler characteristic described in Theorem 4.16 is extremely flexible. Various factorizations of the zeta function are likely to be useful, but particularly those which produce lots of zeros and for which many terms are known to have the same sign. Further explorations of the possibilities are certainly needed.

4.3 Conformal CAT(0) and sectional curvature

In this final section I discuss two uses of the 2-dimensional version of the combinatorial Gauss-Bonnet theorem (Theorem 4.15). The first involves the notion of conformally CAT(0) 2-complexes as developed by Gersten [47] and investigated further by Pride [75], Corson [42], Huck and Rosebrock [53] and others. The second application uses a more recent notion due to Dani Wise [84] that he calls non-positive section curvature. After showing that these two notions are equivalent for the class of special polyhedra, the section concludes with some speculations about a possible relationship between these notions and a long-standing open question about one-relator groups.

Definition 4.17 (Conformally CAT(0) 2-complexes) *Let X be an angled 2-complex as defined in the previous section and notice that the links* *of the vertices in X are metric graphs. If the face curvatures are non-positive and the vertex links are* CAT(1)*, then X is called* conformally CAT(0)*. Concretely, the second condition means that every simple closed curve in a vertex links should have internal angles which add up to at least* 1 *(or* 2π *in the unnormalized version). If the face curvatures are negative and the vertex links are* CAT(1) *then X is called* conformally CAT(−1).

Conformally CAT(0) 2-complexes have a number of nice properties. Steve Gersten has shown, for example, that conformally CAT(0) complexes are aspherical [47] and Jon Corson has shown that conformally CAT(−1) complexes (with all internal angles positive) are word-hyperbolic [42]. It is an interesting question whether either of these results can be extended to higher dimensions.

Problem 4.18 *Is there a higher dimensional analogue of the notion of a conformally* CAT(0) *or* CAT(−1) *2-complex? More concretely, are there any reasonably general conditions on the angles assigned in an angled n-complex X which would allow one to conclude that X is aspherical or that $\pi_1 X$ is word hyperbolic?*

The distinction between 2-dimensional CAT(0) spaces and 2-dimensional conformal CAT(0) spaces is witnessed by the standard 2-complex for the Baumslag-Solitar groups $BS(n, m) = \langle a, b | a^m b = ba^n \rangle$. Assigning internal and external angles so that the defining relator looks like a rectangle with vertical b's and horizontal a's easily shows that this complex is conformally CAT(0). See Figure 33 for an illustration. On the other hand, an easy argument using translation length shows that $BS(n, m)$ is a CAT(0) group if and only if $n = m$.

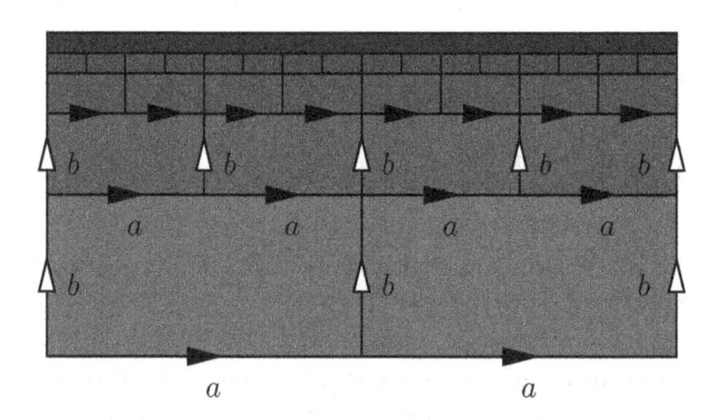

Figure 33: A portion of universal cover of the Baumslag-Solitar group $BS(1, 2)$.

The second concept is of more recent vintage: non-positive sectional curvature.

Definition 4.19 (Non-positive sectional curvature) *Let X be an angled 2-complex. A* section *of a vertex v in X is a based immersion $(S, s) \to (X, v)$ so that the image of the inclusion of the link of s into the link of v is a non-trivial finite spurless connected subgraph. Non-trivial means the link contains at least one edge and spursless means that the link contains no vertices of degree 1. Given a section $(S, s) \to (X, v)$ we can pull back the angle assignments and speak of the vertex curvature at s. If every face of X has non-positive face curvature and every section $(S, s) \to (X, v)$ has non-positive vertex curvature (calculated at s), then X has is said to have* non-positive sectional curvature.

Notice that given a simple closed curve in the link of v, it is easy to construct a section with this loop as the image of its vertex link. Thus, it is immediate that non-positive sectional curvature implies conformally CAT(0). It is also easy to construct examples which distinquish between the two since conformally CAT(0) only checks simple closed loops and non-positive sectional curvature adds restrictions on many additional subgraphs of the vertex link. One situation where the distintion between the two concepts disappears is when X is a special polyhedron.

Definition 4.20 (Special polyhedra) *Let X be a 2-complex in which the link of every point is either a circle, a theta graph, or the complete graph on 4 vertices. See Figure 34. The collection of points with each of these link types define the intrinsic 2-skeleton, 1-skeleton and 0-skeleton of X, respectively. Complexes with this property are called* closed fake surfaces. *If, in addition, the components of the intrinsic 2-skeleton are discs, then X is called a* special polyhedron.

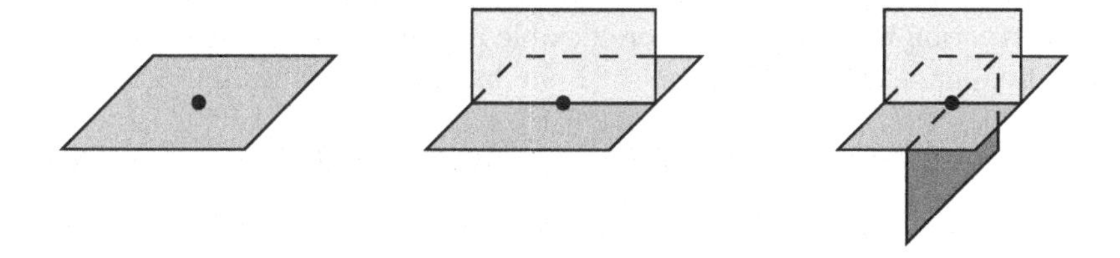

Figure 34: The three possible local structures in a special polyhedron.

Closed fake surfaces and special polyhedra are ubiquitous in at least one sense: every compact 2-complex is simple homotopy equivalent to a special polyhedron. See [50] for an in-depth discussion of these ideas. The simplicity of the vertex links in special polyhedra makes it easy to show the following.

Lemma 4.21 *If X is an angled 2-dimensional special polyhedron, then X is conformally* CAT(0) *if and only if X has non-positive sectional curvature.*

One direction is immediate. In the other direction, note that the only section of a vertex v not tested by the conformal CAT(0) condition is the one derived from the entire link. Adding up the known inequalities for each of the simple closed curves in the link of length 3 one finds that twice the sum of all six angles at v is at least twice the value required to make the vertex curvature non-positive. This completes the proof.

Dani Wise has been able to show that non-positive sectional curvature has strong group-theoretic consequences [84]. Recall that a group is *coherent* if every finitely generated subgroup is also finitely presented. Examples of coherent groups include free groups, surface groups and fundamental groups of compact 3-manifolds. The standard example of a group which is not coherent is the direct product of two non-abelian free groups. Wise proves the following.

Theorem 4.22 (Non-positive sectional curvature and coherence) *If X is an angled 2-complex with non-positive sectional curvature, then $\pi_1 X$ is coherent.*

The proof uses Howie towers [51] and the 2-dimensional combinatorial Gauss-Bonnet theorem. The types of sublinks used to define sectional curvature are precisely those that need to be considered as one lifts various complexes and subcomplexes through the tower. Using the 2-dimensional combinatorial Gauss-Bonnet theorem, one can then show that certain intermediate complexes have non-positive Euler characteristics which, moreover, are strictly increasing as the process continues. It is precisely this tension between increasing Euler characteristics and non-positivity which forces the process to terminate at a complex which then witnesses the finite presentability of the subgroup. The general strategy is similar to the earlier work by myself and Wise on coherence using the idea of perimeter reduction [64], but the construction here are much more flexible in spirit. As an immediate corollary of Lemma 4.21 and Theorem 4.22 one has the following.

Corollary 4.23 *If X is a 2-dimensional special polyhedron with a conformal* CAT(0) *structure, then $\pi_1 X$ is coherent.*

Finally, this section concludes with some speculations about a possible relationship between the general results list above and the coherence of one-relator groups. Consider the following two conjectures.

Conjecture 4.24 (Coherent) *Every one-relator group is coherent.*

Conjecture 4.25 (Conformally CAT(0)) *Every one-relator group is either the fundamental group of a 2-dimensional special polyhedron with a conformal* CAT(0) *structure, or it can be built from such a one-relator group using very simple constructions.*

The first conjecture was asked as a question by Gilbert Baumslag as early as his 1974 survey of one-relator groups [6]. The second is being proposed here as a natural geometric conjecture. Although it is being stated in a slightly vague manner, the idea is that in light of Corollary 4.23, Conjecture 4.25 can be used as a possible line of attack on Conjecture 4.24. There is a much more detailed and precise version of Conjecture 4.25 which I have not stated formally since its development is not suitable for a survey article.

To give at least some idea of the general strategy, consider the following observations. First, it is well-known that every one-relator group is a subgroup of a 2-generator one-relator group so that in order to prove the coherence of all one-relator groups, it is sufficient to consider only those with 2-generators. Next, these groups are particularly nice from the perspective of conformally CAT(0) structures on special polyhedra, since their Euler characteristics are zero. As a result, the only feasible conformal CAT(0) structures are those where the face curvatures are 0 and the vertex curvatures are 0. In other words, all the inequalities become equalities. Another less obvious consequence of the tightness of the constraints is that the pairs of non-adjacent angles in the link of a vertex must have the exact same angles assigned. Thus, if we let n denote the number of vertices in the complex, there are exactly $3n$ variables which determine the angle values. There are n linear equations associated with the face curvatures and n linear equations associated with the curvature of the vertex links. From a linear algebra perspective, so long as the system of $2n$ linear equations in $3n$ variables is consistent, there should be at least an n-dimensional family of solutions. Finally, the process of converting the standard 2-complex for a one-relator group into a special polyhedron is quite flexible. It is possible to insist that the result has no polygons with one or two sides and no "untwisted" triangles. Even at this point there are a number of moves which can be done which only slightly change the complex while significantly changing the system of linear equations. The net result is that the n vertex equations are always consistent with each other and the n face equations are almost always consistent each other.

Concretely, Noel Brady and I have also tried our hand at finding conformal CAT(0) structures for a half-a-dozen examples of 2-generator one-relator groups chosen at random. Starting from a randomly chosen relator of length 12 or so, we haphazardly converted it to a special polyhedron using the pineapple and banana tricks among others [50]. Once we had a special polyhedron, we then worked to remove all monogons and bigons. At this point we stopped and found the linear system of equations which needed to be satisfied and we tested whether this system had a solution. In all cases, it did, and in all but one case, the resulting angle assignment could be extended to a piecewise Euclidean metric on the 2-complex. In other words, the results were typically CAT(0) in addition to being conformally CAT(0). The positive evidence is, admittedly, rather meager at the moment. A computer program to check all of the 2-generator one-relator groups out to a modest size would be more convincing and is currently under development. A general program for investigating conformal CAT(0) structures on special polyhedra for 2-generator one-relator groups is also being actively pursued. In addition to being a reasonable approach to the coherence question for one-relator groups, even partial progress on this geometric conjecture would certainly help to explain why one-relator groups have such a tendency to act like non-positively curved groups, even though it is well-known that they are not non-positively curved in general.

References

[1] J. M. Alonso and et al. Notes on word hyperbolic groups. In *Group theory from a geometrical viewpoint (Trieste, 1990)*, pages 3–63. World Sci. Publishing, River Edge, NJ, 1991. Edited by H. Short.

[2] John C. Baez. The octonions. *Bull. Amer. Math. Soc. (N.S.)*, 39(2):145–205 (electronic), 2002.

[3] W. Ballmann and S. Buyalo. Nonpositively curved metrics on 2-polyhedra. *Math. Z.*, 222(1):97–134, 1996.

[4] Werner Ballmann. *Lectures on spaces of nonpositive curvature*, volume 25 of *DMV Seminar*. Birkhäuser Verlag, Basel, 1995. With an appendix by Misha Brin.

[5] Werner Ballmann, Mikhael Gromov, and Viktor Schroeder. *Manifolds of nonpositive curvature*, volume 61 of *Progress in Mathematics*. Birkhäuser Boston Inc., Boston, MA, 1985.

[6] Gilbert Baumslag. Some problems on one-relator groups. In *Proceedings of the Second International Conference on the Theory of Groups (Australian Nat. Univ., Canberra, 1973)*, pages 75–81. Lecture Notes in Math., Vol. 372. Springer, Berlin, 1974.

[7] David Bessis. The dual braid monoid. *Ann. Sci. École Norm. Sup. (4)*, 36(5):647–683, 2003.

[8] Mladen Bestvina. Non-positively curved aspects of Artin groups of finite type. *Geom. Topol.*, 3:269–302 (electronic), 1999.

[9] Louis J. Billera, Susan P. Holmes, and Karen Vogtmann. Geometry of the space of phylogenetic trees. *Adv. in Appl. Math.*, 27(4):733–767, 2001.

[10] Nicolas Bourbaki. *Éléments de mathématique.* Masson, Paris, 1981. Groupes et algèbres de Lie. Chapitres 4, 5 et 6. [Lie groups and Lie algebras. Chapters 4, 5 and 6].

[11] B. H. Bowditch. Notes on locally cat(1) spaces. In *Geometric group theory (Columbus, OH, 1992)*, pages 1–48. de Gruyter, Berlin, 1995.

[12] Noel Brady and John Crisp. On the dimensions of CAT(0) and CAT(−1) complexes with the same hyperbolic group action. Preprint available at `http://aftermath.math.ou.edu/~nbrady/papers/`.

[13] Noel Brady and Jon McCammond. Nonpositive curvature and small cancellation groups. In preparation.

[14] Noel Brady and Jon McCammond. The length spectrum of a compact constant curvature complex is discrete. *Geom. Dedicata*, 119:159–167, 2006.

[15] Noel Brady, Jon McCammond, and John Meier. Bounding edge degrees in triangulated 3-manifolds. *Proc. Amer. Math. Soc.*, 132(1):291–298 (electronic), 2004.

[16] Thomas Brady. Artin groups of finite type with three generators. *Michigan Math. J.*, 47(2):313–324, 2000.

[17] Thomas Brady. A partial order on the symmetric group and new $K(\pi,1)$'s for the braid groups. *Adv. Math.*, 161(1):20–40, 2001.

[18] Thomas Brady and Jonathan P. McCammond. Four-generator artin groups of finite-type. In preparation.

[19] Thomas Brady and Jonathan P. McCammond. Three-generator Artin groups of large type are biautomatic. *J. Pure Appl. Algebra*, 151(1):1–9, 2000.

[20] Thomas Brady and Colum Watt. $K(\pi,1)$'s for Artin groups of finite type. In *Proceedings of the Conference on Geometric and Combinatorial Group Theory, Part I (Haifa, 2000)*, volume 94, pages 225–250, 2002.

[21] Martin R. Bridson. Geodesics and curvature in metric simplicial complexes. In *Group theory from a geometrical viewpoint (Trieste, 1990)*, pages 373–463. World Sci. Publishing, River Edge, NJ, 1991.

[22] Martin R. Bridson. On the existence of flat planes in spaces of nonpositive curvature. *Proc. Amer. Math. Soc.*, 123(1):223–235, 1995.

[23] Martin R. Bridson. Non-positive curvature in group theory. In *Groups St. Andrews 1997 in Bath, I*, volume 260 of *London Math. Soc. Lecture Note Ser.*, pages 124–175. Cambridge Univ. Press, Cambridge, 1999.

[24] Martin R. Bridson. On the semisimplicity of polyhedral isometries. *Proc. Amer. Math. Soc.*, 127(7):2143–2146, 1999.

[25] Martin R. Bridson. Length functions, curvature and the dimension of discrete groups. *Math. Res. Lett.*, 8(4):557–567, 2001.

[26] Martin R. Bridson and André Haefliger. *Metric spaces of non-positive curvature*, volume 319 of *Grundlehren der Mathematischen Wissenschaften [Fundamental Principles of Mathematical Sciences]*. Springer-Verlag, Berlin, 1999.

[27] M. Burger and A. Iozzi. Boundary maps in bounded cohomology. Appendix to: "Continuous bounded cohomology and applications to rigidity theory" [Geom. Funct. Anal. **12** (2002), no. 2, 219–280; MR 2003d:53065a] by Burger and N. Monod. *Geom. Funct. Anal.*, 12(2):281–292, 2002.

[28] M. Burger and N. Monod. Continuous bounded cohomology and applications to rigidity theory. *Geom. Funct. Anal.*, 12(2):219–280, 2002.

[29] J. W. Cannon, W. J. Floyd, and W. R. Parry. Ample twisted face-pairing 3-manifolds. Available at `http://www.math.vt.edu/people/floyd/research/`.

[30] J. W. Cannon, W. J. Floyd, and W. R. Parry. Introduction to twisted face-pairings. *Math. Res. Lett.*, 7(4):477–491, 2000.

[31] J. W. Cannon, W. J. Floyd, and W. R. Parry. Twisted face-pairing 3-manifolds. *Trans. Amer. Math. Soc.*, 354(6):2369–2397 (electronic), 2002.

[32] J. W. Cannon, W. J. Floyd, and W. R. Parry. Heegaard diagrams and surgery descriptions for twisted face-pairing 3-manifolds. *Algebr. Geom. Topol.*, 3:235–285 (electronic), 2003.

[33] James W. Cannon. The theory of negatively curved spaces and groups. In Tim Bedford, Michael Keane, and Caroline Series, editors, *Ergodic theory, symbolic dynamics, and hyperbolic spaces*, Oxford Science Publications, pages 315–369, New York, 1991. Oxford University Press.

[34] R. Charney, M. Davis, and G. Moussong. Nonpositively curved, piecewise Euclidean structures on hyperbolic manifolds. *Michigan Math. J.*, 44(1):201–208, 1997.

[35] Ruth Charney. Metric geometry: connections with combinatorics. In *Formal power series and algebraic combinatorics (New Brunswick, NJ, 1994)*, volume 24 of *DIMACS Ser. Discrete Math. Theoret. Comput. Sci.*, pages 55–69. Amer. Math. Soc., Providence, RI, 1996.

[36] Ruth Charney and Michael Davis. The Euler characteristic of a nonpositively curved, piecewise Euclidean manifold. *Pacific J. Math.*, 171(1):117–137, 1995.

[37] Indira Chatterji and Graham Niblo. From wall spaces to CAT(0) cube complexes. *Internat. J. Algebra Comput.*, 15(5-6):875–885, 2005.

[38] Jeff Cheeger, Werner Müller, and Robert Schrader. On the curvature of piecewise flat spaces. *Comm. Math. Phys.*, 92(3):405–454, 1984.

[39] Woonjung Choi. Software package `coxeter.g`. Available to download from `http://www.math.ucsb.edu/~mccammon/software/`.

[40] Woonjung Choi. *The existence of metrics of nonpositive curvature on the Brady-Krammer complexes for finite-type Artin groups.* PhD thesis, Texas A&M University, May 2004.

[41] John H. Conway and Derek A. Smith. *On quaternions and octonions: their geometry, arithmetic, and symmetry.* A K Peters Ltd., Natick, MA, 2003.

[42] Jon Michael Corson. Conformally nonspherical 2-complexes. *Math. Z.*, 214(3):511–519, 1993.

[43] Murray Elder and Jon McCammond. Software packages `cat.g` and `cat.gp`. Available at `http://www.math.ucsb.edu/~mccammon/software/`.

[44] Murray Elder and Jon McCammond. Curvature testing in 3-dimensional metric polyhedral complexes. *Experiment. Math.*, 11(1):143–158, 2002.

[45] Murray Elder and Jon McCammond. CAT(0) is an algorithmic property. *Geom. Dedicata*, 107:25–46, 2004.

[46] Murray Elder, Jon McCammond, and John Meier. Combinatorial conditions that imply word-hyperbolicity for 3-manifolds. *Topology*, 42(6):1241–1259, 2003.

[47] S. M. Gersten. Reducible diagrams and equations over groups. In *Essays in group theory*, volume 8 of *Math. Sci. Res. Inst. Publ.*, pages 15–73. Springer, New York, 1987.

[48] M. Gromov. Hyperbolic groups. In *Essays in group theory*, volume 8 of *Math. Sci. Res. Inst. Publ.*, pages 75–263. Springer, New York, 1987.

[49] Branko Grünbaum. *Convex polytopes*, volume 221 of *Graduate Texts in Mathematics*. Springer-Verlag, New York, second edition, 2003. Prepared and with a preface by Volker Kaibel, Victor Klee and Günter M. Ziegler.

[50] Cynthia Hog-Angeloni and Wolfgang Metzler, editors. *Two-dimensional homotopy and combinatorial group theory*, volume 197 of *London Mathematical Society Lecture Note Series*. Cambridge University Press, Cambridge, 1993.

[51] James Howie. On pairs of 2-complexes and systems of equations over groups. *J. Reine Angew. Math.*, 324:165–174, 1981.

[52] G. Christopher Hruska. *Nonpositively curved spaces with isolated flats.* PhD thesis, Cornell University, 2002.

[53] Günther Huck and Stephan Rosebrock. Weight tests and hyperbolic groups. In *Combinatorial and geometric group theory (Edinburgh, 1993)*, volume 204 of *London Math. Soc. Lecture Note Ser.*, pages 174–183. Cambridge Univ. Press, Cambridge, 1995.

[54] James E. Humphreys. *Reflection groups and Coxeter groups.* Cambridge University Press, Cambridge, 1990.

[55] S. V. Ivanov and P. E. Schupp. On the hyperbolicity of small cancellation groups and one-relator groups. *Trans. Amer. Math. Soc.*, 350(5):1851–1894, 1998.

[56] Richard Kane. *Reflection groups and invariant theory.* CMS Books in Mathematics/Ouvrages de Mathématiques de la SMC, 5. Springer-Verlag, New York, 2001.

[57] M. Kapovich and B. Leeb. On asympototic cones and quasi-isometry classes of fundamental groups of 3-manifolds. *Geom. Funct. Anal.*, 5(3):582–603, 1995.

[58] Misha Kapovich. An example of a 2-dimensional hyperbolic group which can't act on 2-dimensional negatively curved complexes. Preprint, Utah, 1994.

[59] Bruce Kleiner. Review of ballmann, bridson-haefliger and eberlein. *Bulletin of the American Mathematical Society*, 39(2):273–279, 2002.

[60] Daan Krammer. The braid group B_4 is linear. *Invent. Math.*, 142(3):451–486, 2000.

[61] Daan Krammer. Braid groups are linear. *Ann. of Math. (2)*, 155(1):131–156, 2002.

[62] Wolfgang Lueck. L^2-Invariants from the Algebraic Point of View.

[63] Roger C. Lyndon and Paul E. Schupp. *Combinatorial group theory.* Springer-Verlag, Berlin, 1977. Ergebnisse der Mathematik und ihrer Grenzgebiete, Band 89.

[64] J. P. McCammond and D. T. Wise. Coherence, local quasiconvexity, and the perimeter of 2-complexes. *Geom. Funct. Anal.*, 15(4):859–927, 2005.

[65] Jon McCammond. Noncrossing partitions in surprising locations. *Amer. Math. Monthly*, 113(7):598–610, 2006.

[66] Jon McCammond and John Meier. The hypertree poset and the l^2-Betti numbers of the motion group of the trivial link. *Math. Ann.*, 328(4):633–652, 2004.

[67] Jonathan P. McCammond and Daniel T. Wise. Fans and ladders in small cancellation theory. *Proc. London Math. Soc. (3)*, 84(3):599–644, 2002.

[68] P. McMullen. Non-linear angle-sum relations for polyhedral cones and polytopes. *Math. Proc. Cambridge Philos. Soc.*, 78(2):247–261, 1975.

[69] Karl Menger. New foundations of euclidean geometry. *American J. of Math.*, 53:721–745, 1931.

[70] Igor Mineyev and Guoliang Yu. The Baum-Connes conjecture for hyperbolic groups. *Invent. Math.*, 149(1):97–122, 2002.

[71] G. Moussong. *Hyperbolic Coxeter Groups.* PhD thesis, The Ohio State University, 1988.

[72] G. A. Niblo and L. D. Reeves. The geometry of cube complexes and the complexity of their fundamental groups. *Topology*, 37(3):621–633, 1998.

[73] G. A. Niblo and L. D. Reeves. Coxeter groups act on CAT(0) cube complexes. *J. Group Theory*, 6(3):399–413, 2003.

[74] Graham Niblo and Lawrence Reeves. Groups acting on CAT(0) cube complexes. *Geom. Topol.*, 1:approx. 7 pp. (electronic), 1997.

[75] S. J. Pride. Star-complexes, and the dependence problems for hyperbolic complexes. *Glasgow Math. J.*, 30(2):155–170, 1988.

[76] Simon Rentzmann. *The Melzak problem for triangulated convex polytopes.* PhD thesis, Texas A&M University, 2000.

[77] Hyam Rubinstein and Michah Sageev. Intersection patterns of essential surfaces in 3-manifolds. *Topology*, 38(6):1281–1291, 1999.

[78] Michah Sageev. Codimension-1 subgroups and splittings of groups. *J. Algebra*, 189(2):377–389, 1997.

[79] I. J. Schoenberg. Remarks to Maurice Fréchet's article "Sur la définition axiomatique d'une classe d'espace distanciés vectoriellement applicable sur l'espace de Hilbert" [MR1503246]. *Ann. of Math. (2)*, 36(3):724–732, 1935.

[80] D. M. Y. Sommerville. *An introduction to the geometry of n dimensions.* Dover Publications Inc., New York, 1958.

[81] John Sullivan. New tetrahedrally close-packed structures. Preprint 2000.

[82] B. T. Williams. *Two topics in geometric group theory.* PhD thesis, University of Southampton, 1999.

[83] D. T. Wise. Cubulating small cancellation groups. *Geom. Funct. Anal.*, 14(1):150–214, 2004.

[84] D. T. Wise. Sectional curvature, compact cores, and local quasiconvexity. *Geom. Funct. Anal.*, 14(2):433–468, 2004.

[85] Daniel T. Wise. *Non-positively curved squared complexes, aperiodic tilings, and non-residually finite groups.* PhD thesis, Princeton University, 1996.

[86] Daniel T. Wise. A flat plane that is not the limit of periodic flat planes. *Algebr. Geom. Topol.*, 3:147–154 (electronic), 2003.

Homology and dynamics in quasi-isometric rigidity of once-punctured mapping class groups

Lee Mosher

Abstract

Combining recent homological methods of Kevin Whyte with older dynamical methods developed by Benson Farb and myself, we obtain a new quasi-isometric rigidity theorem for the mapping class group $\mathcal{MCG}(S^1_g)$ of a once punctured surface S^1_g: if K is a finitely generated group quasi-isometric to $\mathcal{MCG}(S^1_g)$ then there is a homomorphism $K \to \mathcal{MCG}(S^1_g)$ with finite kernel and finite index image. This theorem is joint with Kevin Whyte.

Gromov proposed the program of classifying finitely generated groups according to their large scale geometric behavior. The goal of this paper is a new quasi-isometric rigidity theorem for mapping class groups of once punctured surfaces:

Theorem 1 (Mosher-Whyte). *If S^1_g is an oriented, once-punctured surface of genus $g \geq 2$ with mapping class group $\mathcal{MCG}(S^1_g)$, and if K is a finitely generated group quasi-isometric to $\mathcal{MCG}(S^1_g)$, then there exists a homomorphism $K \to \mathcal{MCG}(S^1_g)$ with finite kernel and finite index image.*

This theorem will be restated later with a more quantitatively precise conclusion; see Theorem 9.

Whyte is also able to apply his techniques to obtain a strong quasi-isometric rigidity theorem for the group $\mathbf{Z}^n \rtimes \mathrm{GL}(n, \mathbf{Z})$, which we will not state here.

Our theorem about $\mathcal{MCG}(S^1_g)$, answers a special case of:

Conjecture 2. *If S is a nonexceptional surface of finite type then for any finitely generated group K quasi-isometric to $\mathcal{MCG}(S)$ there exists a homomorphism $K \to \mathcal{MCG}(S)$ with finite kernel and finite index image.*

The exceptional surfaces that should be ruled out include several for which we already have quasi-isometric rigidity theorems of a different type: the sphere with ≤ 3 punctures whose mapping class groups are finite; the once-punctured torus and the four punctured sphere whose mapping class groups are commensurable to a free group of rank ≥ 2. Probably the techniques used for the once-punctured case may not be too useful in the general case.

The theorem about $\mathcal{MCG}(S_g^1)$, and Whyte's results about $\mathbf{Z}^n \rtimes \mathrm{GL}(n, \mathbf{Z})$, are both about "universal extension" groups of certain $\mathrm{PD}(n)$ groups: more specifically, $\mathcal{MCG}(S_g^1) \approx \mathrm{Aut}(\pi_1 S_g)$ is the universal extension of the PD(2) group $\pi_1 S_g$; and $\mathbf{Z}^n \rtimes \mathrm{GL}(n, \mathbf{Z})$ is the universal extension of the $\mathrm{PD}(n)$ group $\mathbf{Z}^n$. If one wishes to pursue quasi-isometric rigidity for the group $\mathrm{Aut}(F_n)$, where F_n is the free group of rank ≥ 2, noting that $\mathrm{Aut}(F_n)$ is the universal extension of F_n, the difficulty is that the homological techniques we shall use do not apply: the Poincaré duality groups $\pi_1 S_g$ and $\mathbf{Z}^n$ each have a fundamental class in uniformly finite homology, which F_n does not have.

Contents: This paper is based on LaTeX slides that were prepared for lectures given at the LMS Durham Symposium on Geometry and Cohomology in Group Theory, July 2003. Here is an outline of the paper, based approximately on my four lectures at the conference:

1. Survey of results and techniques in quasi-isometric rigidity.
2. Whyte's techniques: uniformly finite homology applied to extension groups.
3. Surface group extensions and Mess subgroups.
4. Dynamical techniques: extensions of surface groups by pseudo-Anosov homeomorphisms.

Acknowledgements. Supported in part by NSF grant DMS-0103208.

1 Results and techniques in QI-rigidity

A map $f\colon X \to Y$ of metric spaces is a *quasi-isometric embedding* if there exists $K \geq 1, C \geq 0$ such that

$$\frac{1}{K} \cdot d_X(x,y) - C \leq d_Y(fx, fy) \leq K \cdot d_X(x,y) + C$$

A *coarse inverse* for f is a quasi-isometry $\bar{f}\colon Y \to X$ s.t.

$$d_{\sup}(\bar{f} \circ f, \mathrm{Id}_X), \qquad d_{\sup}(f \circ \bar{f}, \mathrm{Id}_Y) < \infty$$

A coarse inverse exists if and only if there exists $C' \geq 0$ such that for all $y \in Y$ there exists $x \in X$ such that

$$d_Y(fx, y) \leq C'$$

If this happens then $f\colon X \to Y$ is a *quasi-isometry*, and X, Y are *quasi-isometric* metric spaces. We will use the abbreviation "QI" to stand for "quasi-isometric".

Given a finitely generated group G, a *model space* for G is a metric space X on which G acts by isometries such that:

- X is *proper*, meaning that closed balls are compact.
- X is *geodesic*, meaning that any $x, y \in X$ are connected by a rectifiable path γ such that $\text{Length}(\gamma) = d(x, y)$.
- The action is properly discontinuous and cobounded.

Here are some examples of model spaces:

- The Cayley graph of G with respect to a finite generating set.
- $X = \widetilde{Y}$ where Y is a compact, connected Riemannian manifold or piecewise Riemannian cell complex, and $G = \pi_1 Y$.

Fact: If X, Y are two model spaces for G then X, Y are quasi-isometric. Also, any model space is quasi-isometric to G with its word metric.

As a consequence, a finitely generated group G has a notion of geometry that is well-defined up to quasi-isometry, namely the geometry of any model space, or of G itself with a word metric.

Definition: Two finitely generated groups are *quasi-isometric* if, with their word metrics, they are quasi-isometric as metric spaces; equivalently, their Cayley graphs are quasi-isometric.

Notation: Given $\mathcal{G}$ a collection of finitely generated groups, let $\langle \mathcal{G} \rangle$ be the class of all groups quasi-isometric to some group in $\mathcal{G}$. More generally, given $\mathcal{X}$ a collection of metric spaces, let $\langle \mathcal{X} \rangle$ be the class of all groups quasi-isometric to some metric space in $\mathcal{X}$.

Examples of QI-rigidity theorems. To reformulate Gromov's program in a practical way: given a collection of metric spaces $\mathcal{X}$, describe the collection of groups $\langle \mathcal{X} \rangle$, preferably in simple algebraic or geometric terms that do not invoke the concept of quasi-isometry. Also, describe all of the quasi-isometry classes within $\langle \mathcal{X} \rangle$. In particular, identify interesting classes of groups $\mathcal{G}$ that are *QI-rigid*, meaning $\mathcal{G} = \langle \mathcal{G} \rangle$. There are many theorems describing interesting QI-rigid classes of groups, proved using an incredibly broad range of mathematical tools.

Example: Gromov's polynomial growth theorem [Gro81] implies:

Theorem 3. *The class of virtually nilpotent groups is quasi-isometrically rigid. The class of virtually abelian groups is quasi-isometrically rigid, with one QI-class for each rank.*

Within the class of virtually nilpotent groups, there are many interesting QI-invariants:

- The Hirsch rank is a QI-invariant.
- The sequence of ranks of the abelian subquotients is a QI-invariant, which is finer than the Hirsch rank.
- There is an even finer QI-invariant of a virtually nilpotent Lie group G: Pansu proved that the asymptotic cone of G is a graded Lie group, whose associated graded Lie algebra is a quasi-isometry invariant that subsumes the previous invariants [Pan83].
- This is still not the end of the story: recently Yehuda Shalom produced two finitely generated nilpotent groups which are not quasi-isometric but whose associated graded Lie algebras are isomorphic [Sha04].

The full QI-classification of virtually nilpotent groups remains unknown.

Example: Stallings' ends theorem [Sta68] implies:

Theorem 4. *The class of groups which splits over a finite group is quasi-isometrically rigid. For each $n \geq 2$, the class $\langle F_n \rangle$ consists of all groups that are virtually free of rank ≥ 2.*

Closely related to this example is recent work of Papasoglu and Whyte [PW02] which, combined with Dunwoody's accessibility theorem [Dun85], completely reduces the QI classification of finitely presented groups to the QI classification of finitely presented one ended groups.

Example: Sullivan proved [Sul81] that any uniformly quasiconformal action on S^2 is quasiconformally conjugate to a conformal action. This implies:

Theorem 5 (Sullivan–Gromov). *$\langle \mathbf{H}^3 \rangle$ consists of all groups H for which there exists a homomorphism $H \to \mathrm{Isom}(\mathbf{H}^3)$ with finite kernel and whose image is a cocompact lattice.*

This theorem is prototypical of a broad range of QI-rigidity theorems, including our theorem about $\mathcal{MCG}(S^1_g)$. However, the conclusion of our theorem should be contrasted with the Sullivan–Gromov theorem: the latter gives only a "topological" characterization of $\langle \mathbf{H}^3 \rangle$, which does not serve to give us an effective list of those groups in $\langle \mathbf{H}^3 \rangle$. There is still no effective listing of the cocompact lattices acting on $\mathbf{H}^3$. The conclusion of our theorem gives an "algebraic" characterization of $\langle \mathcal{MCG}(S^1_g) \rangle$, allowing an effective listing.

Example: Rich Schwartz proved a strong quasi-isometric rigidity theorem for noncocompact lattices in $\mathrm{Isom}(\mathbf{H}^3)$ [Sch96]. To state the theorem we need some definitions.

The commensurator group: Given two groups G, H, a *commensuration* from G to H is an isomorphism from a finite index subgroup of G to a finite index subgroup of H. Two commensurations are *equivalent* if they agree upon restriction to another finite index subgroup. The *commensurator group* $\mathrm{Comm}(G)$ is the set of self-commensurations of G up to equivalence, with the following group law: given commensurations $\phi\colon A \to B, \psi\colon C \to D$ restrict the range of ϕ and the domain of ψ to the finite index subgroup $B \cap C$, and then compose $\psi \circ \phi$.

The left action of G on itself by conjugation induces a homomorphism $G \to \mathrm{Comm}(G)$, whose kernel is the *virtual center* of G, consisting of all elements $g \in G$ such that the centralizer of g has finite index in G.

Two groups G, H are *abstractly commensurable* if there exists a commensuration from G to H. Any abstract commensuration from G to H induces an isomorphism from $\mathrm{Comm}(G)$ to $\mathrm{Comm}(H)$.

Theorem 6 (Schwartz). *If G is a noncocompact, nonarithmetic lattice in $\mathrm{Isom}(\mathbf{H}^3)$, then $\langle G \rangle$ consists of those finitely generated groups H which are abstractly commensurable to G. More precisely, the homomorphism $G \to \mathrm{Comm}(G)$ is an injection with finite index image, and $\langle G \rangle$ consists of those finitely generated groups H for which there exists a homomorphism $H' \to \mathrm{Comm}(G)$ with finite kernel and finite index image.*

This theorem gives a very precise and effective enumeration of $\langle \mathcal{G} \rangle$, similar to the conclusion of our main theorems. Schwartz' theorem also can be formulated in the arithmetic case, although there the homomorphism $G \to \mathrm{Comm}(G)$ has infinite index image.

The general techniques of Sullivan-Gromov theorem and of Schwartz' theorem give models for the proof of our main theorem, as we now explain.

Technique: the quasi-isometry group of a group. Consider a metric space X, for example a model space for a finitely generated group. Let $\widehat{\mathrm{QI}}(X)$ be the set of self quasi-isometries of X, equipped with the operation of composition. Define an equivalence relation on $\widehat{\mathrm{QI}}(X)$, where $f \sim g$ if $d_{\sup}(f, g) = \sup\{fx, gx\} < \infty$. Composition descends to a group operation on the set of equivalence classes, giving a group

$$\mathrm{QI}(X) = \text{the } \textit{quasi-isometry group} \text{ of } X$$

Notation: let $[f]$ denote the equivalence class of f in $\mathrm{QI}(X)$. Note that $[f]^{-1} = [\bar{f}]$ for any coarse inverse $\bar{f}$ to f.

For any quasi-isometry $f\colon X \to Y$ there is an isomorphism $\mathrm{ad}_f\colon \mathrm{QI}(X) \to \mathrm{QI}(Y)$ defined by $\mathrm{ad}_f[g] = [f \circ g \circ \bar{f}]$, where $\bar{f}$ is any coarse inverse for f.

It follows that if G is a finitely generated group then the *quasi-isometry group* of G is well-defined up to isomorphism by taking it to be $\mathrm{QI}(X)$ for any model space X of G.

The group $\mathrm{QI}(G)$ is an important quasi-isometry invariant of a group G, and it is often important to be able to compute it. Here are some properties of $\mathrm{QI}(G)$, followed a little later by some examples of computations.

The left action of G on itself by multiplication, defined by $L_g(h) = gh$, induces a homomorphism $G \to \mathrm{QI}(G)$ whose kernel is the virtual center. The left action of G on itself by conjugation, given by $C_g(h) = ghg^{-1}$, induces the same homomorphism $G \to \mathrm{QI}(G)$, because $d_{\sup}(L_g, C_g)$ is finite, equal to the word length of g.

Every commensuration defines a natural quasi-isometry of G, well defined in $\mathrm{QI}(G)$ up to equivalence of commensurations, thereby defining a homomorphism $\mathrm{Comm}(G) \to \mathrm{QI}(G)$. The homomorphism $G \to \mathrm{QI}(G)$ factors as

$$G \to \mathrm{Comm}(G) \to \mathrm{QI}(G)$$

Technique: quasi-actions Let G be a finitely generated group, X a model space for G, and H a finitely generated group quasi-isometric to G. Fix a quasi-isometry $\Phi\colon H \to X$ and a coarse inverse $\bar{\Phi}\colon X \to H$. Define $A\colon H \to \mathrm{QI}(X)$ by the formula

$$A(h) = \Phi \circ L_h \circ \bar{\Phi}$$

This map has the following properties:

A is a quasi-action: There exists constants $K \geq 1$, $C \geq 0$ such that

- The maps $A(h)$ are K, C quasi-isometries for all $h \in H$
- $d_{\sup}(A(hh'), A(h) \circ A(h')) \leq C$ for all $h, h' \in H$
- $d_{\sup}(A(\mathrm{Id}), \mathrm{Id}) \leq C$

and so we obtain a homomorphism $A\colon H \to \mathrm{QI}(X)$.

A is proper: For all $r \geq 0$ there exists n such that if $B, B' \subset X$ have diameter $\leq r$ then

$$\left|\{h \in H \mid (A(h) \cdot B) \cap B' \neq \emptyset\}\right| \leq n$$

A is cobounded: there exists $s \geq 0$ such that for all $x, y \in X$ there exists $h \in H$ such that $d(A(h) \cdot x, y) \leq s$.

Given a group G and a model space X, a common strategy in investigating quasi-isometric rigidity of G is: compute $\mathrm{QI}(X)$; and then describe those homomorphisms $H \to \mathrm{QI}(X)$ arising from quasi-actions, called "uniform" homomorphism. If necessary, restrict to proper, cobounded quasi-actions. Try to "straighten" any such quasi-action.

Examples of QI-rigidity. Here are some examples of how this strategy is carried out, taken from the above examples:

Proof sketch (Sullivan-Gromov rigidity theorem). For all groups in the quasi-isometry class of $\mathbf{H}^3$, the boundary is $\partial\mathbf{H}^3 = S^2$.

First, one calculates $\mathrm{QI}(\mathbf{H}^3) = \mathrm{QC}(S^2)$, the group of quasi-conformal homeomorphisms of S^2; this is a classical result in quasiconformal geometry.

Second, the isometry group $\mathrm{Isom}(\mathbf{H}^3) = \mathrm{Conf}(S^2)$ is a uniform subgroup of $\mathrm{QC}(S^2)$, and one proves that every uniform subgroup can be conjugated into $\mathrm{Conf}(S^2)$. In other words, every quasi-action on $\mathbf{H}^3$ is quasiconjugate to an action. This result, due to Sullivan, is the heart of the proof.

The properties of "properness" and "coboundedness" are invariant under quasiconjugacy. It follows that if H is a finitely generated group quasi-isometric to $\mathbf{H}^3$ then H has a proper, cobounded action on $\mathbf{H}^3$. In other words, there is a homomorphism $H \to \mathrm{Isom}(\mathbf{H}^3)$ with finite kernel and discrete, cocompact image. ◊

By constrast we now give:

Proof sketch for the Schwartz rigidity theorem. Let G be a noncocompact lattice in $\mathbf{H}^3$.

The heart of the proof is essentially a calculation

$$\mathrm{QI}(G) \approx \mathrm{Comm}(G)$$

This calculation holds in both the arithmetic case and the nonarithmetic case, the difference being that the induced map $G \to \mathrm{Comm}(G)$ has finite index image if and only if G is nonarithmetic. Assuming this to be the case, it follows that the homomorphism $\mathrm{Comm}(G) \to \mathrm{QI}(G)$ is an isomorphism and that the map $G \to \mathrm{Comm}(G) \approx \mathrm{QI}(G)$ is an injection with finite index image.

Schwartz' proof is actually a bit more quantitative, as follows. If G is nonarithmetic then there exists an embedding

$$\mathrm{Comm}(G) \hookrightarrow \mathrm{Isom}(\mathbf{H}^3)$$

whose image Γ is a noncocompact lattice containing G with finite index, so that the injection $G \to \mathrm{Comm}(G)$ agrees with the inclusion $G \hookrightarrow \Gamma$. The hard part of Schwartz' proof is to show the following:

- For all $K \geq 1$, $C \geq 0$ there exists $A \geq 0$ such that if $\Phi\colon G \to G$ is a K, C quasi-isometry then there exists $\gamma \in \Gamma$ such that

$$d_{\sup}(\Phi, L_\gamma) \leq A$$

To be more precise, the sup distance on the left is a comparison of two different functions from G into Γ, one being $G \xrightarrow{\Phi} G \hookrightarrow \Gamma$, and the other being $G \hookrightarrow \Gamma \xrightarrow{L_\gamma} \Gamma$.

Noting that any quasi-isometry of G extends to the finite index supergroup Γ, and that $\mathrm{QI}(G) \approx \mathrm{QI}(\Gamma)$, we can abstract this discussion as follows.

Consider a finitely generated group Γ, and suppose that the following holds:

Strong QI-rigidity: For all $K \ge 1$, $C \ge 0$ there exists $A \ge 0$ such that if $\Phi\colon \Gamma \to \Gamma$ is a K, C quasi-isometry then there exists $\gamma \in \Gamma$ such that

$$d_{\sup}(\Phi, L_\gamma) \le A$$

This property, coupled with triviality of the virtual center (true for lattices in $\mathrm{Isom}(\mathbf{H}^3)$ as well as for $\mathcal{MCG}(S^1_g)$), immediately imply that the homomorphism $\Gamma \to \mathrm{QI}(\Gamma)$ is an isomorphism.

To complete the proof of Schwartz' Theorem, we now apply the following fact:

Proposition 7. *If Γ is a strongly QI-rigid group whose virtual center is trivial, then for any finitely generated group H quasi-isometric to Γ there exists a homomorphism $H \to \Gamma$ with finite kernel and finite index image.*

Proof. As explained earlier, the left action of H on itself by translation can be quasiconjugated to a proper, cobounded quasi-action of H on Γ, which induces a homomorphism $\phi\colon H \to \mathrm{QI}(\Gamma) = \Gamma$.

Let $K \ge 1$, $C \ge 0$ be uniform constants for the quasi-action of H on Γ.

Applying strong QI-rigidity of Γ, we obtain a constant A such that the (quasi-)action of each $h \in H$ on Γ is within sup distance A of left multiplication by $\phi(h)$. It immediately follows that the kernel of ϕ is finite, because the quasi-action of H is proper and so there are only finitely many elements $h \in H$ for which $\phi(h)$ is within distance A of the identity on Γ.

It also follows that the image of ϕ has finite index, because the quasi-action of H on Γ is cobounded, whereas the left action on Γ of any infinite index subgroup of Γ is not cobounded. $\diamondsuit$

This completes the proof of Schwartz' Theorem. $\diamondsuit$

This proof immediately yields an interesting corollary:

Corollary 8. *If Γ is strongly QI-rigid with trivial virtual center, then every commensuration of Γ is the restriction of an inner automorphism of Γ. It follows that the homomorphisms $\Gamma \to \mathrm{Comm}(\Gamma)$ and $\Gamma \to \mathrm{Out}(\Gamma)$ are isomorphisms.*

Now we can give the more quantitative statement of the main theorem about $\mathcal{MCG}(S_g^1)$:

Theorem 9 (Mosher-Whyte). *The group $\mathcal{MCG}(S_g^1)$ is strongly QI-rigid: for all $K \geq 1$, $C \geq 0$ there exists $A \geq 0$ such that for any K, C quasi-isometry $\Phi \colon \mathcal{MCG}(S_g^1) \to \mathcal{MCG}(S_g^1)$ there exists $\gamma \in \mathcal{MCG}(S_g^1)$ for which $d_{\sup}(\Phi, L_\gamma) < A$.*

Combining Corollary 8 and Theorem 9 we get a new proof of a result of Ivanov [Iva97]:

Corollary 10. *The homomorphisms*

$$\mathcal{MCG}(S_g^1) \to \mathrm{Comm}(S_g^1) \quad \textit{and} \quad \mathcal{MCG}(S_g^1) \to \mathrm{Out}(\mathcal{MCG}(S_g^1))$$

are isomorphisms.

We remark that Ivanov's proof of this result ultimately depends on his theorem that the action of the mapping class group on the curve complex of a finite type surface induces an isomorphism between the mapping class group and the automorphism group of the curve complex. Our proofs of Theorem 9 and Corollary 10 make no use of the curve complex and its automorphism group.

2 Fiber preserving quasi-isometries

In this section we explain Kevin Whyte's methods for using uniformly finite homology classes and their supports to investigate quasi-isometric rigidity problems for certain fiber bundles. Bruce Kleiner also outlined, at the AMS Ann Arbor conference in 2000, how to use support sets to study quasi-isometric rigidity, using a coarse version of the Künneth formula applied to fiber bundles.

Suppose one wants to investigate quasi-isometric rigidity for the fundamental group of a graph of groups where each vertex and edge group is (virtually) π_1 of an aspherical n-manifold for fixed integer $n \geq 0$. Let us focus on the example $F_2 \times \mathbf{Z}^n$.

To start the proof, pick a nice model space for $F_2 \times \mathbf{Z}^n$, namely, $T \times \mathbf{R}^n$ where T is a Cayley tree of F_2. As described earlier, the technique will be to study quasi-actions on $T \times \mathbf{R}^n$. The first step, carried out by Farb and myself, is to prove that each quasi-isometry of $T \times \mathbf{R}^n$ coarsely preserves the $\mathbf{R}^n$ fibers:

Theorem 11 ([FM00]). *For every $K \geq 1$, $C \geq 0$ there exists $A \geq 0$ such that if $\Phi \colon T \times \mathbf{R}^n \to T \times \mathbf{R}^n$ is a K, C quasi-isometry, then for all $t \in T$ there exists $t' \in T$ such that $d_{\mathcal{H}}(\Phi(t \times \mathbf{R}^n), t' \times \mathbf{R}^n) < A$.*

The notation $d_{\mathcal{H}}(\cdot,\cdot)$ means Hausdorff distance between subsets of a metric space.

More generally, this theorem is true when the product $T \times \mathbf{R}^n$ is replaced by a "coarse fibration" over a tree whose fiber is a uniformly contractible manifold, or even more generally by the Bass-Serre tree of spaces that arise from a finite graph of groups whose vertex and edge groups are coarse PD(n) groups of fixed dimension n [MSW03].

In this section we shall give Kevin Whyte's proof of Theorem 11. This proof also applies to situations where the base space T of the fibration is replaced by certain higher dimensional complexes, for example:

- Thick buildings.
- The model space for $\mathcal{MCG}(S_g)$, over which a model space for $\mathcal{MCG}(S_g^1)$ fibers, with fiber $\mathbf{H}^2$.
- A model space for $\mathrm{SL}(n,\mathbf{Z})$, over which a model space for $\mathbf{Z}^n \rtimes \mathrm{GL}(n,\mathbf{Z})$ fibers, with fiber $\mathbf{R}^n$.

For the example $T \times \mathbf{R}^n$, the idea of the proof is that a subset of the form (line in T) $\times\, \mathbf{R}^n \approx \mathbf{R}^{n+1}$ is the support of a "top dimensional uniformly finite homology class". A quasi-isometry of $T \times \mathbf{R}^n$ acts on such classes, coarsely preserving their supports. Each fiber is the intersection of some finite number of these supports, and so the fibers are preserved.

In this proof it is necessary that the fiber be a uniformly contractible *manifold*, on which there is a "uniformly finite" fundamental class of full support. (For those who live in outer space, that's why the proof does not apply to the extension $1 \to F_n \to \mathrm{Aut}(F_n) \to \mathrm{Out}(F_n) \to 1$, whose fiber F_n is not a manifold and does not have a uniformly finite fundamental class of full support).

In order to make this proof rigorous, we have to discuss:

1. Uniformly finite homology
2. Top dimensional supports
3. Application to fiber bundles

2.1 Uniformly finite homology

Let X be a simplicial complex. Fix a geodesic metric in which each simplex is a regular Euclidean simplex with side length 1. We say that X is *uniformly locally finite* or ULF if there exists $A \ge 0$ such that the link of each simplex contains at most A simplices. We say that X is *uniformly contractible* or UC if for all $r > 0$ there exists $s(r) \ge 0$ such thateach subset $A \subset X$ with $\mathrm{diam}(A) \le r$ is contractible to a point inside $N_s(A)$. The function $s(r)$ is called a *gauge* of uniform contractibility.

For example, any tree is uniformly contractible. Also, if X is a contractible simplicial complex, and if there is a cocompact, simplicial group action on X (for example if X is the universal cover of a compact, aspherical simplicial complex), the X is UC and ULF.

Simplicial uniformly finite homology. If X is ULF, define $H_n^{\text{suf}}(X)$: a chain in $C_n^{\text{suf}}(X)$ is a uniformly bounded assignments of integers to n-simplices. Since X is ULF, the boundary map $\partial\colon C_n^{\text{suf}}(X) \to C_{n-1}^{\text{suf}}(X)$ is defined, and clearly $\partial\partial = 0$.

Theorem 12. *$H_*^{suf}(X)$ is a quasi-isometry invariant among UC, ULF simplicial complexes.*

Proof. We start the proof of the theorem by defining *uniformly finite homology* $H_*^{\text{uf}}(X)$ which is a large scale version of simplicial uniformly finite homology $H_*^{\text{suf}}(X)$, then proving that $H_*^{\text{uf}}(X)$ and $H_*^{\text{suf}}(X)$ are isomorphic for UC, ULF simplicial complexes, and then proving that $H_*^{\text{uf}}(X)$ is a QI-invariant.

Step 1: Uniformly finite homology. For each $n \geq 1$, define the nth Rips complex, $R^n(X)$, with one k-simplex for each ordered $k+1$-tuple of vertices with diameter $\leq n$. Note that $R^1(X) = X$.

Since X is ULF, it follows that $R^n(X)$ is ULF, and the sequence of inclusions

$$X = R^1(X) \subset R^2(X) \subset \cdots$$

therefore induces homomorphisms

$$H_*^{\text{suf}}(R^1(X)) \to H_*^{\text{suf}}(R^2(X)) \to H_*^{\text{suf}}(R^3(X)) \cdots$$

Define the uniformly finite homology to be the direct limit

$$H_*^{\text{uf}}(X) = \lim_{k\to\infty} H_*^{\text{suf}}(R^k(X))$$

We can also describe $H_*^{\text{uf}}(X)$ as the homology of a chain complex. We have a direct system

$$C_*^{\text{suf}}(R^1(X)) \to C_*^{\text{suf}}(R^2(X)) \to C_*^{\text{suf}}(R^3(X)) \to \cdots$$

so we can take the direct limit

$$C_*^{\text{uf}}(X) = \lim_{k\to\infty} C_*^{\text{suf}}(R^k(X))$$

The boundary homomorphism is defined, and the homology of this chain complex is canonically isomorphic to $H_*^{\text{uf}}(X)$.

Step 2: $H_*^{\mathrm{uf}}(X) = H_*^{\mathrm{suf}}(X)$.

The identity map $i\colon X \to X = R^1(X)$ induces a chain map $i\colon C_*^{\mathrm{suf}}(X) \to C_*^{\mathrm{uf}}(X)$. Using that X is UC, we'll define a chain map

$$j\colon C_*^{\mathrm{uf}}(X) \to C_*^{\mathrm{suf}}(X)$$

which will turn out to be a uniform chain homotopy inverse to the inclusion $C_*^{\mathrm{suf}}(X) \to C_*^{\mathrm{uf}}(X)$.

The idea for defining j is: connect the dots.

Consider a 1-simplex in $C_1^{\mathrm{uf}}(X)$, which means a 1-simplex σ in $R^k(X)$ for some k, which means $\sigma = (u,v)$ with $d(u,v) \le k$. Connecting the dots, we get a 1-chain $j(\sigma)$ in X with boundary $v-u$, consisting of at most k different 1-simplices. Given now a simplicial uniformly finite 1-chain $\sum a_\sigma \sigma$ in $R^k(X)$, the infinite sum

$$j(\sum a_\sigma \sigma) = \sum a_\sigma j(\sigma)$$

is defined because it is locally finite: for each simplex τ in X there are finitely many terms of the sum $\sum a_\sigma j(\sigma)$ which assign a nonzero coefficient to τ. This finishes the definition of $j\colon C_1^{\mathrm{uf}}(X) \to C_1^{\mathrm{suf}}(X)$.

Next consider a 2-simplex σ in $C_2^{\mathrm{uf}}(X)$, which means $\sigma = (u,v,w)$, where u,v,w have pairwise distances at most k. Now connect the 2-dimensional dots: the 1-chain

$$j(u,v) + j(v,w) + j(w,u)$$

is a cycle. Its support is a subset of diameter at most $3k/2$, and so

$$j(u,v) + j(v,w) + j(w,u) = \partial j(\sigma)$$

for some 2-chain $j(\sigma)$ supported on a subset of diameter at most $s(3k/2)$, where s is a gauge of uniform contractibility. More generally, the boundary of a simplicially uniformly finite 2-chain in $R^2(X)$, is again defined as a locally finite infinite sum. This finishes the definition of $j\colon C_2^{\mathrm{uf}}(X) \to C_2^{\mathrm{suf}}(X)$, and the chain map condition is obvious.

Now continue the definition of the chain map j by induction, using connect-the-dots.

Similarly, using connect the dots and induction, we can construct a chain homotopy between identity and ji, and similarly for ij. This finishes Step 2.

Step 3: QI-invariance of uniformly finite homology. Consider a quasi-isometry $\Phi\colon X \to Y$ with coarse inverse $\bar{\Phi}\colon Y \to X$, both K,C quasi-isometries, and C-coarse inverses of each other. Moving a bounded distance, we may assume $\Phi, \bar{\Phi}$ take vertices to vertices.

If $d(u,v) = 1$ then $d(\Phi u, \Phi v) \le p = K + C$. We therefore obtain an induced simplicial map $X = R^1(X) \to R^p(X)$, inducing a chain map

$$\Phi_\# : C_*^{\mathrm{suf}}(R^1(X)) \to C_*^{\mathrm{suf}}(R^p(X))$$

In the backwards direction, if $d(u', v') \le p$ in Y, then $d(\bar\Phi u', \bar\Phi v') \le p' = Kp + C$, and so we get an induced simplicial map $R^p(Y) \to R^{p'}(X)$ inducing a chain map

$$\bar\Phi_\# : C_*^{\mathrm{suf}}(R^p(Y)) \to C_*^{\mathrm{suf}}(R^{p'}(X))$$

The composition $\bar\Phi \circ \Phi$ induces a chain map

$$\bar\Phi_\# \circ \bar\Phi_\# : C_*^{\mathrm{suf}}(R^1(X)) \to C_*^{\mathrm{suf}}(R^{p'}(X))$$

but $\bar\Phi \circ \Phi$ is C-close to the identity map on vertices of X, and so a connect-the-dots argument shows that this chain map is chain homotopic to the inclusion map.

A similar argument applies to the composition $\Phi \circ \bar\Phi$.

This finishes the proof that H_*^{suf} is a QI invariant. ◇

2.2 Top dimensional supports.

Suppose now that X is a UC, ULF simplicial complex of dimension d. There are no simplices of dimension $d+1$, and so each class $c \in H_d^{\mathrm{uf}}(X)$ is represented by a unique d-cycle in $C_d^{\mathrm{suf}}(X)$, also denoted c. Its support $\mathrm{supp}(c)$ is therefore a well-defined subset of X.

Proposition 13. *With X as above, every quasi-isometry of X coarsely respects supports of classes in $H_d^{uf}(X)$. More precisely: for all K, C there exists A such that if $\Phi\colon X \to X$ is a K, C quasi-isometry, and if $c \in H_d^{uf}(X)$, then*

$$d_{\mathcal{H}}\big(\Phi(\mathrm{supp}(c)), \mathrm{supp}(\Phi_*(c))\big) < A$$

Most QI-rigidity theorems have a similar step: find some collection $\mathcal{C}$ of objects in the model space which are coarsely respected by quasi-isometries: for all K, C there exists A such that for each K, C quasi-isometry $\Phi\colon X \to X$, and for each object $c \in \mathcal{C}$ there exists an object $c' \in \mathcal{C}$ such that

$$d_{\mathcal{H}}(\Phi(c), c') < A$$

Proof. Moving Φ a bounded distance, we may assume that Φ takes vertices to vertices. We get an induced chain map

$$\Phi_\# : C^{\mathrm{suf}}(X) \to C^{\mathrm{suf}}(R^p(X))$$

Note first that for all $c \in C^{\mathrm{suf}}(X)$, the subset $\mathrm{supp}(\Phi_\#(c))$ is contained in a uniformly bounded neighborhood of $\Phi_\#(\mathrm{supp}(c))$, where the support of a chain in $C^{\mathrm{suf}}(R^p(X))$ is simply the set of vertices occurring among the summands in the chain. In other words, $\Phi_\#$ induces coarse inclusion of supports.

Compose with the connect-the-dots map

$$C^{\mathrm{suf}}(R^p(X)) \to C^{\mathrm{suf}}(X)$$

which also induces coarse inclusion of supports. By composition we obtain an induced map

$$\Phi_{\#\#} : C^{\text{suf}}(X) \to C^{\text{suf}}(X)$$

which also induces coarse inclusion of supports. Since top dimensional supports are unique, it follows that

$$\Phi(\text{supp}(c)) \subset N_A(\text{supp}(\Phi_{\#\#}(c)) = N_A(\text{supp}(\Phi_*(c)))$$

for some uniform constant A.

To get the inverse inclusion, applying the same argument to a coarse inverse $\bar{\Phi}$ we have

$$\bar{\Phi}(\text{supp}(\Phi_*(c))) \subset N_A(\text{supp}(\bar{\Phi}_*\Phi_*(c))) = N_A(\text{supp}(c))$$

where the last equation follows from uniqueness of supports. Now apply Φ to both sides of this equation:

$$\begin{aligned} \text{supp}(\Phi_*(c)) &\subset N_{A'}(\Phi\bar{\Phi}(\text{supp}(\Phi_*(c))) \\ &\subset N'_A(\Phi(N_A(\text{supp}(c))) \\ &\subset N_{A''}(\Phi(\text{supp}(c))) \end{aligned}$$

2.3 Application to fiber bundles.

Consider now a fiber bundle $\pi\colon E \to B$ with fiber F_x over each $x \in B$. We assume that E, B are UC, ULF simplicial complexes, π is a simplicial map, each fiber $F_x = \pi^{-1}(x)$ is a manifold of dimension n, and for each vertex x the subcomplex F_x is UC, with gauge independent of x. It follows that for each k-simplex σ, the $k+n$ simplices of $\pi^{-1}(\sigma)$ that are not contained in $\pi^{-1}(\partial\sigma)$, intersected with the fiber $F_\sigma = \pi^{-1}(\text{barycenter}(\sigma))$, define a cellular structure on F_σ which is UC, with gauge independent of σ.

Let $d = \dim(B)$, $n = \dim(F)$, $d + n = \dim(E)$.

Make the following assumption about the top dimensional, uniformly finite homology $H_d^{\text{uf}}(B)$:

Top dimensional classes in B separate points: There exists $r > 0$ so that for all $s > 0$ there exists $D > 0$ so that for any $x, y \in B$ with $d(x, y) > D$, there is a top dimensional class $c \in H_d^{\text{uf}}(B)$ such that

$$d(\text{supp}(c), x) \leq r \quad \text{and} \quad d(\text{supp}(c), y) > s$$

Example: $T \times \mathbf{R}^n$, where T = Cayley tree of F_2. In the base space T, each bi-infinite line is the support of a top dimensional class, and lines in T clearly separate points.

Example to come: later we shall construct a model space for $\mathcal{MCG}(S_g^1)$, fibering over model space for $\mathcal{MCG}(S_g)$, with fiber $\mathbf{H}^2$. The dimension of the base space will equal $4g-5$, which is the virtual cohomological dimension of $\mathcal{MCG}(S_g)$. We shall prove that the top dimensional classes of uniformly finite homology separate points of $\mathcal{MCG}(S_g)$.

The main result for this section is that every quasi-isometry of the total space E coarsely preserves fibers:

Theorem 14 (Whyte). *Consider the fibration $F \to E \to B$ as above, and assume that top dimensional classes in $H_d^{uf}(B)$ coarsely separate points. For all K, C there exists A such that if $\Phi \colon E \to E$ is a K, C quasi-isometry, then for each $x \in B$ there exists $x' \in B'$ such that*

$$d_{\mathcal{H}}(\Phi(F_x), F_{x'}) \leq A$$

Proof. The key observation of the proof is that the support of every top dimensional class in E is saturated by fibers. To be precise: for every top dimensional class $c \in H_{d+n}^{\mathrm{uf}}(E)$, there exists a unique top dimensional class $c' = \pi(c) \in H_d^{\mathrm{uf}}(B)$, such that

$$\mathrm{supp}(c) = \pi^{-1}(\mathrm{supp}(c'))$$

To see why, for each d-simplex $\sigma \subset B$, $\pi^{-1}(B)$ is a manifold with boundary of dimension $d+n$. So, for any class c of dimension $d+n$, if $\mathrm{supp}(c)$ contains some $d+n$ simplex in $\pi^{-1}(\sigma)$, it follows that $\mathrm{supp}(c)$ contains all of $\pi^{-1}(\sigma)$.

The converse is also true: for each top dimensional class $c' \in H_d^{\mathrm{uf}}(B)$ there exists a top dimensional class in E, denoted $c = \pi^{-1}(c') \in H_{d+n}^{\mathrm{uf}}(E)$ such that $\mathrm{supp}(c) = \pi^{-1}(\mathrm{supp}(c'))$: over each simplex $\sigma \subset \mathrm{supp}(c')$, weight all the simplices in $\pi^{-1}(\sigma)$ with the same weight as σ, using a coherent orientation of fibers to choose the sign.

Thus, the projection π induces an isomorphism

$$H_{d+n}^{\mathrm{uf}}(E) \to H_d^{\mathrm{uf}}(B)$$

so that a $d+n$-cycle c in E, and the corresponding d-cycle c' in B, are related by

$$\mathrm{supp}(c) = \pi^{-1}(\mathrm{supp}(c'))$$

Now we use the property that supports of top dimensional classes in B separate points. Up to changing constants, it follows that:

Supports of top dimensional classes in E separate fibers: There exists $r > 0$ such that for all $s > 0$ there exists $D > 0$ such that given fibers F_x, F_y with $d_{\mathcal{H}}(F_x, F_y) > D$, there is a top dimensional class $c \in H_{d+n}^{\mathrm{uf}}(E)$ so that $F_x \subset N_r(\mathrm{supp}(c))$ but $F_y \cap N_r(\mathrm{supp}(c)) = \emptyset$.

From this it follows that fibers in E are coarsely respected by a quasi-isometry. Here are the details.

Fix a K, C quasi-isometry $\Phi \colon E \to T$ and a fiber F_x, $x \in G$. We want to show that $\Phi(F_x)$ is uniformly Hausdorff close to some fiber $F_{x'}$.

Fix some $R > r$, to be chosen later, and let $\mathcal{C}_x$ denote the collection of classes $c \in H^{\text{uf}}_{d+n}(E)$ such that $F_x \subset N_R(\text{supp}(c))$. From the fact that top dimensional classes in E separate fibers, it follows that F_x has (uniformly) finite Hausdorff distance from the set of points $\xi \in E$ such that $\xi \in N_R(\text{supp}(c))$ for all $c \in \mathcal{C}_x$.

Notation: let $\hat{\mathcal{C}}_x = \{\Phi_\#(c) \mid c \in \mathcal{C}_x\}$, and let $\hat{c} = \Phi_\#(c)$.

By applying Proposition 13, the Hausdorff distance between $\Phi(\text{supp}(c))$ and $\text{supp}(\hat{c})$ is at most a constant A, for any $c \in \mathcal{C}_x$. Thus for any $\hat{c} \in \hat{\mathcal{C}}_x$ we have

$$\Phi(F_x) \subset N_{R'}(\text{supp}(\hat{c}))$$

where $R' = KR + C + A$. Now we say how large to choose R, namely, so that $R' > r$.

It now follows that $\Phi(F_x)$ has (uniformly) finite Hausdorff distance from the set $\mathcal{F}$ of points $\eta \in E$ such that $\eta \in N_{R'}(\text{supp}(\hat{c}))$ for all $\hat{c} \in \hat{\mathcal{C}}_x$. But the set $\mathcal{F}$ is clearly a union of fibers of E. Pick one fiber $F_{x'}$ in $\mathcal{F}$. Taking $s = R'$ in the definition of coarse separation of fibers, there is a resulting D. If $F_{y'}$ is a fiber whose distance from $F_{x'}$ is more than D, it follows that $F_{y'}$ is not contained in $\mathcal{F}$, because that would violate coarse separation of fibers. This shows that the set $\mathcal{F}$ contains $F_{x'}$ and is contained in the D-neighborhood of $F_{x'}$, that is, $\mathcal{F}$ has Hausdorff distance at most D from $F_{x'}$. But $\mathcal{F}$ also has (uniformly) finite Hausdorff distance from $\Phi(F_x)$, and so $\Phi(F_x)$ has (uniformly) finite Hausdorff distance from $F_{x'}$.

This finishes the proof of Theorem 14. $\diamond$

3 Extensions of surface groups and Mess subgroups

Let S_g be a closed, oriented surface of genus $g \geq 2$, and let S^1_g be S_g minus a single base point p. There is a short exact sequence

$$1 \to \pi_1(S_g) \to \mathcal{MCG}(S^1_g) \to \mathcal{MCG}(S_g) \to 1$$

The homomorphism $\mathcal{MCG}(S^1_g) \to \mathcal{MCG}(S_g)$ is the map that "fills in the puncture". The homomorphism $\pi_1(S_g) \to \mathcal{MCG}(S^1_g)$ is the "push" map, which isotopes the base point p around a loop, at the end of the isotopy defining a map of S_g taking p to itself; then remove the base point to define a mapping class on S^1_g.

The main theorem of this section is:

Theorem 15. *Every quasi-isometry of $\mathcal{MCG}(S_g^1)$ coarsely preserves the system of cosets of $\pi_1(S_g)$.*

The meaning of this theorem is that for each $K \geq 1$, $C \geq 0$ there exists $A \geq 0$ such that if $\Phi\colon \mathcal{MCG}(S_g^1) \to \mathcal{MCG}(S_g^1)$ is a K, C quasi-isometry, then for each coset C of $\pi_1(S_g)$ there exists a coset C' such that $d_{\mathcal{H}}(\Phi(C), C') \leq A$.

The setup of Theorem 15 is to represent the short exact sequence by a fibration

$$\mathbf{H}^2 \to E \to B$$

as above, where E is a model space for $\mathcal{MCG}(S_g^1)$, and B is a model space for $\mathcal{MCG}(S_g)$. The theorem can then be translated into geometric terms by saying that every quasi-isometry of E coarsely preserves the fibers.

Whyte's idea for proving Theorem 15 is to apply Theorem 14, by using Mess subgroups of $\mathcal{MCG}(S_g)$ to provide top dimensional classes in the uniformly finite homology of B that coarsely separate points.

3.1 Dimension of $\mathcal{MCG}(S_g)$.

Given a contractible model space B for a group G, one would expect that the top dimension in which $H_n^{\mathrm{uf}}(B)$ is nontrivial would be $n = \mathrm{vcd}(G)$. So we need the following formula of John Harer [Har86]:

Theorem 16.

$$\mathrm{vcd}(\mathcal{MCG}(S_g)) = 4g - 5.$$

Proof. We will sketch Harer's proof of the upper bound $\mathrm{vcd}(\mathcal{MCG}(S_g)) \leq 4g - 5$, and then give Geoff Mess' proof of the lower bound $\mathrm{vcd}(\mathcal{MCG}(S_g)) \geq 4g - 5$. Mess' proof will provide the basic ingredients we need to investigate top dimensional uniformly finite homology classes in a model space for $\mathcal{MCG}(S_g)$.

To prove $\mathrm{vcd}(\mathcal{MCG}(S_g)) \leq 4g - 5$, by using the short exact sequence, in which

$$\mathrm{vcd}(\mathrm{kernel}) = \dim(\mathbf{H}^2) = 2$$

it suffices to prove

$$\mathrm{vcd}(\mathcal{MCG}(S_g^1)) \leq 4g - 3$$

Harer constructs a contractible complex K of dimension $4g - 3$ which is a model space for $\mathcal{MCG}(S_g^1)$: K is the complex of "filling arc systems" of the once punctured surface. Fixing a base point $p \in S_g$, a filling arc system is a system of arcs $\mathcal{A} = \{A_i\}$ whose interiors are pairwise disjoint, whose ends are all located at the base point p, so that for each component C of $S_g - \cup\{A_i\}$, regarding C as the interior of a polygon whose sides are arcs of $\{A_i\}$, the number of sides of C is at least 3. Each filling arc system $\mathcal{A}$ can be refined by adding more arcs until it is triangulated, and the number of such arcs is called the *defect* of $\mathcal{A}$. The complex K that Harer uses has one cell $C_{\mathcal{A}}$ of dimension d for each isotopy class of filling arc systems $\mathcal{A}$ of defect d, and the boundary of $C_{\mathcal{A}}$ consists of all cells $C_{\mathcal{A}'}$ such that $\mathcal{A} \subset \mathcal{A}'$. Harer used Strebel differentials to prove contradictiblity of K, but a purely combinatorial proof was given by Hatcher [Hat91].

Remarks: Harer does *not* directly construct a $4g-3$ dimensional model space for $\mathcal{MCG}(S_g)$. Thurston, in his three page 1986 preprint "A spine for the Teichmüller space of a closed surface" [Thu86], does construct a model space for $\mathcal{MCG}(S_g)$. While we can prove, with some work, that the dimension of Thurston's spine in genus $g = 2$ is indeed equal to $4g-5 = 8-5 = 3$, we are unable to prove that Thurston's spine in genus $g \geq 3$ has dimension $4g-5$, and it may be false. For present purposes we will ultimately depend on the Eilenberg-Ganea-Wall theorem [Bro82] to obtain an appropriate model space for $\mathcal{MCG}(S_g)$, which is why we need to compute the vcd.

3.2 Mess subgroups

Now we give Geoff Mess' proof of the lower bound: $\text{vcd}(\mathcal{MCG}(S_g)) \geq 4g-5$.

The proof exhibits a subgroup $M_g < \mathcal{MCG}(S_g)$ which is a Poincaré duality group of dimension $4g-5$, in fact M_g is the fundamental group of a compact, aspherical $4g-5$ manifold. The group M_g is called a *Mess subgroup* of $\mathcal{MCG}(S_g)$.

Mess subgroups are constructed by induction on genus.

Base case: Genus 2. With $g = 2$, we have $4g-5 = 3$, so we need a 3-dimensional subgroup of $\mathcal{MCG}(S_2)$. Take a curve family $\{c_1, c_2, c_3\} \subset S_2$ consisting of three pairwise disjoint, pairwise nonisotopic curves. The Dehn twists about c_1, c_2, c_3 generate a rank 3 free abelian group, and we are done.

Up to the action of $\mathcal{MCG}(S_2)$, there are two orbits of curve families $\{c_1, c_2, c_3\}$, depending on whether or not some curve in the family separates. So, there are two conjugacy classes of Mess subgroups in $\mathcal{MCG}(S_2)$.

Induction step: Let M_{g-1} be a Mess subgroup in $\mathcal{MCG}(S_{g-1})$, and so M_{g-1} is a Poincaré duality group of dimension $4(g-1)-5$.

Consider the short exact sequence

$$1 \to \pi_1(S_{g-1}) \to \mathcal{MCG}(S^1_{g-1}) \to \mathcal{MCG}(S_{g-1}) \to 1$$

Let M'_{g-1} = preimage of M_{g-1}, so we get

$$1 \to \pi_1(S_{g-1}) \to M'_{g-1} \to M_{g-1} \to 1$$

and it follows that M'_{g-1} is Poincaré duality of dimension $4(g-1)-5+2$.

Let $S_{g,1}$ be the surface S_g with a hole removed, and with one boundary component. There is a central extension

$$1 \to \mathbf{Z} \to \mathcal{MCG}(S_{g,1}) \to \mathcal{MCG}(S^1_g) \to 1$$

obtained by collapsing the hole to a punctuure; here, the group $\mathcal{MCG}(S_{g,1})$ is defined as the group of homeomorphisms constant on the boundary, modulo isotopies that are stationary on the boundary.

Let M''_{g-1} be the preimage of M'_{g-1} in $\mathcal{MCG}(S_{g,1})$, and we get

$$1 \to \mathbf{Z} \to M''_{g-1} \to M'_{g-1} \to 1$$

from which it follows that M''_{g-1} is Poincaré duality of dimension $4(g-1)-5+3$.

Now attach a handle (a one-holed torus) to $S_{g,1}$ to get S_{g+1}, so we get an embedding

$$\mathcal{MCG}(S_{g,1}) \to \mathcal{MCG}(S_{g+1})$$

Pick a simple closed curve c contained in the handle and not isotopic to the boundary. The Dehn twist τ_c commutes with $\mathcal{MCG}(S_{g,1})$, and in fact the subgroup of $\mathcal{MCG}(S_{g+1})$ generated by τ_c and $\mathcal{MCG}(S_{g,1})$ is isomorphic to the product $\mathcal{MCG}(S_{g,1}) \times \tau_c$. We can therefore define

$$M_g = M''_{g-1} \times \langle \tau_c \rangle$$

which is a Poincaré duality group of dimension $4(g-1) - 5 + 4 = 4g - 5$ contained in $\mathcal{MCG}(S_g)$.

This finishes Mess' proof that $\mathrm{vcd}(\mathcal{MCG}(S_g)) \geq 4g - 5$. ◇

Remarks: The construction of M_g is completely determined by the isotopy type of a certain filtration of S_g by subsurfaces. There are only finitely many such isotopy types up to the action of $\mathcal{MCG}$, and so there are only finitely many conjugacy classes of Mess subgroups. In fact, a little thought shows that there are exactly two conjugacy classes of Mess subgroups, distinguished by whether the original curve system $\{c_1, c_2, c_3\}$ chosen in a genus 2 subsurface with one hole contains a separating curve.

Letting $\mathrm{Stab}(c) < \mathcal{MCG}(S_g)$ be the stabilizer group of the closed curve c picked in the last step, we have

$$M_g \subset \mathrm{Stab}(c)$$

This fact will be significant later on.

3.3 Model spaces and Mess cycles

We will use a small trick: for the moment, we won't actually work with a model space for $\mathcal{MCG}(S_g)$, instead we'll work with a model space for a finite index, torsion free subgroup $\Gamma_g < \mathcal{MCG}(S_g)$. This is OK because the inclusion $\Gamma_g \hookrightarrow \mathcal{MCG}(S_g)$ is a quasi-isometry. Since $\mathrm{vcd}(\mathcal{MCG}(S_g)) = 4g - 5$ it follows that $\mathrm{cd}(\Gamma_g) = 4g - 5$. We can therefore apply the Eilenberg-Ganea-Wall theorem, to obtain a model space B for Γ_g of dimension $4g - 5$.

The reason for this trick is that for a group with torsion such as $\mathcal{MCG}(S_g)$, the construction of a model space of dimension equal to the vcd is problematical.

Next we obtain a model space for E and a fibration $\mathbf{H}^2 \to E \to B$ as follows. Start with the canonical $\mathbf{H}^2$ bundle over Teichmüller space $\mathcal{T}$, map B to $\mathcal{T}$ by a Γ_g-equivariant map, and pull the bundle back to get E. This space E is then a model space for the canonical $\pi_1(S_g)$ extension of Γ_g, which is quasi-isometric to $\mathcal{MCG}(S^1_g)$. To prove that quasi-isometries of $\mathcal{MCG}(S^1_g)$ coarsely respect cosets of $\pi_1(S_g)$, it suffices to prove the same for the $\pi_1(S_g)$ extension of Γ_g, and for this it suffices to prove that quasi-isometries of E coarsely respect the $\mathbf{H}^2$ fibers. Applying Theorem 14, it remains to construct top dimensional uniformly finite cycles in B which coarsely separate points.

Given a Mess subgroup $M < \mathcal{MCG}(S_g)$, the intersection $M' = M \cap \Gamma_g$ has finite index in M, and so M' is still a Poincaré duality group of dimension $4g-5$. The complex B/M' is therefore a $K(M',1)$ space of dimension $4g-5$. Since M' is a Poincaré duality group, the homology $H_{4g-5}(M')$ is infinite cyclic generated by the fundamental class $[M']$, and this class is represented by a unique $4g-5$ cycle in B/M'. This cycle lifts to a $4g-5$ dimensional, uniformly finite cycle in B; call this a *Mess cycle* in B.

We will prove that the Mess cycles in B separate points.

3.4 Passage to cosets of curve stabilizers

We now pass from Mess cycles to left cosets of Mess subgroups to left cosets of curve stabilizers, as follows. Although $\mathcal{MCG}$ does not act on B, it does quasi-act, which is good enough. The quasi-action of $\mathcal{MCG}$ permutes the Mess cycles. There is a bijection between Mess subgroups and Mess cycles: each Mess subgroup M corresponds to a unique Mess cycle c such that M that (coarsely) stabilizes c. If M (coarsely) stabilizes c and if $\Phi \in \mathcal{MCG}(S_g)$ then $\Phi M \Phi^{-1}$ (coarsely) stabilizes $\Phi(c)$.

Pick representatives $M_1, \ldots, M_k$ of the finitely many conjugacy classes of Mess subgroups in $\mathcal{MCG}(S_g)$. It follows that, under the quasi-isometry $B \to \mathcal{MCG}(S_g)$, Mess cycles correspond to left cosets in $\mathcal{MCG}(S_g)$ of $M_1, \ldots, M_k$. So, it suffices to show that left cosets of $M_1, \ldots, M_k$ coarsely separate points in $\mathcal{MCG}(S_g)$.

Each Mess subgroup M_i fixes some curve c_i, and so $M_i < \mathrm{Stab}(c_i)$. Thus, each left coset of M_i is contained in a left coset of $\mathrm{Stab}(c_i)$. So, choosing curves $c_0, \ldots, c_n$ representing the orbits of simple closed curves, it suffices to prove that the left cosets of the groups $\mathrm{Stab}(c_i)$ coarsely separate points in $\mathcal{MCG}$.

3.5 New model space

We now switch to a new model space Γ for $\mathcal{MCG}(S_g)$, no longer contractible. We will pass from left cosets of the groups $\mathrm{Stab}(c_i)$ to subsets of the new model space Γ.

Γ is a graph whose vertices are pairs (C, D) where each of C, D is a pairwise disjoint curve system, the systems C, D jointly fill the surface, and each component of $S - (C \cup D)$ is a hexagon. This implies that $\mathcal{MCG}$ acts on the vertex set with finitely many orbits. Since $\mathcal{MCG}$ is finitely generated, and since there are finitely many orbits of vertices, it follows that we can attach edges in an $\mathcal{MCG}$-equivariant way so that the graph Γ is connected and has finitely many orbits of edges. There's probably some nice scheme for attaching edges, based on low intersection numbers, but it's not necessary. The graph Γ is now quasi-isometric to $\mathcal{MCG}$. Given a curve c, define Γ_c to be the subgraph of Γ spanned by vertices (C, D) such that $c \in C \cup D$.

3.6 The subgraphs Γ_c coarsely separate points

We can now pass from left cosets of curve stabilizers to the sets Γ_c. Our ultimate goal is to show that the system of subgraphs Γ_c, one for each curve c, coarsely separates points in Γ.

Given vertices (C, D) and (C', D') which are very far from each other, we shall pick a curve c in $C \cup D$ and show that (C', D') is far from Γ_c. This is enough, because (C, D) is contained in Γ_c.

Since (C, D) and (C', D') are very far from each other, there exists $c \in C \cup D$ and $c' \in C' \cup D'$ such that the intersection number $< c, c' >$ is very large. Proof: fixing (C, D), if all such intersection numbers $< c, c' >$ are uniformly small, then there is a uniform cardinality to the number of possible (C', D'), so the distance from (C, D) to (C', D') is uniformly bounded.

Consider now any curve system (C_1, D_1) in Γ_c, meaning that (C_1, D_1) contains c. The curve $c \in C_1 \cup D_1$ has very large intersection number with the curve $c' \in C' \cup D'$. It follows that (C_1, D_1) and (C', D') are far from each other. Proof: if (C_1, D_1) and (C', D') are close, there is a uniform bound to the intersection number of a curve in $C_1 \cup D_1$ with a curve in (C', D').

This completes the proof that quasi-isometries of $\mathcal{MCG}(S_g^1)$ coarsely preserve fibers.

4 Dynamical techniques: extensions of surface groups by pseudo-Anosov homeomorphisms.

In this section we give the proof of Theorem 1, by proving Strong QI-Rigidity of $\mathcal{MCG}(S_g^1)$ in the sense of Section 1. By applying Theorem 15, we are reduced to showing the following:

Fibered QI-rigidity For all $K \geq 1$, $C \geq 0$, $R \geq 0$ there exists $A \geq 0$ such that if Φ is a K, C quasi-isometry of $\mathcal{MCG}(S_g^1)$, and if Φ takes each coset of $\pi_1 S_g$ to within a Hausdorff distance R of some other coset of $\pi_1 S_g$, then there exists $h \in \mathcal{MCG}(S_g^1)$ such that

$$d_{\sup}(\Phi(f), hf) < A \quad \text{for all} \quad f \in \mathcal{MCG}(S_g^1)$$

The methods of proof are very similar to the following result of Farb and myself:

Theorem 17 ([FM02b]). *If F is a Schottky subgroup of $\mathcal{MCG}(S_g)$ with extension*

$$1 \to \pi_1(S_g) \to \Gamma_F \to F \to 1$$

then the injection $\Gamma_F \to \mathrm{QI}(\Gamma_F)$ has finite index image. Moreover, if H is a group quasi-isometric to Γ_F then there exists a homomorphism $H \to \mathrm{QI}(\Gamma_F)$ with finite kernel and finite index image.

In that proof, using a tree as a model space for F, we proved that every quasi-isometry of Γ_F coarsely preserves fibers. Then we used pseudo-Anosov dynamics (as we will here) to prove "Fibered QI-rigidity", from which Theorem 17 follows.

4.1 Teichmüller space and its canonical $\mathbf{H}^2$ bundle.

We have already briefly mentioned these objects; here they are in more detail.

The Teichmüller space $\mathcal{T}_g$ of S_g is the space of hyperbolic structures on S_g modulo isotopy, or equivalently the space of conformal structures on S_g modulo isotopy. We also need the Teichmüller space $\mathcal{T}_g^1$ of S_g^1, defined similarly using finite area complete hyperbolic structures on S_g^1, or equivalently conformal structures with a removable singularity at the puncture. It follows that each element of $\mathcal{T}_g^1$ can be expressed, up to isotopy, as a pair (σ, p) where σ is a hyperbolic structure on S_g representing an element of $\mathcal{T}_g$, and p is a point in the universal cover $\widetilde{\sigma} \approx \mathbf{H}^2$. We therefore obtain a fiber bundle structure

$$\mathbf{H}^2 \to \mathcal{T}_g^1 \to \mathcal{T}_g$$

on which the short exact sequence

$$1 \to \pi_1(S_g) \to \mathcal{MCG}(S_g^1) \to \mathcal{MCG}(S_g) \to 1$$

acts. For each $x \in \mathcal{T}_g$ we use Σ_x to denote the fiber of $\mathcal{T}_g^1$ over x, and so Σ_x is an isometric copy of $\mathbf{H}^2$.

Note that the action of $\mathcal{MCG}(S_g^1)$ on $\mathcal{T}_g^1$ is *not* cocompact, and so we cannot regard $\mathcal{T}_g^1$ as a model space of $\mathcal{MCG}(S_g^1)$, and similarly for $\mathcal{MCG}(S_g)$ acting on $\mathcal{T}_g$. There is, however, a cocompact equivariant spine $Y_g \subset \mathcal{T}_g$

whose inverse image is a cocompact equivariant spine $Y_g^1 \subset \mathcal{T}_g^1$ and we have a fibration

$$\mathbf{H}^2 \to Y_g^1 \to Y_g$$

on which the short exact sequence acts. By cocompactness, the Y's are model spaces for the $\mathcal{MCG}$'s.

The action of $\mathcal{MCG}(S_g^1)$ on Y_g^1 respects the $\mathbf{H}^2$ fibers, and the stabilizer of each fiber is $\pi_1 S_g$. It follows that there is a quasi-isometry $\mathcal{MCG}(S_g^1) \to Y_g^1$ taking cosets of $\pi_1 S_g$ to $\mathbf{H}^2$ fibers.

We can translate the result of Theorem 15 to the language of Y_g^1. The translation says: every quasi-isometry $\Phi\colon Y_g^1 \to Y_g^1$ coarsely respects fibers, that is, there exists a constant $A \geq 0$ such that for each $x \in Y_g$ there exists $x' \in Y_g$ such that $d_{\mathcal{H}}(\Phi(\Sigma_x), \Sigma_{x'}) \leq A$. Choosing an x' for each x, we obtain an induced map $\phi\colon Y_g \to Y_g$ with $\phi(x) = x'$, and ϕ is a quasi-isometry.

This reduces the proof of Fibered QI-rigidity for $\mathcal{MCG}(S_g^1)$ to the analogous statement for quasi-isometries of Y_g^1: with Φ as above, we must find $h \in \mathcal{MCG}(S_g^1)$ so that the actions of Φ and h on S_g^1 agree to within bounded distance.

$\mathbf{H}^2$ bundles over lines. Consider bi-infinite, proper paths $\ell\colon \mathbf{R} \to \mathcal{T}_g$, with image often in Y_g. The path ℓ is always assumed to be piecewise smooth and Lipschitz. Let

$$\Sigma_\ell = \pi^{-1}(\ell) \subset \mathcal{T}_g^1$$

so we have an $\mathbf{H}^2$ bundle over the line ℓ:

$$\mathbf{H}^2 \to \Sigma_\ell \to \ell$$

There is a reasonably natural metric on Σ_ℓ, obtained by combining the $\mathbf{H}^2$ metric on fibers with the metric on the $\mathbf{R}$ factor. There are some choices, but the metric is natural up to quasi-isometry. If ℓ is piecewise geodesic then for the metric on Σ_ℓ we can take the pullback of the metric on $\mathcal{T}_g^1$.

Given a quasi-isometry $\Phi\colon Y_g^1 \to Y_g^1$ with induced quasi-isometry $\phi\colon Y_g \to Y_g$, for any bi-infinite proper path $\ell\colon \mathbf{R} \to \mathcal{T}_g$ the map Φ restricts to a fiber respecting quasi-isometry

$$\Sigma_\ell \to \Sigma_{\phi(\ell)}$$

Also, if $\ell, \ell'\colon \mathbf{R} \to Y_g$ are "fellow travellers", then there is an induced $\pi_1 S$-equivariant map $\Sigma_\ell \to \Sigma_{\ell'}$ which is a quasi-isometry.

Example: Suppose that ℓ is the axis of a pseudo-Anosov diffeomorphism, or more generally, that ℓ fellow travels such an axis. Thurston's hyperbolization theorem for fibered 3-manifolds implies that Σ_ℓ is quasi-isometric to $\mathbf{H}^3$, and so Σ_ℓ is a Gromov hyperbolic metric space.

Teichmüller geodesics and their singular solv spaces. A *quadratic differential* on S_g is a transverse pair of measured foliations:

$$q = (\mathcal{F}_u, \mathcal{F}_s)$$

Each quadratic differential q determines a singular Euclidean metric, which determines conformal structure with removable singularities, which determines a point $\sigma(q) \in \mathcal{T}_g$. For each $t \in \mathbf{R}$ define

$$q_t = (e^{-t}\mathcal{F}_u, e_t\mathcal{F}_s)$$

The path

$$\gamma_q = \{t \mapsto \sigma(q_t) \bigm| t \in \mathbf{R}\}$$

in $\mathcal{T}_g$ is the *Teichmüller geodesic* corresponding to q.

Let Σ_q^{SOLV} denote the hyperbolic plane bundle Σ_{γ_q} with the *singular* SOLV *metric*, defined by

$$e^{-2t}\, d\mathcal{F}_u^2 + e^{2t}\, d\mathcal{F}_s^2 + dt^2$$

Fact: if γ_q is cobounded in $\mathcal{T}_g$, meaning that it is contained in the $\mathcal{MCG}(S_g)$ orbit of some bounded subset B of $\mathcal{T}_g$, then the the identity map $\Sigma_{\gamma_q} \to \Sigma_q^{\text{SOLV}}$ is a quasi-isometry between the "natural" metric and the singular SOLV metric, with quasi-isometry constants depending only on B.

Pseudo-Anosov homeomorphisms, their axes, and their hidden symmetries.

A homeomorphism $f \colon S_g \to S_g$ is *pseudo-Anosov* if there exists a quadratic differential $q_f = (\mathcal{F}_u, \mathcal{F}_s)$ and $\lambda > 1$ such that

$$f(\mathcal{F}_u, \mathcal{F}_s) = (\lambda^{-1}\mathcal{F}_u, \lambda\mathcal{F}_s)$$

It follows that the path γ_{q_f} is invariant under f in $\mathcal{T}$, and in fact γ_{q_f} is the set of points $\sigma \in \mathcal{T}$ at which $d(f\sigma, \sigma)$ is minimized; we call γ_{q_f} the *axis* of f in $\mathcal{T}$. Conversely, a mapping class $\Phi \in \mathcal{MCG}(S_g)$ is represented by a pseudo-Anosov homeomorphism only if $d(\Phi\sigma, \sigma)$ has a positive minimum in $\mathcal{T}$. These facts were proved by Bers [Ber78].

Let Σ_f denote Σ_{q_f}, with a superscript "SOLV" added to the notation is we wish to denote the singular SOLV metric. The group $J_f = \pi_1(S_g) \rtimes_f \mathbf{Z}$ acts by isometries on Σ_f^{SOLV}, that is,

$$J_f < I_f = \text{Isom}(\Sigma_f^{\text{SOLV}}))$$

It is possible that J_f is properly contained in I_f. We can think of the elements of $I_f - J_f$ as "hidden symmetries" of f. One possibility for hidden symmetries occurs when f is a proper power. Another possibility occurs when f or some power of f is conjugate to its own inverse. In general, I_f is a virtually cyclic group, and like all such groups there is an epimorphism $I_f \to C$ whose image C is either infinite cyclic or infinite dihedral, and whose kernel is finite; a nontrivial kernel provides a further source of hidden symmetries of f.

We shall need an alternate description of I_f. There exists a maximal index orbifold subcover $S_g \to O_f$ such that f descends to a pseudo-Anosov homeomorphism of the orbifold O_f which we shall denote f'. Let $\mathrm{VN}_{f'}$ be the virtual normalizer of f' in $\mathcal{MCG}(\mathcal{O}_f)$, consisting of all $g \in \mathcal{MCG}(\mathcal{O}_f)$ such that $g^{-1}\langle f'\rangle g \cap \langle f'\rangle$ has finite index in each of the infinite cyclic subgroups $g^{-1}\langle f'\rangle g$ and $\langle f'\rangle$.

Fact 18. *There is a natural extension*

$$1 \to \pi_1(\mathcal{O}_f) \to I_f \to \mathrm{VN}_{f'} \to 1$$

Since the virtual normalizer of a pseudo-Anosov homeomorphism is virtually cyclic, it follows that I_f contains J_f with finite index.

A direct construction can be used to show:

Fact 19. *There exist pseudo-Anosov homeomorphisms with no hidden symmetries, that is, so that $I_f = J_f$.*

For example, consider the fact the dimension of the measured foliation space of S_g is $6g-6$, and the transition matrix of a train track representative of every pseudo-Anosov homeomorphism f is an $n \times n$ matrix with $n \leq 6g-6$; it follows that the algebraic degree of the expansion factor $\lambda(f)$ is at most $6g-6$. One can construct primitive pseudo-Anosov homeomorphisms f of S_g such that $\lambda(f)$ has maximal algebraic degree $6g-6$. For such an f, the kernel K of the epimorphism $I_f \to C$ must be trivial, for if K is nontrivial then f or some power of f commutes with K and so descends through an orbifold covering map $S_g \to \mathcal{O} = S_g/K$ where the measured foliation space for $\mathcal{O}$ has strictly smaller dimension than for S_g; it follows that the degree of $\lambda(f)$ is strictly smaller than $6g-6$.

This still leaves open the possibility that I_f is infinite dihedral, implying that f is conjugate to its own inverse, but a random example will fail to have this property.

4.2 Proof of fibered QI-rigidity

Consider a quasi-isometry $\Phi\colon Y_g^1 \to Y_g^1$. As noted in Section 4.1, Φ coarsely respects the fibers of the fibration $Y_g^1 \to Y_g$, and so Φ induces a quasi-isometry $\phi\colon Y_g \to Y_g$. For any bi-infinite, proper path $\ell\colon \mathbf{R} \to Y_g$, the quasi-isometry Φ induces a coarse fiber respecting quasi-isometry from Σ_ℓ to $\Sigma_{\phi(\ell)}$. We shall apply various fiber respecting quasi-isometry invariants to the metric spaces Σ_ℓ.

For example, we say that ℓ and its $\mathbf{H}^2$ bundle Σ_ℓ are *hyperbolic* if Σ_ℓ is a δ-hyperbolic metric space for some $\delta \geq 0$. Since hyperbolicity is a quasi-isometry invariant, we obtain:

Fact 20 (Hyperbolic spaces are preserved). *Every quasi-isometry Φ of the space $\mathcal{MCG}(S_g^1)$ coarsely respects the hyperbolic spaces Σ_ℓ, and the induced quasi-isometry ϕ of $\mathcal{MCG}(S_g)$ coarsely respects the hyperbolic paths ℓ.*

Much more surprising is that a quasi-isometry Φ coarsely respects the *periodic* hyperbolic spaces. To be precise, a Teichmüller geodesic $\gamma \subset \mathcal{T}_g$, or its $\mathbf{H}^2$ bundle Σ_γ, is *periodic* if γ is the axis of some pseudo-Anosov mapping class in $\mathcal{MCG}(S_g)$. Equivalently, by Thurston's hyperbolization theorem, Σ_γ is the universal cover of a fibered hyperbolic 3-manifold. A bi-infinite path $\ell \colon \mathbf{R} \to Y_g$ is *coarsely periodic* if there exists an axis γ in $\mathcal{T}$ such that ℓ and γ are fellow travellers, meaning that $d(\ell(h(t)), \gamma(t))$ is uniformly bounded where $h \colon \mathbf{R} \to \mathbf{R}$ is some quasi-isometry of the real line.

Here is the heart of the matter:

Theorem 21. *Every quasi-isometry of $\mathcal{MCG}(S_g^1)$ coarsely respects the periodic hyperbolic 3-manifolds Σ_γ. To be precise, given: a quasi-isometry $\Phi \colon Y_g^1 \to Y_g^1$ inducing a quasi-isometry $\phi \colon Y_g \to Y_g$, given a periodic axis $\gamma \subset \mathcal{T}_g$, and given a coarsely periodic path ℓ in Y_g that fellow travels γ, there exists a periodic axis $\gamma' \subset \mathcal{T}_g$ such that $\phi \circ \ell$ fellow travels γ'. Moreover, $\Phi \colon \Sigma_\ell \to \Sigma_{\phi(\ell)}$ is a bounded distance from an isometry $H \colon \Sigma_\gamma^{\text{SOLV}} \to \Sigma_{\gamma'}^{\text{SOLV}}$.*

The meaning of the final sentence of this theorem is that there is a commutative diagram of fiber respecting quasi-isometries

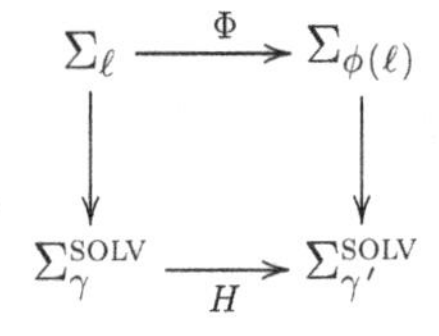

where the vertical arrows are maps that move points a uniformly bounded distance in $\mathcal{T}_g^1$.

In this theorem, the bounds in the conclusion depend only on the quasi-isometry constants of Φ and on the fellow traveller constants for ℓ and γ.

Before proving this theorem we first apply it to:

Proof of QI-rigidity of $\mathcal{MCG}(S_g^1)$.

Fix a quasi-isometry $\Phi \colon Y_g^1 \to Y_g^1$.

Consider a pseudo-Anosov homeomorphism $f \colon S_g \to S_g$ without hidden symmetries: the group $J_f = \pi_1(S_g) \rtimes_f \mathbf{Z}$ is the entire isometry group of the singular SOLV manifold Σ_f^{SOLV}. It follows that each conjugate gfg^{-1} has no hidden symmetries. By Theorem 21, there is a pseudo-Anosov $f' \colon S_g \to S_g$ and an isometry $H_f \colon \Sigma_f^{\text{SOLV}} \to \Sigma_{f'}^{\text{SOLV}}$ such that Φ takes Σ_f to $\Sigma_{f'}$ by a map which is a bounded distance from H_f. The isometry H_f conjugates $I_f = \text{Isom}(\Sigma_f^{\text{SOLV}})$ to $I_{f'} = \text{Isom}(\Sigma_{f'}^{\text{SOLV}})$, implying that H_f conjugates $\pi_1(\mathcal{O}_f)$ to $\pi_1(\mathcal{O}_{f'})$. However, since f has no hidden symmetries, $\mathcal{O}_f = S_g$, and so

$\pi_1(\mathcal{O}_{f'})$ must also equal S_g; replacing f' if necessary by some root, it follows that f' has no hidden symmetries. The isometry H_f therefore agrees with the action of some automorphism of $\pi_1(S_g)$, which is identified with a mapping class $h_f \in \mathcal{MCG}(S^1_g)$, and conjugation by h_f takes f to f'.

Next we must show that h_f is independent of f. For this we use the well known fact that $\mathcal{MCG}(S^1_g)$ acts faithfully on the circle at infinity S^1_∞ of $\mathbf{H}^2 = \widetilde{S}_g$; this fact is the basis of Nielsen theory.

The quasi-isometry Φ also acts on S^1_∞. To see why, fix a fiber Σ_x and identify this fiber isometrically with $\mathbf{H}^2$, so that the boundary of Σ_x is identified with S^1_∞. Since $\Phi(\Sigma_x)$ is (uniformly) coarsely equivalent to some fiber $\Sigma_{x'}$, and since $\Sigma_{x'}$ is (nonuniformly) coarsely equivalent to Σ_x, the action of Φ induces a self quasi-isometry of Σ_x, thereby inducing a homeomorphism of S^1_∞.

By construction of h_f, the actions of Φ and of h_f on S^1_∞ agree, that is to say, the action of h_f on S^1_∞ is independent of f. By faithfulness of the action of $\mathcal{MCG}(S^1_g)$ on S^1_∞, it follows that $h = h_f \in \mathcal{MCG}(S^1_g)$ is independent of f. We have thus constructed the desired mapping class $h \in \mathcal{MCG}(S^1_g)$, and from the construction it is evident that $\Phi\colon Y^1_g \to Y^1_g$ is within bounded distance of the action of h.

4.3 Proof of Theorem 21

This theorem reduces quickly to results from [Mos03] (see also [Bow02]) and from [FM02b], for each of which we will sketch a proof in broad strokes.

Step 1: Hyperbolic lines in $\mathcal{T}_g$. First we need the following theorem, proved independently by Bowditch and by myself:

Theorem 22 ([Bow02]; [Mos03]). *A line $\ell\colon \mathbf{R} \to \mathcal{T}_g$ is hyperbolic if and only if there exists a cobounded Teichmüller geodesic $\gamma\colon \mathbf{R} \to \mathcal{T}_g$ that fellow travels ℓ.*

For example, a pseudo-Anosov axis is a cobounded Teichmüller geodesic, forming a countable family. In toto, there are uncountably many cobounded Teichmüller geodesics, making Theorem 21 all the more surprising.

A one minute proof. (Every theorem should have a one minute proof, a five minute proof, a twenty minute proof...)

We must construct a quadratic differential $q = (\mathcal{F}_s, \mathcal{F}_u)$.

"Hyperbolicity" means "exponential divergence of geodesics" [Can91]. Ordinarily this applies to geodesic rays passing transversely through spheres, but it also applies to geodesics in Σ_ℓ passing transversely through fibers the $\Sigma_t = \ell^{-1}(t)$ [FM02a]. It follows that every geodesic contained in a fiber Σ_t is stretched exponentially in either the forward or backward direction, as $t \to +\infty$ or as $t \to -\infty$. Some geodesics are stretched exponentially in both directions; indeed, this is true of a random geodesic in a fiber. Certain geodesics contained in the fibers Σ_t are stretched exponentially as $t \to \infty$, but not as $t \to -\infty$; these geodesics form the unstable foliation $\mathcal{F}_u$. Certain other geodesics in Σ_t are stretched exponentially as $t \to -\infty$ but not as $t \to +\infty$, and these form the stable foliation $\mathcal{F}_s$.

Taking $q = (\mathcal{F}_u, \mathcal{F}_s)$, a compactness argument shows that ℓ fellow travels the Teichmüller geodesic γ_q. ◊

Step 2: Periodic hyperbolic lines. The key fact, from which Theorem 21 quickly follows, is:

Theorem 23 ([FM02b]). *Let γ, γ' be cobounded geodesics in $\mathcal{T}_g$, and suppose that γ is periodic. If there exists a fiber respecting quasi-isometry $\Phi\colon \Sigma^{\text{SOLV}}_\gamma \to \Sigma^{\text{SOLV}}_{\gamma'}$ then γ' is periodic and Φ is a bounded distance from an isometry.*

Before sketching the proof, we apply it to:

Proof of Theorem 21. We replace the map $\mathcal{MCG}(S^1_g) \to \mathcal{MCG}(S_g)$, fibered by cosets of $\pi_1 S_g$, with the map $Y^1_g \to Y_g$, fibered by copies of $\mathbf{H}^2$. Let $\Phi\colon Y^1_g \to Y^1_g$ be a quasi-isometry, inducing a quasi-isometry $\phi\colon Y_g \to Y_g$. Let $\gamma \subset \mathcal{T}_g$ be a periodic axis fellow travelling a coarsely periodic path ℓ. We obtain a fiber respecting quasi-isometry $\Sigma_\ell \to \Sigma^{\text{SOLV}}_\gamma$. The quotient $\Sigma^{\text{SOLV}}_\gamma$ modulo its isometry group is a hyperbolic 3–orbifold, by Thurston's hyperbolization theorem, and so $\Sigma^{\text{SOLV}}_\gamma$ is quasi-isometric to $\mathbf{H}^3$. It follows that Σ_ℓ is a hyperbolic metric space, that is, ℓ is a hyperbolic line. By Fact 20, $\phi(\ell)$ is a hyperbolic line and $\Sigma_{\phi(\ell)}$ is a hyperbolic metric space. Applying Theorem 22, there is a cobounded geodesic γ' in $\mathcal{T}$ that fellow travels $\phi(\ell)$. By combining the fiber respecting quasi-isometries $\Sigma^{\text{SOLV}}_\gamma \to \Sigma_\ell \to \Sigma_{\Phi(\ell)} \to \Sigma^{\text{SOLV}}_{\gamma'}$, we obtain a fiber respecting quasi-isometry $\Sigma^{\text{SOLV}}_\gamma \to \Sigma^{\text{SOLV}}_{\gamma'}$. Applying Theorem 23, γ' is periodic and the latter quasi-isometry is a bounded distance from an isometry. ◊

Step 3: Proof of Theorem 23. The proof in [FM02b] uses Thurston's hyperbolization theorem for fibered 3-manifolds [Ota96] together with Rich Schwartz' geodesic pattern rigidity theorem [Sch97]. It would be extremely nice to have a proof which uses only pseudo-Anosov dynamics, but in lieu of that, here is a broad sketch of the proof in [FM02b].

Given a cobounded Teichmüller geodesic γ, define $\mathrm{QI}_f(\Sigma^{\text{SOLV}}_\gamma)$ to be the group of "fiber respecting quasi-isometries" of $\Sigma^{\text{SOLV}}_\gamma$. We have an injection $\mathrm{Isom}(\Sigma^{\text{SOLV}}_\gamma) = I_\gamma \hookrightarrow \mathrm{QI}_f(\Sigma^{\text{SOLV}}_\gamma)$, and the question arises whether there is anything else in $\mathrm{QI}_f(\Sigma^{\text{SOLV}}_\gamma)$.

First we prove, when γ is a periodic geodesic, that the injection $I_\gamma \to \mathrm{QI}_f(\Sigma^{\text{SOLV}}_\gamma)$ is an isomorphism, that is, every self quasi-isometry of $\Sigma^{\text{SOLV}}_\gamma$ that coarsely respects fibers is a bounded distance from an isometry.

The singular lines of $\Sigma_\gamma^{\text{SOLV}}$ form a collection of singular SOLV geodesics intersecting the fibers at right angles; let Ω denote this collection of geodesics. If Φ is a fiber respecting quasi-isometry of $\Sigma_\gamma^{\text{SOLV}}$, then Φ coarsely respects leaves of f^s and f^u as noted earlier, and in fact Φ coarsely respects the suspensions of these leaves. It follows that Φ coarsely respects Ω, because the singular lines in Ω are precisely the sets which, coarsely, are intersections of three or more suspensions of leaves of f^s (or of f^u) whose pairwise intersections are unbounded. We may then move values of Φ by a bounded amount so that Φ is a homeomorphism that strictly respects leaves of f^s, leaves of f^u, and Ω.

By Thurston's hyperbolization theorem, there is an I_γ-equivariant $\mathbf{H}^3$ metric on $\Sigma_\gamma^{\text{SOLV}}$. The lines in Ω can be straightened to hyperbolic geodesics, which are evidently invariant under I_γ. This is exactly the setup of Schwartz' theorem, whose conclusion is that the group of quasi-isometries of $\mathbf{H}^3$ that coarsely respects Ω contains I_γ with finite index, that is, $\text{QI}_f(\Sigma_\gamma^{\text{SOLV}})$ contains I_γ with finite index. But then an easy argument shows that I_γ must actually be all of $\text{QI}_f(\Sigma_\gamma^{\text{SOLV}})$.

Using the isomorphism $I_\gamma \to \text{QI}_f(\Sigma_\gamma^{\text{SOLV}})$ and the coarse fiber respecting quasi-isometry $\Phi\colon \Sigma_\gamma^{\text{SOLV}} \to \Sigma_{\gamma'}^{\text{SOLV}}$, we now show that Φ is a bounded distance from an isometry. As above, we may first move values of Φ by a bounded amount so that Φ is a homeomorphism that respects the fibers and the stable and unstable foliations. Let Σ_x denote a fiber of $\Sigma_\gamma^{\text{SOLV}}$ whose image under Φ is a fiber $\Sigma_{x'}$ of $\Sigma_{\gamma'}^{\text{SOLV}}$. Let I_x be the subgroup of $\text{Isom}(\Sigma_x)$ that preserves the stable foliation and the unstable foliation, and similarly for $I_{x'}$. The conjugate action $\Phi^{-1} \circ I_{x'} \circ \Phi$ is an action on $\Sigma_\gamma^{\text{SOLV}}$ by quasi-isometries that preserves each fiber. By the computation of $\text{QI}_f(\Sigma_\gamma^{\text{SOLV}})$ just given, we obtain $\Phi^{-1} \circ I_{x'} \circ \Phi \subset I_x$, and in particular $\Phi^{-1} \circ I_{x'} \circ \Phi$ preserves the invariant measures on the stable and unstable foliations of Σ_x. Conjugating back now to $\Sigma_{x'}$, it follows that $I_{x'}$ preserves two sets of invariant measures on the stable and unstable foliations of $\Sigma_{x'}$: the ones coming from the singular SOLV structure on $\Sigma_{\gamma'}^{\text{SOLV}}$, and the ones pushed forward via Φ. But the stable and unstable foliations on $\Sigma_{x'}/I_{x'}$ are uniquely ergodic: this follows from a theorem of Masur [Mas80], which says that the stable and unstable foliations associated to a cobounded geodesic in Teichmüller space are uniquely ergodic. Up to rescaling, therefore, the map that Φ induces from Σ_x to $\Sigma_{x'}$ is an isometry; and the rescaling may be ignored by moving the point x' up or down. But this immediately implies that $\Sigma_\gamma^{\text{SOLV}}$ and $\Sigma_{\gamma'}^{\text{SOLV}}$ are isometric, by an isometry that agrees with Φ on $\Sigma_x \to \Sigma_{x'}$ and that moves other fibers up or down by uniformly bounded adjustments.

References

[Ber78] L. Bers, *An extremal problem for quasiconformal mappings and a theorem by Thurston*, Acta Math. **141** (1978), 73–98.

[Bow02] B. Bowditch, *Stacks of hyperbolic spaces and ends of 3-manifolds*, preprint, 2002.

[Bro82] K. Brown, *Cohomology of groups*, Graduate Texts in Math., vol. 87, Springer, 1982.

[Can91] J. Cannon, *The theory of negatively curved spaces and groups*, Ergodic theory, symbolic dynamics, and hyperbolic spaces (C. Series T. Bedford, M. Keane, ed.), Oxford Univ. Press, 1991.

[Dun85] M. J. Dunwoody, *The accessibility of finitely presented groups*, Invent. Math. **81** (1985), 449–457.

[FM00] B. Farb and L. Mosher, *On the asymptotic geometry of abelian-by-cyclic groups*, Acta Math. **184** (2000), no. 2, 145–202.

[FM02a] ______, *Convex cocompact subgroups of mapping class groups*, Geometry and Topology **6** (2002), 91–152.

[FM02b] ______, *The geometry of surface-by-free groups*, Geom. Funct. Anal. **12** (2002), 915–963.

[Gro81] M. Gromov, *Groups of polynomial growth and expanding maps*, IHES Sci. Publ. Math. **53** (1981), 53–73.

[Har86] J. L. Harer, *The virtual cohomological dimension of the mapping class group of an orientable surface*, Invent. Math. **84** (1986), 157–176.

[Hat91] Allen Hatcher, *On triangulations of surfaces*, Topology Appl. **40** (1991), no. 2, 189–194.

[Iva97] N. V. Ivanov, *Automorphism of complexes of curves and of Teichmüller spaces*, Internat. Math. Res. Notices (1997), no. 14, 651–666.

[Mas80] H. Masur, *Uniquly ergodic quadratic differentials*, Comment. Math. Helv. **55** (1980), 255–266.

[Mos03] L. Mosher, *Stable quasigeodesics in Teichmüller space and ending laminations*, Geometry and Topology **7** (2003), 33–90.

[MSW03] L. Mosher, M. Sageev, and K. Whyte, *Quasi-actions on trees I: Bounded valence*, Ann. of Math. (2003), 116–154, ARXIV:MATH.GR/0010136.

[Ota96] J.-P. Otal, *Le théorème d'hyperbolisation pour les variétés fibrées de dimension 3*, Astérisque, no. 235, Société Mathématique de France, 1996.

[Pan83] P. Pansu, *Croissance des boules et des géodésiques fermées dan les nilvariétés*, Ergodic Th. & Dyn. Sys. **3** (1983), 415–455.

[PW02] P. Papasoglu and K. Whyte, *Quasi-isometries between groups with infinitely many ends*, Comment. Math. Helv. **77** (2002), no. 1, 133–144.

[Sch96] R. Schwartz, *The quasi-isometry classification of rank one lattices*, IHES Sci. Publ. Math. **82** (1996), 133–168.

[Sch97] ———, *Symmetric patterns of geodesics and automorphisms of surface groups*, Invent. Math. **128** (1997), 177–199.

[Sha04] Y. Shalom, *Harmonic analysis, cohomology, and the large-scale geometry of amenable groups*, Acta Math. **192** (2004), no. 2, 119–185.

[Sta68] J. Stallings, *On torsion free groups with infinitely many ends*, Ann. of Math. **88** (1968), 312–334.

[Sul81] D. Sullivan, *On the ergodic theory at infinity of an arbitrary discrete group of hyperbolic motions*, Riemann surfaces and related topics, Proceedings of the 1978 Stony Brook Conference, Ann. Math. Studies, vol. 97, Princeton University Press, 1981, pp. 465–496.

[Thu86] W. P. Thurston, *A spine for Teichmüller space*, preprint, 1986.

Lee Mosher:
Department of Mathematics, Rutgers University, Newark
Newark, NJ 07102
E-mail: mosher@andromeda.rutgers.edu

Hattori-Stallings trace and Euler characteristics for groups

Indira Chatterji*and Guido Mislin

Introduction

For G a group and P a finitely generated projective module over the integral group ring, Bass conjectured in [2] that the Hattori-Stallings rank of P should vanish on elements different from $1 \in G$, and proved it in many cases such as torsion-free linear groups. Later, this conjecture has been proved for many more groups, notably by Eckmann [13], Emmanouil [15] and Linnell [22]. The latest advances are given in [3] and the first section of the present paper is a quick survey of the Bass conjecture, together with an outline of the proof of the main result of [3].

In most cases, one proves a stronger conjecture, which asserts that the Hattori-Stallings rank of a finitely generated projective module over the complex group ring should vanish on elements of infinite order (that this conjecture is indeed stronger follows from Linnell's work [22]). Given a group G of type FP over $\mathbb{C}$, its complete Euler characteristic $E(G)$ is the Hattori-Stallings rank of an alternating sum of finitely generated projective modules over $\mathbb{C}G$, and on the elements of finite order, one could then ask of what the values do depend. It is Brown in [7] who first studied that question, proving a formula in many cases. In Section 2 we shall explain the basics to understand Brown's formula and propose Conjecture 1 below as a generalization. Our generalization amounts to putting Brown's work in the context of L^2-homology (not available at the time when [7] was written), and applies to cases where Brown's formula is not available.

Conjecture 1 *Let G be a group of type FP over $\mathbb{C}$ such that the centralizer of every element of finite order in G has finite L^2-Betti numbers. Then for every $s \in G$*

$$E(G)(s) = \chi^{(2)}(C_G(s)), \tag{1}$$

where $E(G)(s)$ is the s-component of the complete Euler characteristic of G and $\chi^{(2)}(C_G(s))$ is the L^2-Euler characteristic of the centralizer of s in G.

*Partially supported by the Swiss National Science Foundation

This formula, as opposed to the Bass conjecture, has nice stability properties that we discuss in Section 3. We describe in Section 4 a class of groups containing all G with cocompact $\underline{E}G$ for which Formula (1) holds. It is straightforward (see Lemma 2.1 below) that Formula (1) always holds for $s = 1$. If we write $\chi(G)$ for the naive Euler characteristic $\sum(-1)^i \dim_{\mathbb{C}} H_i(G;\mathbb{C})$ then for G satisfying Conjecture 1, we find (cf. Corollary 4.5)

$$\chi(G) = \sum_{[s]\in[G]} \chi^{(2)}(C_G(s)). \tag{2}$$

If $K(G,1)$ is a finite complex, then G satisfies Conjecture 1 and G is necessarily torsion-free so that Formula (2) reduces to Atiyah's celebrated theorem: $\chi(G) = \chi^{(2)}(G)$.

1 Review of the Bass conjecture

For a group G we denote by $\mathrm{HS} : K_0(\mathbb{C}G) \to HH_0(\mathbb{C}G) = \bigoplus_{[G]} \mathbb{C}$ the *Hattori-Stallings Trace*; $[G]$ stands for the set of conjugacy classes of G. If P denotes a finitely generated projective $\mathbb{C}G$-module and $[P] \in K_0(\mathbb{C}G)$ the corresponding element, we write

$$\mathrm{HS}(P) := \mathrm{HS}([P]) = \sum_{[s]\in[G]} \mathrm{HS}(P)(s) \cdot [s] \in \bigoplus_{[G]} \mathbb{C}$$

with $\mathrm{HS}(P)(s)$ depending on the conjugacy class $[s]$ of $s \in G$ only. Therefore, we can think of $\mathrm{HS}(P) : G \to \mathbb{C}$ as a class function. It is well-known that for $s \in G$ a *central* element of infinite order, one has $\mathrm{HS}(P)(s) = 0$. More generally, if $C_G(s)$ denotes the centralizer of $s \in G$, and $[G : C_G(s)]$ is finite and s has infinite order, then $\mathrm{HS}(P)(s) = 0$, because $\mathrm{HS}(P|_{C_G(s)})(s) = \mathrm{HS}(P)(s)$ (in general, if $H < G$ is a subgroup of finite index and $s \in H$, then $\mathrm{HS}(P|_H)(s)$ $= [C_G(s) : C_H(s)]\,\mathrm{HS}(P)(s)$, see [2, Corollary 6.3] or Chiswell's notes [11]).

Another very useful result in this context goes back to Bass (cf. [2], Proposition 9.2), and states that if $\mathrm{HS}(P)(s) \neq 0$ then there is an $N > 0$ such that for almost all primes p, the elements s^{p^N} are conjugate to s; note that in case s has infinite order, this implies that for almost all primes p, s is contained in a subgroup of G which is isomorphic to $\mathbb{Z}[1/p]$. According to Bass in [2], the following general vanishing theorem ought to be true:

Conjecture 2 (Bass Conjecture over $\mathbb{C}$) *For P a finitely generated projective $\mathbb{C}G$-module and $s \in G$ an element of infinite order,* $\mathrm{HS}(P)(s) = 0$.

The Bass Conjecture over $\mathbb{C}$ is known to hold for many groups, including:

- linear groups (cf. Bass [2]), which includes Braid Groups (they are linear: [19] and [5])

- groups with $\mathrm{cd}_{\mathbb{C}} \leq 2$ (cf. Eckmann [13]; see also Emmanouil [15], as well as [16] for more general results using techniques of cyclic homology), which includes one-relator groups and knot groups
- subgroups of semihyperbolic groups (this follows from results of Alonso and Bridson [1], see also [14] or the discussion in [24], following Corollary 7.17 of Part 1; for the definition of semihyperbolic groups the reader is referred to [6]); these include (subgroups of) word hyperbolic groups and cocompact CAT(0)-groups
- mapping class groups Γ_g of closed surfaces of genus g (cf. Corollary 7.17 (Part 1) of [24])
- amenable groups, more generally groups satisfying the Bost Conjecture (for the Bost Conjecture, see [20] and [26]) have been shown to satisfy the Bass Conjecture over $\mathbb{C}$ in [3]. For instance groups which have the *Haagerup Property* (also called *a-T-menable groups*). We recall that a group is said to have the Haagerup Property if it admits an isometric, metrically proper affine action on some Hilbert space (for a discussion of such groups, see [10]); the class of groups having the Haagerup property contains all countable groups which are extensions of amenable groups with free kernel, and is closed under subgroups, finite products, passing to the fundamental group of a countable, locally finite graph of groups with finite edge stabilizers (vertex stabilizers are assumed to have the Haagerup property), countable increasing unions, amalgamations $A*_B C$ with A and C both countable amenable and B central in A and C (use Propositions 4.2.12 and 6.2.3 of [10]) and passing to finite index supergroups. Groups which act metrically properly and isometrically on a uniformly locally finite, weakly δ-geodesic and strongly δ-bolic space (see [18] and [20]); examples of groups satisfying these conditions are word hyperbolic groups (see [25]) and cocompact CAT(0)-groups.

A prominent class of groups for which Conjecture 2 is not known in general, is the class of profinite groups. However, if G is any group and Q a finitely generated projective $\mathbb{Z}G$-module, then according to Linnell [22], if $s \neq 1$ is such that $\mathrm{HS}(\mathbb{C}G \otimes_{\mathbb{Z}G} Q)(s) \neq 0$, then s is contained in a subgroup of G isomorphic to the additive group $\mathbb{Q}$ of rationals. This in particular implies that for G profinite one has $\mathrm{HS}(\mathbb{C}G \otimes_{\mathbb{Z}G} Q)(s) = 0$ for all $s \in G \setminus \{1\}$, because $\mathbb{Q}$ cannot be a subgroup of a profinite group.

We give an outline of the strategy for proving the main result of [3], which states that the Bost Conjecture implies the Bass Conjecture over $\mathbb{C}$ (see Theorem 1.1 below). The Bost Conjecture asserts that the *Bost assembly map*

$$\beta_0^G : K_0^G(\underline{E}G) \to K_0(\ell^1 G)$$

is an isomorphism (see [20] and [26]). Here, the left hand side denotes the equivariant K-homology of the classifying space for proper actions of G, and the right hand side is the projective class group of the Banach algebra $\ell^1 G$ of summable complex valued functions on G.

Theorem 1.1 *Suppose that G satisfies the Bost Conjecture. Then G satisfies the Bass Conjecture over $\mathbb{C}$.*

Before outlining the proof of Theorem 1.1, we need to address some auxiliary constructions. We can extend $\mathrm{HS} : K_0(\mathbb{C}G) \to \bigoplus_{[G]} \mathbb{C}$ to a trace $\mathrm{HS}^1 : K_0(\ell^1 G) \to \prod_{[G]} \mathbb{C}$ as follows. If $[Q] \in K_0(\ell^1 G)$, with Q a finitely generated projective $\ell^1 G$-module, we choose an idempotent (n,n)-matrix $M = (m_{ij})$ with entries in $\ell^1 G$ representing Q (i.e. $(\ell^1 G)^n \cdot M \cong Q$ as left $\ell^1 G$-modules), then we put

$$\mathrm{HS}^1(Q) := \mathrm{HS}^1([Q]) := \{\sum_{i=1}^{n} \sum_{t \in [s]} m_{ii}(t)\}_{[s] \in [G]} \in \prod_{[G]} \mathbb{C}.$$

The $m_{ii}(t)$'s stand for the t-coefficients of $m_{ii} \in \ell^1 G$, $1 \leq i \leq n$. We will write $\mathrm{HS}^1(x)(s)$ for the $[s]$-component of $\mathrm{HS}^1(x)$, $x \in K_0(\ell^1 G)$. One checks that HS^1 is well-defined and compatible with the ususal Hattori-Stallings trace. To get informations on HS^1 via the Bost assembly map, we embed G into an acyclic group of a very special kind. Recall that a group G is called *acyclic*, if $H_i(G; \mathbb{Z}) = 0$ for $i > 0$. As proved in [3], every group G admits a functorial embedding into an acyclic group $A = A(G)$, which we call the *pervasively acyclic hull of* G, satisfying the following:

- For every finitely generated abelian subgroup $B < A$ the centralizer $C_A(B)$ is acyclic (such a group is called *pervasively acyclic*)
- A is countable if G is and the induced map on conjugacy classes $[G] \to [A]$ is injective.

In this context, the important feature of a pervasively acyclic group A is that its classifying space for proper actions is $K_0^A \otimes \mathbb{Q}$-*discrete*, meaning that the inclusion $\underline{E}A^0 \to \underline{E}A$ of the 0-skeleton induces a surjective map

$$K_0^A(\underline{E}A^0) \otimes \mathbb{Q} \to K_0^A(\underline{E}A) \otimes \mathbb{Q},$$

see Corollary 3.9 of [3]. In other words, all the information of the A-CW-complex $\underline{E}A$ captured by the equivariant K-homology is contained in its 0-skeleton $\underline{E}A^0 = \coprod_\alpha A/A_\alpha$, where $A_\alpha < A$ stands for a finite subgroup, corresponding to the stabilizer of some 0-cell of $\underline{E}A$. The equivariant K-homology we use here is the one defined by Davis and Lück (cf. [12]), arising from a spectrum over the orbit category of G. It is defined on the category of *all* G-CW-complexes; on proper, cocompact G-CW-complexes, this *representable* equivariant K-homology agrees with the one used in the original version of the Baum-Connes or Bost conjectures (see [17]) so that if

X is a proper, not necessarily cocompact G-CW-complex, then $K_*^G(X) = \operatorname{colim}_{Y \subset X, Y/G \text{ compact}} K_*^G(Y)$, in accordance with the classical setup for the Baum-Connes and Bost conjectures. It follows that K_0^G is fully additive so that

$$K_0^A(\underline{E}A^0) = \bigoplus_\alpha K_0^A(A/A_\alpha) = \bigoplus_\alpha K_0^{A_\alpha}(\{pt\}) = \bigoplus_\alpha K_0(\ell^1 A_\alpha)$$

and $K_0(\ell^1 A_\alpha) \cong R_{\mathbb{C}}(A_\alpha)$, the additive group of the complex representation ring of the finite group A_α.

Outline of the proof of Theorem 1.1. Let P be a finitely generated projective $\mathbb{C}G$-module and assume that G satisfies the Bost Conjecture. Then $x := [\ell^1 G \otimes_{\mathbb{C}G} P]$ lies in the image of the Bost assembly map β_0^G and $\mathrm{HS}^1(x)$ captures the information contained in $\mathrm{HS}(P)$. We embed G into its pervasively acyclic hull $A(G) =: A$ and together with the standard embedding $\underline{E}A^0 \to \underline{E}A$ of the 0-skeleton this yields a commutative diagram

$$\begin{array}{ccccccc}
 & & [P] \in K_0(\mathbb{C}G) & \xrightarrow{\mathrm{HS}} & \bigoplus_{[G]} \mathbb{C} \\
 & & \downarrow & & \downarrow \\
K_0^G(\underline{E}G) & \xrightarrow[\cong]{\beta_0^G} & K_0(\ell^1 G) & \xrightarrow{\mathrm{HS}^1} & \prod_{[G]} \mathbb{C} \\
\downarrow & & \downarrow & & \downarrow \\
K_0^A(\underline{E}A) & \xrightarrow{\beta_0^A} & K_0(\ell^1 A) & \xrightarrow{\mathrm{HS}^1} & \prod_{[A]} \mathbb{C} \\
\uparrow & & \uparrow & & \uparrow \\
K_0^A(\underline{E}A^0) & \xrightarrow{\oplus \beta_0^{A_\alpha}} & \bigoplus_\alpha K_0(\ell^1 A_\alpha) & \xrightarrow{\oplus_\alpha \mathrm{HS}} & \bigoplus_{\alpha, [A_\alpha]} \mathbb{C}.
\end{array}$$

Using the facts that $K_0^A(\underline{E}A^0) \to K_0^A(\underline{E}A)$ is rationally surjective (since $\underline{E}A$ is $K_0^A \otimes \mathbb{Q}$-discrete), and that the induced map $\prod_{[G]} \mathbb{C} \to \prod_{[A]} \mathbb{C}$ is injective, we conclude by diagram chasing that $\mathrm{HS}^1(x)$ lies in the subspace of functions $[G] \to \mathbb{C}$, whose support is contained in the subset of those conjugacy classes of G, which are represented by elements of finite order. Therefore $\mathrm{HS}^1(x)(s) = 0$ for $s \in G$ of infinite order, which implies that $\mathrm{HS}(P)(s) = 0$ too, establishing the Bass conjecture over $\mathbb{C}$ for the group G.

QED

2 Euler characteristics

In this section we shall explain the basics to discuss Conjecture 1. Let G be a group of type FP over $\mathbb{C}$, meaning that there exists a resolution

$$P_* : 0 \to P_n \to P_{n-1} \to \cdots \to P_0 \to \mathbb{C}$$

with each P_i finitely generated projective over $\mathbb{C}G$; in case the P_i's may be chosen to be finitely generated and free over $\mathbb{C}G$, the G is termed of type FF over $\mathbb{C}$. The element $W(G) := \sum_i (-1)^i [P_i] \in K_0(\mathbb{C}G)$ depends on G only and we call it the *Wall element*. Under the Hattori-Stallings trace, the Wall element $W(G)$ is mapped to $E(G) = \sum_{[s]\in[G]} E(G)(s)[s]$, the sum being taken over the set $[G]$ of conjugacy classes $[s]$ of elements $s \in G$. This is the *complete Euler characteristic of* G (see [27]). If G has a cocompact $\underline{E}G$, Conjecture 1 is true as it reduces to Brown's formula [7] that we shall now discuss. For G of type FP over $\mathbb{C}$, the *Euler characteristic of* G (in the sense of Bass [2] and Chiswell [11]) is given by $e(G) = E(G)(1)$. Note also that $W(G) = 0$ if and only if $e(G) = 0$ and G is of type FF over $\mathbb{C}$. Brown conjectures under suitable finiteness conditions for G the following formula:

$$E(G)(s) = \begin{cases} e(C_G(s)) & \text{if } s \text{ has finite order} \\ 0 & \text{otherwise} \end{cases} \tag{3}$$

and proves it in many cases, including groups with cocompact $\underline{E}(G)$. Brown's assumptions always require in particular $C_G(s)$ to be of type FP over $\mathbb{C}$, and in this case we will show that Formula (1) reduces to Brown's formula (3). To do this, we first recall the definition of L^2-Euler characteristic. For $i \in \mathbb{N}$, the i-th L^2-Betti number is defined as the von Neumann dimension of the $\mathcal{N}(G)$-module $H_i(G; \mathcal{N}(G))$

$$\beta_i(G) = \dim_G H_i(G; \mathcal{N}(G)) \in [0, \infty],$$

where $\mathcal{N}(G)$ is the group von Neumann algebra of G (see Lück's book [23]). If $\sum (-1)^i \beta_i(G)$ converges, the L^2-*Euler characteristic* is defined as

$$\chi^{(2)}(G) = \sum_{i\in\mathbb{N}} (-1)^i \beta_i(G) \in \mathbb{R}. \tag{4}$$

In case G is finite, $\chi^{(2)}(C_G(s)) = 1/|C_G(s)|$ and Formulae (1) and (2) reduce to well-known results. With no finiteness restrictions imposed on G, one can find for any $r \in \mathbb{R}$ a group G with $\chi^{(2)}(G) = r$. However, if G is of type FP over $\mathbb{C}$ then $\chi^{(2)}(G) \in \mathbb{Q}$, as shown by the following.

Lemma 2.1 *Suppose that a group G is of type FP over $\mathbb{C}$. Then $\chi^{(2)}(G) = e(G)$ and $e(G)$ is a rational number.*

Proof. Let $P_* : 0 \to P_n \to P_{n-1} \to \cdots \to P_0 \to \mathbb{C}$ be a projective resolution of type FP for G. Then

$$\begin{aligned} \chi^{(2)}(G) &= \sum_{i\in\mathbb{N}} (-1)^i \beta_i(G) = \sum_{i\in\mathbb{N}} (-1)^i \dim_G \mathcal{N}(G) \otimes_{\mathbb{C}G} P_i \\ &= \sum_{i\in\mathbb{N}} (-1)^i HS(P_i)(1) = e(G). \end{aligned}$$

Here we used the fact that for a finitely generated projective $\mathbb{C}G$-module P, $\dim_G \mathcal{N}(G) \otimes_{\mathbb{C}G} P = HS(P)(1)$, which is actually just the Kaplansky trace of P; the Kaplansky trace of a finitely generated projective $\mathbb{C}G$-module is a rational number, by Zalesskii's theorem (see [8]). QED

The L^2-Betti numbers turn out to be often 0. In particular we mention the following vanishing result.

Theorem 2.2 (Cheeger-Gromov, [9]) *If G contains an infinite normal amenable subgroup, then $\beta_i(G) = 0$ for all $i \in \mathbb{N}$, and therefore $\chi^{(2)}(G) = 0$.*

This theorem immediately implies that for an arbitrary group G, the L^2-Euler characteristic of $C_G(s)$ is 0 for all $s \in G$ of infinite order, so that one more evidence for Conjecture 1 is the following simple observation: *If the Bass Conjecture over $\mathbb{C}$ holds for G, then Formula (1) holds on elements of infinite order.* Indeed, the Bass conjecture will say that the left hand side vanishes on elements of infinite order. The following fact on L^2-Euler characteristics will be used later, mainly for the case of subgroups $H < G$, with G of type FP over $\mathbb{C}$. Since then $\mathrm{cd}_{\mathbb{C}} H < \infty$, the Euler characteristic $\chi^{(2)}(H)$ is well-defined if and only if all L^2-Betti numbers $\beta_i(H)$ are finite.

Lemma 2.3 *Let H and K be groups with $\sum_i \beta_i(H)$ and $\sum_i \beta_i(K)$ convergent. Then $\chi^{(2)}(H \times K) = \chi^{(2)}(H)\chi^{(2)}(K)$.*

Proof. One uses the Künneth Formula for L^2-Betti numbers [9]: $\beta_n(H \times K) = \sum_{i+j=n} \beta_i(H)\beta_j(K)$, and takes the alternating sum; note that $\sum_n \beta_n(H \times K)$ is convergent so that $\chi^{(2)}(H \times K)$ is well-defined. QED

A 1-dimensional contractible G-CW-complex T with vertex set V and edge set E (for short: a G-tree) is given by a G-push-out

$$\begin{array}{ccc} (\coprod_{\beta \in E/G} G/G_\beta) \times S^0 & \longrightarrow & \coprod_{\alpha \in V/G} G/G_\alpha \\ \downarrow & & \downarrow \\ (\coprod_{\beta \in E/G} G/G_\beta) \times D^1 & \longrightarrow & T \end{array}$$

and the cellular chain complex of T has the form

$$0 \to \bigoplus_{\beta \in E/G} \mathbb{C}[G/G_\beta] \to \bigoplus_{\alpha \in V/G} \mathbb{C}[G/G_\alpha] \to \mathbb{C}.$$

The group G is then the fundamental group of a graph of groups $\{G_\gamma\}_{\gamma \in I}$, $I = V/G \sqcup E/G$; the graph is called *finite*, if I is a finite set (i.e. if the action of G on T is cocompact). If X is an arbitrary G-CW-complex, we write $H_*(X; \mathcal{N}(G)) := H_*(\mathcal{N}(G) \otimes_{\mathbb{Z}G} C_*^{\mathrm{cell}}(X))$ for its L^2-homology so that $H_*(G; \mathcal{N}(G)) = H_*(EG; \mathcal{N}(G))$.

Lemma 2.4 *Let G be the fundamental group of a (not necessarily finite) graph of groups $\{G_\gamma\}_{\gamma\in I}$, where $I = V/G \sqcup E/G$. If for each of the groups G_γ the series $\sum_i \beta_i(G_\gamma)$ is convergent and equals 0 for almost all $\gamma \in I$, then*

$$\chi^{(2)}(G) = \sum_{\alpha\in V/G} \chi^{(2)}(G_\alpha) - \sum_{\beta\in E/G} \chi^{(2)}(G_\beta).$$

Proof. The group G acts on a tree $T = (V, E)$ with chain complex

$$0 \to \bigoplus_{\beta\in E/G} \mathbb{C}[G/G_\beta] \to \bigoplus_{\alpha\in V/G} \mathbb{C}[G/G_\alpha] \to \mathbb{C}.$$

Take a projective resolution of this complex in the category of chain complexes over $\mathbb{C}G$, say $P_* \to Q_* \to R_*$, with P_* a projective resolution for $\bigoplus \mathbb{C}[G/G_\beta]$, Q_* one for $\bigoplus \mathbb{C}[G/G_\alpha]$ and R_* for $\mathbb{C}$. Upon tensoring with $\mathcal{N}(G) \otimes_{\mathbb{C}G} -$ we obtain a short exact sequence of chain complexes

$$\mathcal{N}(G) \otimes_{\mathbb{C}G} P_* \to \mathcal{N}(G) \otimes_{\mathbb{C}G} Q_* \to \mathcal{N}(G) \otimes_{\mathbb{C}G} R_*;$$

the exactness results from the fact that the sequences $P_i \to Q_i \to R_i$ are split exact for all i, because R_i is projective. Taking homology yields a long exact sequence of L^2-homology groups

$$\cdots \to H_{i+1}(G; \mathcal{N}(G)) \to \bigoplus_{\beta\in E/G} H_i(\mathrm{Ind}_{G_\beta}^G EG_\beta; \mathcal{N}(G)) \to$$

$$\bigoplus_{\alpha\in V/G} H_i(\mathrm{Ind}_{G_\alpha}^G EG_\alpha; \mathcal{N}(G)) \to H_i(G; \mathcal{N}(G)) \to \cdots .$$

We used here that a $\mathbb{C}G$-projective resolution of $\mathbb{C}[G/G_\gamma]$ is chain homotopy equivalent to the cellular $\mathbb{C}G$-chain complex of the induced G-CW-complex $\mathrm{Ind}_{G_\gamma}^G EG_\gamma = G \times_{G_\gamma} EG_\gamma$. Therefore

$$H_*(\mathcal{N}(G) \otimes_{\mathbb{C}G} P_*) = \bigoplus_{\beta\in E/G} H_*(\mathrm{Ind}_{G_\beta}^G EG_\beta; \mathcal{N}(G))$$

and similarly for $H_*(\mathcal{N}(G) \otimes_{\mathbb{C}G} Q_*)$. According to [23] (Theorem 6.54, (7)), for any induced G-CW-complex $\mathrm{Ind}_{G_\gamma}^G X$ one has

$$\dim_G H_i(\mathrm{Ind}_{G_\gamma}^G X; \mathcal{N}(G)) = \dim_{G_\gamma} H_i(X; \mathcal{N}(G_\gamma))$$

and it follows that $\dim_G H_i(\mathrm{Ind}_{G_\gamma}^G EG_\gamma; \mathcal{N}(G)) = \beta_i(G_\gamma)$. Thus, by taking the alternating sum of L^2-Betti numbers in the long exact homology sequence above, the desired formula follows. QED

There are groups G of type FP over $\mathbb{C}$ containing centralizers $C_G(s)$ which are not of type FP over $\mathbb{C}$. Such examples have first been constructed by Leary and Nucinkis in [21], and those cannot satisfy Brown's formula, because then $e(C_G(s))$ is not defined. The following group G is a simple example for which Formula (1) holds whereas (3) doesn't apply. Take first a group $\mathcal{G}$ as described by Leary-Nucinkis in [21] with the following property:

$\mathcal{G}$ is of type FP over $\mathbb{C}$ and contains an element $t \in \mathcal{G}$ of finite order such that $C_{\mathcal{G}}(t)$ is not of type FP over $\mathbb{C}$.

Then the right hand side of Brown's formula (3) doesn't make sense for the group $G = \mathcal{G} \times \mathbb{Z}$, which is of type FP over $\mathbb{C}$ but none of the centralizers $C_G((t,n)) = C_{\mathcal{G}}(t) \times \mathbb{Z}$ are; note that $(t,n) \in G$ is of finite order if and only if $n = 0$. But nevertheless, the group G satisfies Conjecture 1 because of the following.

Lemma 2.5 *Let H be a group of type FP over $\mathbb{C}$ and $G := H \times \mathbb{Z}$. Then G is of type FF over $\mathbb{C}$, satisfies Conjecture 1 and $W(G) = 0 \in K_0(\mathbb{C}G)$.*

Proof. Let $P_* : 0 \to P_n \to \cdots \to P_1 \to P_0 \to \mathbb{C} \to 0$ be a resolutions of type FP over $\mathbb{C}$ for H and $D_* : 0 \to \mathbb{C}\langle z\rangle \to \mathbb{C}\langle z\rangle \to \mathbb{C} \to 0$ be the projective resolution for $\mathbb{Z} = \langle z\rangle$ with the map $\mathbb{C}\langle z\rangle \to \mathbb{C}\langle z\rangle$ given by multiplication with $1 - z$. Then $E_* = P_* \otimes D_* \to \mathbb{C} \to 0$ is a resolution of type FP over $\mathbb{C}$ for $G = H \times \mathbb{Z}$, and since $E_i = (P_i \otimes \mathbb{C}\langle z\rangle) \oplus (P_{i-1} \otimes \mathbb{C}\langle z\rangle)$, we see that $W(G) = \sum_{i=0}^{n+1}(-1)^i[E_i] = 0$ (terms cancel pairwise); hence G is of type FF over $\mathbb{C}$ and $E(G) = 0$ so that $E(G)(s) = 0$ for every $s \in G$. On the other hand, the centralizer of $s = (u,v) \in H \times \mathbb{Z}$ contains the normal subgroup $\{1_H\} \times \mathbb{Z}$ so that $\chi^{(2)}(C_G(s)) = 0$ as well. QED.

It follows that Conjecture 1 holds for any group of type FP over $\mathbb{C}$ of the form $H \times \mathbb{Z}$, because both sides are zero; we shall construct non-zero examples later (recall that if $H \times \mathbb{Z}$ is of type FP over $\mathbb{C}$ then so is H by Proposition 2.7 of [4]). We we will show in Section 4 (Theorem 4.6) that for each $\rho \in \mathbb{Q}$ there exists a group $G(\rho)$ of type FP over $\mathbb{C}$ containing an element s of finite order such that $C_{G(\rho)}(s)$ is not of type FP over $\mathbb{C}$ but such that $G(\rho)$ nevertheless satisfies Conjecture 1, with $E(G(\rho))(s) = \rho$.

3 Stability properties of Formula (1)

In this section we shall study some stability properties of Formula (1), starting with the following.

Lemma 3.1 *Let A and B be two groups of type FP over $\mathbb{C}$ such that A satisfies Formula (1) for some $a \in A$, and B satisfies it for some $b \in B$. Then $G = A \times B$ satisfies Formula (1) for the element (a,b).*

Proof. Let $P_* : 0 \to P_n \to \cdots \to P_1 \to P_0 \to \mathbb{C} \to 0$ be a resolution of type FP over $\mathbb{C}$ for A and $Q_* : 0 \to Q_n \to \cdots \to Q_1 \to Q_0 \to \mathbb{C} \to 0$ one for B (by adding trivial modules we can assume that both resolutions have the same length). A projective resolution of type FP over $\mathbb{C}$ for $G = A \times B$ is given by $E_* = P_* \otimes Q_* \to \mathbb{C} \to 0$. For an element $s = (a, b) \in G$ we compute

$$\begin{aligned} HS(W(G))(s) &= \sum_{i=0}^{2n}(-1)^i HS(E_i)(s) = \sum_{i=0}^{2n}(-1)^i \sum_{k+\ell=i} HS(P_k \otimes Q_\ell)(a,b) \\ &= \sum_{i=0}^{2n}\sum_{k+\ell=i}(-1)^{k+\ell} HS(P_k)(a) HS(Q_\ell)(b) \\ &= HS(W(A))(a) HS(W(B))(b) = \chi^{(2)}(C_A(a))\chi^{(2)}(C_B(b)) \\ &= \chi^{(2)}(C_A(a) \times C_B(b)). \end{aligned}$$

Here we used in the second line the fact that for P and Q finitely generated projective modules over $\mathbb{C}A$ and $\mathbb{C}B$ respectively, $HS(P \otimes Q)(a, b) = HS(P)(a)HS(Q)(b)$. We conclude using Lemma 2.3 and the fact that $C_G(a, b) = C_A(a) \times C_B(b)$. Note that the L^2-Betti numbers of $C_A(a)$ are finite, and trivial for large degrees, because $C_A(a)$ is assumed to have a well-defined L^2-Euler characteristic and $\mathrm{cd}_{\mathbb{C}}\, C_A(a)$ is finite; a similar remark applies to $C_B(b)$. QED

Definition 3.2 (Condition (F)) *The fundamental group G of a finite graph of groups $\{G_\gamma\}$ satisfies* Condition (F)*, if the G-action on the associated standard tree T is such that for every element of finite order $s \in G$, the action of $C_G(s)$ on the fixed tree T^s satisfies the hypothesis of Lemma 2.4.*

Remark 3.3 *Condition (F) amounts to say that for any element of finite order $s \in G$ and for each of the stabilizers $H < C_G(s)$ appearing on the fixed tree T^s, the series $\sum_i \beta_i(H)$ is convergent and equals 0 for all but finitely many conjugacy classes (H).*

Lemma 3.4 *Let G be the fundamental group of a finite graph of groups.*

(i) *If all edge and vertex groups satisfy Formula (1) at all elements of infinite order, then so does G.*

(ii) *If G satisfies Condition (F) and all edge and vertex groups satisfy Formula (1) at all elements of finite order, then G satisfies Formula (1) at all elements of finite order.*

Proof. The group G acts cocompactly on a tree $T = (V, E)$, yielding a resolution $0 \to \bigoplus_{\beta \in E/G} \mathbb{C}[G/G_\beta] \to \bigoplus_{\alpha \in V/G} \mathbb{C}[G/G_\alpha] \to \mathbb{C} \to 0$. Each of the groups G_γ (for $\gamma \in V/G \sqcup E/G$) is of type FP over $\mathbb{C}$ (by assumption), so let us denote by $P_*^\gamma : 0 \to P_n^\gamma \to \cdots \to P_0^\gamma \to \mathbb{C} \to 0$ a corresponding resolution of type FP. Tensoring by $\mathbb{C}[G/G_\gamma]$ yields the following resolutions of type FP over $\mathbb{C}$ of induced modules: $\tilde{P}^{\gamma}_{*} : 0 \to \tilde{P}^{\gamma}_{n} \to \cdots \to \tilde{P}^{\gamma}_{1} \to \tilde{P}^{\gamma}_{0} \to \mathbb{C}[G/G_{\gamma}] \to 0$, for $\gamma \in V/G \sqcup E/G$, so that the Wall element for G is given by

$$W(G) = \sum_{\alpha \in V/G} [\mathbb{C}[G/G_{\alpha}]] - \sum_{\beta \in E/G} [\mathbb{C}[G/G_{\beta}]],$$

where $[\mathbb{C}[G/G_{\gamma}]] = \sum_{j=0}^{n}(-1)^{j}[\tilde{P}^{\gamma}_{j}] = i^{\gamma}_{*}W(G_{\gamma}) \in K_0(\mathbb{C}G)$. The complete Euler characteristic of G is then given by

$$E(G) = \sum_{\alpha \in V/G} i^{\alpha}_{*}E(G_{\alpha}) - \sum_{\beta \in E/G} i^{\beta}_{*}E(G_{\beta}).$$

(i) Now let us take $s \in G$ of infinite order. By Cheeger-Gromov's Theorem 2.2 of this note $\chi^{(2)}(C_G(s)) = 0$; on the other hand, $E(G)(s) = 0$ because $E(G_{\gamma})(t) = 0$ for any $\gamma \in V/G \sqcup E/G$ and any t of infinite order, by assumption on the G_{γ}'s.

(ii) If $s \in G$ has finite order, then

$$\begin{aligned} E(G)(s) &= \sum_{[x] \in [s, G_{\alpha}]} E(G_{\alpha})(x) - \sum_{[y] \in [s, G_{\beta}]} E(G_{\beta})(y) \\ &= \sum_{[x] \in [s, G_{\alpha}]} \chi^{(2)}(C_{G_{\alpha}}(x)) - \sum_{[y] \in [s, G_{\beta}]} \chi^{(2)}(C_{G_{\beta}}(y)) \end{aligned}$$

because by assumption the G_{γ}'s satisfy Formula (1) at elements of finite order (we used here the notation $[s, G_{\gamma}]$ for the conjugacy classes of elements in G_{γ} which are G-conjugate to s). So to conclude we need to show that the last line of the above equation is equal to $\chi^{(2)}(C_G(s))$, which we will do now. We think of the G_{γ}'s as representatives for the stabilizers of the G-action on the standard tree T of the given graph of groups so that a general stabilizer will have the form $tG_{\gamma}t^{-1}$. Since s has finite order, $T^s = (V^s, E^s)$ is a non-empty tree, upon which $C_G(s)$ acts via the restriction of the G-action on T. The stabilizer of a vertex or an edge $\in T^s$ has the form $C_G(s) \cap tG_{\gamma}t^{-1}$, where $s \in tG_{\gamma}t^{-1}$, so that $C_G(s) \cap tG_{\gamma}t^{-1} \cong C_{G_{\gamma}}(t^{-1}st)$. Moreover, by assumption G satisfies Condition (F), and hence $\chi^{(2)}(C_{G_{\gamma}}(tst^{-1}))$ is well-defined so that $\chi^{(2)}(C_G(s))$ is well defined too and, by Lemma 2.4 satisfies

$$\chi^{(2)}(C_G(s)) = \sum_{x \in I} \chi^{(2)}(C_{G_{\alpha}}(x)) - \sum_{y \in J} \chi^{(2)}(C_{G_{\beta}}(y))$$

with index set I corresponding to $V^s/C_G(s)$. But this set corresponds bijectively to conjugacy classes of elements x in the $[G_{\alpha}]$'s, which are G-conjugate to s; similarly for J. QED

4 Conjecture 1 and two classes of groups

To begin with, we consider the following class $\mathcal{B}_\infty$ of groups.

Definition 4.1 *Let $\mathcal{B}_\infty$ denote the smallest class of groups which contains all groups of type FF over $\mathbb{C}$, all groups of type FP over $\mathbb{C}$ which satisfy the Bass Conjecture over $\mathbb{C}$, all groups G with cocompact $\underline{E}G$, all groups $G = H \times \mathbb{Z}$ with H of type FP over $\mathbb{C}$ and which is closed under finite products of groups and under passing to the fundamental group of a finite graph of groups.*

Clearly all groups in $\mathcal{B}_\infty$ are of type FP over $\mathbb{C}$. In particular, the Wall element $W(G) \in K_0(\mathbb{C}G)$ is defined for all groups in $\mathcal{B}_\infty$. Examples of groups in $\mathcal{B}_\infty$ include word hyperbolic groups, braid groups, cocompact CAT(0)-groups, Coxeter groups, mapping class groups of surfaces, knot groups, finitely generated one-relator groups, S-arithmetic groups, Artin groups, amenable groups of type FP over $\mathbb{C}$. Many more groups can be obtained using the closure properties mentioned before; the groups thus obtained are in general not known to satisfy the Bass conjecture over $\mathbb{C}$. We do not know of any group of type FP over $\mathbb{C}$ not belonging to $\mathcal{B}_\infty$. As we have seen, there are groups G in $\mathcal{B}_\infty$ containing x of finite (resp. infinite) order, whose centralizer $C_G(x)$ is not of type FP over $\mathbb{C}$ and, therefore, $E(C_G(x))$ is not defined and $C_G(x) \notin \mathcal{B}_\infty$. But nevertheless, the following holds.

Theorem 4.2 *Let G be a group in $\mathcal{B}_\infty$ and $s \in G$ an element of infinite order. Then Formula (1) holds at s:*

$$E(G)(s) = 0 = \chi^{(2)}(C_G(s)).$$

Proof. We have already seen that the right hand side is 0 (cf. Cheeger-Gromov's Theorem 2.2 in this note). The left hand side is certainly 0 in case G is of type FF over $\mathbb{C}$ or if G satisfies the Bass Conjecture over $\mathbb{C}$ or if $\underline{E}G$ is cocompact. Moreover, by Lemmas 3.1 and 3.4 (i), if $G = H \times K$ or G is the fundamental group of a finite graph of groups G_α and if $E(L)(t) = 0$ for all t of infinite order in L, where L is one of the groups H, K, G_α, then $E(G)(s) = 0$ for all elements of infinite order $s \in G$. Finally, $G = H \times \mathbb{Z}$ certainly satisfies $E(G)(s) = 0$ for all s (see Lemma 2.5). QED

We now describe a class of groups $\mathcal{B} \subset \mathcal{B}_\infty$ containing many examples of groups G with $E(G)(s) \neq 0$ for some $s \neq e$ in G satisfying Conjecture 1, but such that the corresponding centralizer $C_G(s)$ is not of type FP over $\mathbb{C}$.

Definition 4.3 *Let $\mathcal{B}$ denote the smallest class of groups which contains all groups G with cocompact $\underline{E}G$, all groups $G = H \times \mathbb{Z}$ with H of type FP over $\mathbb{C}$ and which is closed under finite products of groups and under passing to the fundamental group of a finite graph of groups which satisfy Condition (F).*

Theorem 4.4 *The groups of the class $\mathcal{B}$ satisfy Conjecture 1.*

Proof. This follows by applying Lemmas 2.5, 3.1 and 3.4. QED

Corollary 4.5 *For G satisfying Conjecture 1, $\chi(G) = \sum_{[s]\in[G]} \chi^{(2)}(C_G(s))$.*

Proof. By definition we have that

$$\chi(G) = \sum_i (-1)^i \dim_{\mathbb{C}} H_i(G;\mathbb{C}) = \sum_i (-1)^i \dim_{\mathbb{C}} \mathbb{C} \otimes_{\mathbb{C}G} P_i,$$

where $P_* \to \mathbb{C}$ is a resolution of G of type FP over $\mathbb{C}$. It implies that $\sum_{[s]\in[G]} E(G)(s) = \chi(G)$, because for P a finitely generated projective $\mathbb{C}G$-module, $\sum_{[s]\in[G]} \mathrm{HS}(P)(s) = \dim_{\mathbb{C}} \mathbb{C} \otimes_{\mathbb{C}G} P$. The desired result now follows from Formula (1). QED

We shall now construct explicit non-trivial examples in the class $\mathcal{B}$. More precisely we prove the following.

Theorem 4.6 *Given $\rho \in \mathbb{Q}$ there exists a group $G = G(\rho)$ of type FP over $\mathbb{C}$ with $s \in G$ of order 2 such that G satisfies Conjecture 1, with*

$$E(G)(s) = \chi^{(2)}(C_G(s)) = \rho$$

but with the centralizer $C_G(s)$ not of type FP over $\mathbb{C}$.

Before proceeding with the proof we need the following.

Lemma 4.7 *For $\rho \in \mathbb{Q}$ there exist a group $G_\rho \in \mathcal{B}$ with $\chi^{(2)}(G_\rho) = \rho$.*

Proof. Since a free group F_n of rank n satisfies $\chi^{(2)}(F_n) = 1 - n$, one has for $n, k \geq 0$ that $\chi^{(2)}((F_2 \times F_{n+1}) * F_k) = n - k$, so that for $\ell > 0$

$$\chi^{(2)}(((F_2 \times F_{n+1}) * F_k) \times \mathbb{Z}/\ell\mathbb{Z}) = \frac{n-k}{\ell}.$$

The group $G = ((F_2 \times F_{n+1}) * F_k) \times \mathbb{Z}/\ell\mathbb{Z}$ admits a cocompact $\underline{E}G$ via its obvious quotient action on $E(G/(\mathbb{Z}/\ell\mathbb{Z}))$, with orbit space the finite complex $((\vee^2 S^1) \times (\vee^{n+1} S^1)) \vee (\vee^k S^1)$, thus $G \in \mathcal{B}$. QED

Proof of Theorem 4.6. Let $\mathcal{G}$ be one of the groups described in [21], Example 9, such that $\mathcal{G}$ is of type FP over $\mathbb{C}$, $s \in \mathcal{G}$ is an element of order 2 and $C_{\mathcal{G}}(s)$ is not finitely generated. By definition of $\mathcal{B}$, the group $H := \mathcal{G} \times \mathbb{Z}$ belongs to $\mathcal{B}$, and $C_H((s,0))$ is not finitely generated, because it maps onto $C_{\mathcal{G}}(s)$. Writing t for $(s,0)$, we form $K := H *_{\langle t\rangle} H \in \mathcal{B}$. Thus, K is the fundamental group of a finite graph of groups $\{H, \langle t\rangle\}$, with associated tree T. If $w \in K$ has finite order with w not conjugate to t, the edge stabilizers of the $C_K(w)$ action on T^w are all trivial, and the vertex stabilizers are isomorphic to $C_H(z)$ for some element z of order 2 in H, thus $\beta_i(C_H(z)) = 0$ for all i, because such a centralizer contains a normal infinite cyclic subgroup. The centralizer of $\langle t\rangle$ in K decomposes as a fundamental group of a graph of groups of the form $\{H_\delta, \langle t\rangle\}$ with the H_δ's again isomorphic to groups $C_H(w)$, $w \in H$ of order 2, so that $\beta_i(H_\delta) = 0$ for all i and all δ. It follows that K satisfies Condition (F) and $\chi^{(2)}(C_K(t)) = -\chi^{(2)}(\langle t\rangle) = -1/2$. Note that $C_K(t)/\langle t\rangle$ maps onto $C_H(t)/\langle t\rangle$, which shows that $C_K(t)$ is not finitely generated. Forming $G := K \times G_{-2\rho} \in \mathcal{B}$ where $G_{-2\rho}$ is obtained following Lemma 4.7 above, gives a group with $C_G(t) = C_K(t) \times G_{-2\rho}$ not of type FP over $\mathbb{C}$ (because it is not finitely generated), but

$$\chi^{(2)}(C_G(t)) = \chi^{(2)}(C_K(t)) \cdot \chi^{(2)}(G_{-2\rho}) = -\frac{1}{2} \cdot (-2\rho) = \rho.$$

QED

Acknowledgements. The first named author thanks Ken Brown for friendly discussions on [7].

References

[1] J.M. Alonso and M. Bridson. *Semihyperbolic groups.* Proc. London Math. Soc. **70** (1995), 56–114.

[2] H. Bass. *Euler characteristics and characters of discrete groups.* Invent. Math. **35** (1976), 155–196.

[3] J. Berrick, I. Chatterji and G. Mislin. *From acyclic groups to the Bass conjecture for amenable groups.* Math. Annalen **329** (2004), 597–621.

[4] R. Bieri. *Homological Dimension of Discrete Groups.* Queen Mary College Mathematics Notes, 2nd edition (1981).

[5] S. Bigelow. *Braid groups are linear.* J. Amer. Math. Soc. (electronic) **14** (2001), 471–486.

[6] M. R. Bridson and A. Haefliger. *Metric Spaces of Non-Positive Curvature.* Springer, Grundlehren **391**, 1999.

[7] K. Brown. *Complete Euler characteristics and fixed-point theory.* Journal of Pure and Applied Algebra **24** (1982), 103–121.

[8] M. Burger and A. Valette. *Idempotents in complex group rings: theorems of Zalesskii and Bass revisited.* Journal of Lie Theory **8** (1998), 219–228.

[9] J. Cheeger, M. Gromov. *L_2-cohomology and group cohomology.* Topology **25** (1986), no. 2, 189–215.

[10] P.-A. Cherix, M. Cowling, P. Jolissaint, P. Julg and A. Valette. *Groups with the Haagerup Property (Gromov's a-T-menability).* Birkhäuser, Progress in Math. **197**, 2001.

[11] I. Chiswell. *Euler Characteristics for Groups.* Preprint 2003.

[12] J.F. Davis and W. Lück. *Spaces over a category and assembly maps in isomorphism conjectures in K- and L-theory.* K-Theory **15** (1998), 201–252.

[13] B. Eckmann. *Cyclic homology of groups and the Bass conjecture.* Comment. Math. Helv. **61** (1986), 193–202.

[14] B. Eckmann. *Idempotents in a complex group algebra, projective modules and the von Neumann algebra.* Arch. Math. **76** (2001), 241–249.

[15] I. Emmanouil. *On a class of groups satisfying Bass' conjecture.* Invent. Math. **132** (1998), 307–330.

[16] I. Emmanouil and I.B.S. Passi. *A contribution to Bass' conjecture.* J. Group Theory **7** (2004), no. 3, 409–420.

[17] I. Hambleton and E. Pedersen. *Identifying assembly maps in K- and L-theory.* Math. Ann. **328** (2004), no. 1-2, 27–57.

[18] G. Kasparov and G. Skandalis. *Groups acting properly on "bolic" spaces and the Novikov conjecture.* Ann. of Math.**158** (2003), 165–206.

[19] D. Krammer. *Braid groups are linear.* Ann. of Math. (2) **155** (2002), no. 1, 131–156.

[20] V. Lafforgue. *K-théorie bivariante pour les algèbres de Banach et conjecture de Baum-Connes.* Invent. Math. **149** (2002), 1–95.

[21] I. Leary and B. Nucinkis. *Some groups of type VF.* Invent. Math. **151** (2003), 135-165.

[22] P.A. Linnell. *Decomposition of augmentation ideals and relation modules.* Proc. London Math. Soc. (3) **47** (1983), 83–127.

[23] W. Lück. *L^2-Invariants: Theory and Applications to Geometry and K-Theory.* Ergebnisse der Mathematik und ihrer Grenzgebiete **44**, Springer 2002.

[24] G. Mislin and A. Valette. *Proper Group Actions and the Baum-Connes Conjecture.* Advanced Courses in Mathematics CRM Barcelona, Birkhäuser Verlag 2003.

[25] I. Mineyev, G. Yu. *The Baum-Connes conjecture for hyperbolic groups.* Invent. Math. **149** (2002), 97–122.

[26] G. Skandalis. *Progrès récents sur la conjecture de Baum-Connes. Contribution de Vincent Lafforgue.* Sém. Bourbaki, 1999/2000, exposé **869**; Astérisque **276** (2002), 105–135.

[27] J.R. Stallings. *Centerless groups – an algebraic formulation of Gottlieb's theorem.* Topology **4** (1965), 129 – 134.

Groups of small homological dimension and the Atiyah conjecture

Peter Kropholler, Peter Linnell and Wolfgang Lück

Abstract

A group has homological dimension ≤ 1 if it is locally free. We prove the converse provided that G satisfies the Atiyah Conjecture about L^2-Betti numbers. We also show that a finitely generated elementary amenable group G of cohomological dimension ≤ 2 possesses a finite 2-dimensional model for BG and in particular that G is finitely presented and the trivial $\mathbb{Z}G$-module $\mathbb{Z}$ has a 2-dimensional resolution by finitely generated free $\mathbb{Z}G$-modules.

Key words: (co-)homological dimension, von Neumann dimension, the Atiyah Conjecture.
Mathematics Subject Classification 2000: 18G20, 46L99.

1 Notation

Throughout this paper let G be a (discrete) group. It has *homological dimension* $\leq n$ if $H_p(G; M) = \mathrm{Tor}_p^{\mathbb{Z}G}(\mathbb{Z}, M)$ vanishes for each $\mathbb{Z}G$-module M and each $p > n$. It has *cohomological dimension* $\leq n$ if $H^p(G; M) = \mathrm{Ext}^p_{\mathbb{Z}G}(\mathbb{Z}, M)$ vanishes for each $\mathbb{Z}G$-module M and each $p > n$.

We call G *locally free* if each finitely generated subgroup is free. The *class of elementary amenable groups* is defined as the smallest class of groups, which contains all finite and all abelian groups and is closed under taking subgroups, taking quotient groups, extensions and directed unions. Each elementary amenable group is amenable, but the converse is not true.

2 Review of the Atiyah Conjecture

Denote by $\mathcal{N}(G)$ the group von Neumann algebra associated to G which we will view as a ring (not taking the topology into account) throughout this paper. For a $\mathcal{N}(G)$-module M let $\dim_{\mathcal{N}(G)}(M) \in [0, \infty]$ be its dimension in the sense of [8, Theorem 6.7]. Let $\frac{1}{\mathcal{FIN}(G)}\mathbb{Z} \subseteq \mathbb{Q}$ be the additive abelian subgroup of $\mathbb{Q}$ generated by the inverses $|H|^{-1}$ of the orders $|H|$ of finite subgroups H of G. Notice that $\frac{1}{\mathcal{FIN}(G)}\mathbb{Z}$ agrees with $\mathbb{Z}$ if and only if G is torsion-free.

Conjecture 1 (Atiyah Conjecture). *Consider a ring A with $\mathbb{Z} \subseteq A \subseteq \mathbb{C}$. The* Atiyah Conjecture for A and G *says that for each finitely presented AG-module M we have*

$$\dim_{\mathcal{N}(G)} (\mathcal{N}(G) \otimes_{AG} M) \in \frac{1}{\mathcal{FIN}(G)}\mathbb{Z}.$$

For a discussion of this conjecture and the classes of groups for which it is known we refer for instance to [8, Section 10.1]. It is not clear whether the Atiyah conjecture is subgroup closed; however in the case G is torsion-free, then it certainly is. This can be seen from [8, Theorem 6.29(2)]. We mention Linnell's result [6] that the Atiyah Conjecture is true for $A = \mathbb{C}$ and all groups G which can be written as an extension with a free group as kernel and an elementary amenable group as quotient and possess an upper bound on the orders of its finite subgroups. The Atiyah Conjecture has also been proved by Schick [9] for $A = \mathbb{Q}$ and torsion-free groups G which are residually torsion-free elementary amenable.

3 Results

Theorem 2. *A locally free group G has homological dimension ≤ 1.*

If G is a group of homological dimension ≤ 1 and the Atiyah Conjecture 1 holds for G, then G is locally free.

Theorem 3. *Let G be an elementary amenable group of cohomological dimension ≤ 2. Then*

1. *Suppose that G is finitely generated. Then G possesses a presentation of the form*
$$\langle x, y \mid yxy^{-1} = x^n \rangle.$$
In particular there is a finite 2-dimensional model for BG and the trivial $\mathbb{Z}G$-module $\mathbb{Z}$ possesses a 2-dimensional resolution by finitely generated free $\mathbb{Z}G$-modules;

2. *Suppose that G is countable but not finitely generated. Then G is a non-cyclic subgroup of the additive group $\mathbb{Q}$.*

4 Proofs

Lemma 4. *Let A be a ring with $\mathbb{Z} \subset A \subset \mathbb{C}$. Let P be a projective AG-module such that for some finitely generated AG-submodule $M \subset P$ we have $\dim_{\mathcal{N}(G)}(\mathcal{N}(G) \otimes_{AG} P/M) = 0$. Then P is finitely generated.*

Proof: Choose a free AG-module F and AG-maps $i\colon P \to F$ and $r\colon F \to P$ with $r \circ i = \mathrm{id}$. Since $M \subset P$ is finitely generated, there is a finitely generated free direct summand $F_0 \subset F$ with $i(M) \subset F_0$ and $F_1 := F/F_0$ a free AG-module. Hence i induces a map $f\colon P/M \to F_1$. It suffices to show that f is trivial because then $i(P) \subset F_0$ and the restriction of r to F_0 yields an epimorphism $F_0 \to P$.

Let $g\colon AG \to P/M$ be any AG-map. The map $\mathcal{N}(G) \otimes_{AG} (f \circ g)$ factorizes through $\mathcal{N}(G) \otimes_{AG} P/M$. Hence its image has von Neumann dimension zero because $\dim_{\mathcal{N}(G)}$ is additive [8, Theorem 6.7] and $\dim_{\mathcal{N}(G)}(\mathcal{N}(G) \otimes_{AG} P/M) = 0$ holds by assumption. Since the von Neumann algebra $\mathcal{N}(G)$ is semi-hereditary (see [8, Theorem 6.5 and Theorem 6.7]), the image of $\mathcal{N}(G) \otimes_{AG} (f \circ g)$ is a finitely generated projective $\mathcal{N}(G)$-module, whose von Neumann dimension is zero, and hence is the zero-module. Therefore $\mathcal{N}(G) \otimes_{AG} (f \circ g)$ is the zero map. Since $AG \to \mathcal{N}(G)$ is injective, $f \circ g$ is trivial. This implies that f is trivial since g is any AG-map. □

Lemma 5. *Let A be a ring with $\mathbb{Z} \subset A \subset \mathbb{C}$. Suppose that there is a positive integer d such that the order of any finite subgroup of G divides d and that the Atiyah Conjecture holds for A and G. Let N be a AG-module. Suppose that $\dim_{\mathcal{N}(G)}(\mathcal{N}(G) \otimes_{AG} N) < \infty$. Then there is a finitely generated AG-submodule $M \subset N$ with $\dim_{\mathcal{N}(G)}(\mathcal{N}(G) \otimes_{AG} N/M) = 0$.*

Proof: Since N is the colimit of the directed system of its finitely generated AG-modules $\{M_i \mid i \in I\}$ and tensor products commute with colimits, we get $\mathrm{colim}_{i\in I}\, \mathcal{N}(G) \otimes_{AG} N/M_i = 0$. Additivity (see [8, Theorem 6.7]) implies $\dim_{\mathcal{N}(G)}(\mathcal{N}(G) \otimes_{AG} N/M_i) < \infty$ for all $i \in I$ since $\dim_{\mathcal{N}(G)}(\mathcal{N}(G) \otimes_{AG} N) < \infty$ holds by assumption. We conclude from Additivity and Cofinality (see [8, Theorem 6.7]) and the fact that the functor colimit over a directed system of modules is exact

$$\inf\{\dim_{\mathcal{N}(G)}(\mathcal{N}(G) \otimes_{AG} N/M_i) \mid i \in I\} = 0.$$

The assumption about G implies using [8, Lemma 10.10 (4)]

$$d \cdot \dim_{\mathcal{N}(G)}(\mathcal{N}(G) \otimes_{AG} N/M_i) \in \mathbb{Z}.$$

Hence there must be an index $i \in I$ with $\dim_{\mathcal{N}(G)}(\mathcal{N}(G) \otimes_{AG} N/M_i) = 0$. □

Proof of Theorem 2: A finitely generated free group has obviously homological dimension ≤ 1. Since homology is compatible with colimits over directed systems (in contrast to cohomology), we get for every group G, which is the directed union of the family of subgroups $\{G_i \mid i \in I\}$, and every $\mathbb{Z}G$-module M

$$H_n(G; M) = \mathrm{colim}_{i\in I}\, H_n(G_i; \mathrm{res}_i\, M),$$

where $\mathrm{res}_i\, M$ is the restriction of M to a $\mathbb{Z}G_i$-module. Hence any locally free group has homological dimension ≤ 1.

Suppose that G has homological dimension ≤ 1. Let $H \subset G$ be a finitely generated subgroup. Then the homological dimension of H is ≤ 1. Since each countably presented flat module is of projective dimension ≤ 1 [1, Lemma 4.4], we conclude that the cohomological dimension of H is ≤ 2. We can choose an exact sequence $0 \to P \to \mathbb{Z}H^s \to \mathbb{Z}H \to \mathbb{Z}$, where s is the number of generators and P is projective. Since the homological dimension is ≤ 1, the induced map $\mathcal{N}(H) \otimes_{\mathcal{N}(H)} P \to \mathcal{N}(H) \otimes_{\mathcal{N}(H)} \mathbb{Z}H^s$ is injective and hence $\dim_{\mathcal{N}(H)}(\mathcal{N}(H) \otimes_{\mathcal{N}(H)} P) \leq \dim_{\mathcal{N}(H)}(\mathcal{N}(H) \otimes_{\mathcal{N}(H)} \mathbb{Z}H^s) = s$. Suppose that G satisfies the Atiyah Conjecture. Since G cannot contain a non-trivial finite subgroup, H also satisfies the Atiyah Conjecture, and Lemma 4 and Lemma 5 imply that P is finitely generated. Hence H is of type FP. Since each finitely presented flat module is projective [1, Lemma 4.4], the cohomological dimension and the homological dimension agree for groups of type FP. Hence H has cohomological dimension 1. A result of Stallings [10] implies that H is free. □

In [3] the notion of Hirsch length for an elementary amenable group was defined, generalizing that of the Hirsch length of a solvable group. This was used in the proof of [5, Corollary 2] to show that an elementary amenable group of finite cohomological dimension is virtually solvable with finite Hirsch number, see [4, Theorem 1.11] for further details. We can now state

Lemma 6. *If G is an elementary amenable group of homological dimension ≤ 2, then G is metabelian.*

Proof: A group is metabelian if and only if each finitely generated subgroup is metabelian. Hence we can assume without loss of generality that G is finitely generated. Then by the above remarks and [1, Theorem 7.10(a)], G is virtually solvable of Hirsch length ≤ 2.

If G has Hirsch length 1, then G is infinite cyclic, so we may assume that G has Hirsch length 2. Let N denote the Fitting subgroup of G (so N is generated by the nilpotent normal subgroups of G and is a locally nilpotent normal subgroup).

Suppose that N has finite index in G. Then N is finitely generated and is therefore free abelian of rank 2. Also G/N acts faithfully by conjugation on N (a torsion-free group with a central subgroup of finite index must be abelian). If $g \in G \setminus N$, then $g^r \in N \setminus 1$ for some positive integer r and thus g fixes a nonidentity element of N. We deduce that $|G/N| \leq 2$ and it follows that G is metabelian.

On the other hand if N has infinite index in G, then it has Hirsch length 1. Hence every finitely generated subgroup of N is trivial or isomorphic to $\mathbb{Z}$. This implies that N is abelian and any automorphism of finite order $f\colon N \to N$ has the property that $f(x) \in \{x, -x\}$ holds for $x \in N$. Since the group G/N acts faithfully by conjugation on N and is virtually cyclic, we conclude that G/N is isomorphic to $\mathbb{Z}$ or $\mathbb{Z} \times \mathbb{Z}/2$. Hence G is metabelian. □

Proof of Theorem 3: Since G has cohomological dimension 2, it certainly has homological dimension at most 2 and so by Lemma 6 is metabelian. A result of Gildenhuys [2, Theorem 5], states that a solvable group G of cohomological dimension 2 has a presentation of the form $\langle x, y; y^{-1}xy = x^n \rangle$ for some $n \in \mathbb{Z}$ if G is finitely generated and is a non-cyclic subgroup of the additive group $\mathbb{Q}$ if G is not finitely generated. Given a torsion free finitely generated one-relator group G, the finite two-dimensional CW-complex associated to a presentation with finitely many generators and one non-trivial relation is a model for BG (see [7, Chapter III §§9 -11]). This finishes the proof of Theorem 3. □

References

[1] **Bieri, R**: *"Homological dimension of discrete groups"*, 2-nd edition, Queen Mary College Mathematics Notes, Mathematics Department, Queen Mary College, London (1981).

[2] **Gildenhuys, D.**: *"Classification of soluble groups of cohomological dimension"*, Math. Z. 166, 21–25 (1979).

[3] **Hillman, Jonathan A.**: *"Elementary amenable groups and 4-manifolds with Euler characteristic 0"*, *J. Austral. Math. Soc. Ser. A*, 50(1):160–170, 1991.

[4] **Hillman, J. A.**: *"Four-manifolds, geometries and knots"*, volume 5 of *Geometry & Topology Monographs.* Geometry & Topology Publications, Coventry, 2002.

[5] **Hillman, J. A. and Linnell, P. A.**: *"Elementary amenable groups of finite Hirsch length are locally-finite by virtually-solvable"*, *J. Austral. Math. Soc. Ser. A*, 52(2):237–241, 1992.

[6] **Linnell, P.**: *"Division rings and group von Neumann algebras"*, Forum Math. 5, 561–576 (1993).

[7] **Lyndon, R.C. and Schupp, P. E.**: *"Combinatorial group theory"*, Ergebnisse der Mathematik und ihrer Grenzgebiete 89, Springer (1977).

[8] **Lück, W.**: *"L^2-Invariants: Theory and Applications to Geometry and K-Theory"*, Ergebnisse der Mathematik und ihrer Grenzgebiete 44, Springer (2002).

[9] **Schick, T.**: *"Integrality of L^2-Betti numbers"*, Math. Ann. 317, 727–750 (2000).

[10] **Stallings, J.R.**: *"On torsion-free groups with infinitely many ends"*, Annals of Math. 88, 312–334 (1968).

Addresses:

Peter Kropholler, Department of Mathematics, University at Glasgow, University Garden, Glasgow G12 8QW, Scotland,
p.h.kropholler@maths.gla.ac.uk,
http://www.maths.gla.ac.uk/people/?id=289

Peter Linnell, Department of Mathematics, Virginia Tech, Blacksburg, Virginia, VA 24061-0123, USA,
linnell@math.vt.edu,
http://www.math.vt.edu/people/plinnell/

Wolfgang Lück, FB Mathematik, Universität Münster, Einsteinstr. 62, D-48149 Münster, Germany,
lueck@math.uni-muenster.de,
http://wwwmath.uni-muenster.de/math/u/lueck/

Logarithms and assembly maps on $K_n(\mathbf{Z}_l[G])$

Victor P. Snaith

Abstract

When G is a finite group and l a prime I define a new logarithm or regulator homomorphism on $K_n(\mathbf{Z}_l[G])$, the higher dimensional algebraic K-theory of the l-adic group ring of G. The "torsion free" part of the logarithm is evaluated on the image of the assembly map in terms of a familiar universal coefficient map and the Adams operation ψ^l.

1 Introduction

At the London Mathematical Society Symposium on Geometry and Cohomology in Group Theory, which was held in Durham during July 4-13, 2003 I was particularly taken by lectures which were related to the algebraic K-theory of group-rings. This paper is related to those lectures because in it I will define a new regulator map on the higher algebraic K-groups of l-adic group rings and partially relate this regulator to the K-theory assembly map. Although the tenor of the conference was largely towards infinite discrete groups, the groups that I shall discuss will be finite. On the other hand, the assembly maps and regulators that I will discuss are natural in the group and therefore they have potential applications to residually finite discrete groups or profinite groups.

In Section Two I discuss the recent computation, due to Lars Hesselholt and Ib Madsen, of the l-adic part of the algebraic K-theory of l-adic local fields. In Section Three I use the Explicit Brauer Induction technique of Robert Boltje, Peter Symonds and myself ([17], [18]) to construct a regulator map of the form

$$\rho_{i,G} : K_i(\mathbf{Z}_l[G]) \longrightarrow \mathrm{Hom}_{\Omega_{\mathbf{Q}_l}}(R_{\overline{\mathbf{Q}}_l}(G), K_i(\mathbf{Q}_l(\mu_N)))$$

modelled on my previous [18] construction of the Oliver-Taylor group logarithm in dimension one. In Remark 3.5 and §3.6 I have mentioned a couple of research problems concerning $\rho_{i,G}$. In Section Four I compute the "torsion free" part of the composition of $\rho_{i,G}$ with the assembly map.

I believe that, with a more detailed analysis of $K_i(\mathbf{Z}_l[G])$ when G is a finite l-group, which has been computed using topological cyclic homology methods by Hesselholt and Madsen, it should be possible to compute $\rho_{i,G}$ completely in the l-group case.

Having said this, it is my belief that the usefulness of $\rho_{i,G}$ will be in the context of arithmetic rather than in the context of computations. The first reason for this belief is that it is difficult to imagine doing better than the computations of $K_2(\mathbf{Z}_l[G])$ in [10] and [11] which illustrate the limitations of logarithms. The second reason is that $\rho_{1,G}$ was useful in proving a number of arithmetic results concerning $K_0(\mathbf{Z}[G])$ in [18]. In ([18] §4.5.30) $\rho_{1,G}$ first appeared in a simplification of the construction of the Oliver-Taylor logarithm. From this construction I was able to prove ([18] §4.3) Martin Taylor's determinantal congruence conjecture, posed in [23]. Also I was able to give a simplified proof of the Galois descent for determinants, which is one of the three main steps in Martin Taylor's proof of the Fröhlich conjecture [26]. Finally $\rho_{1,G}$ was used in (see [18] Remark 4.6.4) to give a new proof of the result, due to Ph. Cassou-Noguès and M.J. Taylor ([24], [25]), that the Adams operations preserve the determinantal subgroup – a proof which, for example, avoided the use of the decomposition homomorphism.

2 K-theory of l-adic local fields

2.1. Let l be an odd rational prime and let E be a finite extension of the l-adic numbers $\mathbf{Q}_l$. Let U denote the infinite unitary group with its classical topology and let BU denote its classifying space. Hence the topological K-theory of a space X is represented in the homotopy category by $\mathbf{Z} \times BU$ where $\mathbf{Z}$ denotes the integers with the discrete topology. For each integer r let $\Psi^r : \mathbf{Z} \times BU \longrightarrow \mathbf{Z} \times BU$ denote the r-th Adams operation. Let $F\Psi^r$ denote the homotopy fibre defined by

$$F\Psi^r \longrightarrow \mathbf{Z} \times BU \stackrel{\Psi^r - 1}{\longrightarrow} BU.$$

By ([6] Theorem D), after l-completion, Quillen's K-theory space for E ([5], [14]) is described by a homotopy equivalence of the form

$$\mathbf{Z} \times BGL(E)^+ \simeq F\Psi^{g^{l^{a-1}d}} \times BF\Psi^{g^{l^{a-1}d}} \times U^{[E:\mathbf{Q}_l]}$$

where $d = (l-1)/[E(\mu_l) : E]$, $a = \max\{v \mid \mu_{l^v} \subset E(\mu_l)\}$ and $g \in \mathbf{Z}_l^*$ is a topological generator. Here μ_T denotes the groups of T-th roots of unity. Since $\pi_s(\mathbf{Z} \times BU)$ is isomorphic to the integers when $s \geq 0$ is even and is zero otherwise and since Ψ^r is multiplication by r^{j+1} in dimension $2j+2$ one easily deduces that $\pi_0(F\Psi^r) \cong \mathbf{Z}$, $\pi_{2j+1}(F\Psi^r) \cong \mathbf{Z}/(r^{j+1} - 1)$ for $j \geq 0$ and $\pi_s(F\Psi^r) = 0$ otherwise.

If k is the residue field of E then the algebraic K-theory of E away from the prime l is determined by the isomorphism, if HCF$(l, m) = 1$,

$$K_i(E; \mathbf{Z}/m) \cong K_i(k; \mathbf{Z}/m) \oplus K_{i-1}(k; \mathbf{Z}/m)$$

of Gabber-Suslin [21], where the $K_i(k; \mathbf{Z}/m)$ were computed in [13].

3 A logarithmic construction

3.1. In this section let l be any rational prime, let G be a finite group and let $\overline{\mathbf{Q}}_l$ denote an algebraic closure of $\mathbf{Q}_l$. Let $R_{\overline{\mathbf{Q}}_l}(G) = K_0(\overline{\mathbf{Q}}_l[G])$ denote the representation ring of finite-dimensional $\overline{\mathbf{Q}}_l$-representations of G, endowed with the canonical action by the profinite Galois group $\Omega_{\mathbf{Q}_l} = \mathrm{Gal}(\overline{\mathbf{Q}}_l/\mathbf{Q}_l)$. Suppose that N is a positive integer such that $\mathbf{Q}_l(\mu_N)$ is a splitting field for G ($N = |G|$, for example). We may consider the group of Galois equivariant homomorphisms of the form

$$\mathrm{Hom}_{\Omega_{\mathbf{Q}_l}}(R_{\overline{\mathbf{Q}}_l}(G), K_i(\mathbf{Q}_l(\mu_N))).$$

In this section I shall construct a natural homomorphism – a *logarithm* or *regulator* map – of the form

$$\rho_{i,G} : K_i(\mathbf{Z}_l[G]) \longrightarrow \mathrm{Hom}_{\Omega_{\mathbf{Q}_l}}(R_{\overline{\mathbf{Q}}_l}(G), K_i(\mathbf{Q}_l(\mu_N))).$$

The homomorphism $\rho_{1,G}$ is closely related to the group-ring logarithm first constructed by Martin Taylor and (independently) Bob Oliver ([9], [24], [25]). The relation is the following. Using the Explicit Brauer Induction technique in the representation theory of groups, as developed by Robert Boltje, Peter Symonds and me ([3],[17],[18], [22]) I was able to give a simple construction of the Oliver-Taylor logarithm for all finite groups (see [18] §4.3.25 and §4.5.30). In this section we are going to rephrase that construction and apply it to the higher-dimensional K-groups of the group-ring $\mathbf{Z}_l[G]$.

Explicit Brauer Induction is a canonical method of reducing from arbitrary l-adic representations of G to one-dimensional representations $\phi : H \longrightarrow \overline{\mathbf{Q}}_l^*$ of subgroups $H \subseteq G$. Let $\mathbf{Z}_l(\phi)$ denote the integral closure of the l-adic integers in the cyclotomic extension $\mathbf{Q}_l(\phi)$ of the l-adics given by adjoining the values of ϕ.

Lemma 3.2.

In the situation of §3.1 the two maps associated to a subgroup $H \subseteq G$ of index m and a homomorphism ϕ on H

$$GL_n(\mathbf{Z}_l[G]) \stackrel{Ind_H^G(\phi)}{\longrightarrow} GL_{nm}(\mathbf{Z}_l(\phi)) \stackrel{det}{\longrightarrow} \mathbf{Z}_l(\phi)^*$$

and

$$GL_n(\mathbf{Z}_l[G]) \stackrel{Ind_H^G}{\longrightarrow} GL_{nm}(\mathbf{Z}_l[H]) \stackrel{\phi}{\longrightarrow} GL_{nm}(\mathbf{Z}_l(\phi)) \stackrel{det}{\longrightarrow} \mathbf{Z}_l(\phi)^*$$

are equal.

Proof

Let $g \in G$ and let $g_1, \ldots, g_m$ be coset representatives for G/H. Then $gg_i = g_{\sigma(g)(i)}h(i,g)$ with $h(i,g) \in H$ where $\sigma : G \longrightarrow \Sigma_m$ is a homomorphism to the symmetric group. Then, in $GL_m(\mathbf{Z}_l(\phi))$,

$$Ind_H^G(\phi)(g) = \sigma(g)diag(\phi(h(1,g)), \phi(h(2,g)), \ldots, \phi(h(m,g)))$$

and so

$$\sum_{g \in G} X(g)g \in GL_n(\mathbf{Z}_l[G])$$

with $X(g) \in GL_n(\mathbf{Z}_l)$ is sent by the first homomorphism to

$$det(\sum_{g \in G} X(g) \otimes Ind_H^G(\phi)(g))$$

while the second homomorphism sends it to

$$\sum_{g \in G} X(g) \otimes \sigma(g)diag(h(1,g), h(2,g), \ldots, h(m,g)).$$

We apply ϕ and take the determinant to arrive at

$$det(\textstyle\sum_{g \in G} X(g) \otimes \sigma(g)diag(\phi(h(1,g)), \phi(h(2,g)), \ldots, \phi(h(m,g))))$$

$$= det(\textstyle\sum_{g \in G} X(g) \otimes Ind_H^G(\phi)(g))$$

as required. □

3.3. Explicit Brauer Induction and the logarithm

Let $R_+(G)$ denote the free abelian group on G-conjugacy classes of pairs (H, ϕ) where H is a subgroup of G and $\phi : H \longrightarrow \overline{\mathbf{Q}}_l^*$ is a one-dimensional l-adic representation of H. Write $(H, \phi)^G \in R_+(G)$ for the generator corresponding to (H, ϕ). Then $R_+(G)$ is a Mackey functor from groups to commutative rings, as explained in [18], and there is a homomorphism

$$b_G : R_+(G) \longrightarrow R_{\overline{\mathbf{Q}}_l}(G)$$

given by $b_G((H, \phi)^G) = Ind_H^G(\phi)$. The Explicit Brauer Induction homomorphism is a natural right inverse to the (split) surjection b_G

$$a_G : R_{\overline{\mathbf{Q}}_l}(G) \longrightarrow R_+(G).$$

For each integer t there is an additive endomorphism $\tilde{\psi}^t$ of $R_+(G)$ given by $\tilde{\psi}^t((H, \phi)^G) = ((H, \phi^t)^G)$ and, by naturality, the composite

$$R_{\overline{\mathbf{Q}}_l}(G) \xrightarrow{a_G} R_+(G) \xrightarrow{\tilde{\psi}^t} R_+(G) \xrightarrow{b_G} R_{\overline{\mathbf{Q}}_l}(G)$$

is equal to the Adams operation ψ^t, which is given on character-values by $\psi^t(V)(g) = V(g^t)$ for all $g \in G$.

Now suppose that $\rho : G \longrightarrow GL_s(\overline{\mathbf{Q}}_l)$ is a matrix representation corresponding to V such that, for some integers n_i and some pairs (H_i, ϕ_i),

$$a_G(\rho) = \sum_i n_i (H_i, \phi_i)^G \in R_+(G).$$

Let N be a multiple of $|G|$. Then the map associated to V in the group-ring logarithm context from $K_1(\mathbf{Z}_l[G]) = GL_\infty(\mathbf{Z}_l[G])^{ab}$ to $K_1(\mathbf{Z}_l(\mu_N)) \cong \mathbf{Z}_l(\mu_N)^*$ is given, in the construction of ([18] §4.3.25 and §4.5.30), by sending the class of

$$Z = \sum_{g \in G} X(g) g \in GL_n(\mathbf{Z}_l[G])$$

to

$$\prod_i det(Ind_{H_i}^G(\phi_i^l)(Z))^{n_i} det(Ind_{H_i}^G(\phi_i)(Z))^{-ln_i} \in \mathbf{Z}_l(\mu_N)^*$$

which corresponds, by Lemma 3.2 and writing everything in additive notation, to

$$\sum_i n_i Ind_{H_i}^G(\phi_i^l - l\phi_i)_*(Z) = \sum_i n_i \cdot (\phi_i^l - l\phi_i)_*(Ind_{H_i}^G(Z)) \in K_1(\mathbf{Z}_l(\mu_N))$$

where

$$Ind_{H_i}^G : K_1(\mathbf{Z}_l[G]) \longrightarrow K_1(\mathbf{Z}_l[H_i]) \quad \text{and} \quad (\phi_i)_* : K_1(\mathbf{Z}_l[H_i]) \longrightarrow K_1(\mathbf{Z}_l(\mu_N))$$

are the canonical homomorphisms.

Therefore, for $\epsilon = 1, 2$ and $t \geq 0$, on $K_{2t+\epsilon}(\mathbf{Z}_l[G])$ we define

$$\rho_{2t+\epsilon,G}(V) : K_{2t+\epsilon}(\mathbf{Z}_l[G]) \longrightarrow K_{2t+\epsilon}(\mathbf{Z}_l(\mu_N))$$

by

$$\rho_{2t+\epsilon,G}(V)(Z) = \sum_i n_i \cdot (\phi_i^l - l^{t+1}\phi_i)_*(Ind_{H_i}^G(Z)) \in K_{2t+\epsilon}(\mathbf{Z}_l(\mu_N)).$$

This construction is natural with respect to the Galois action and yields a homomorphism of the form

$$\rho_{2t+\epsilon,G} : K_{2t+\epsilon}(\mathbf{Z}_l[G]) \longrightarrow \mathrm{Hom}_{\Omega_{\mathbf{Q}_l}}(R_{\overline{\mathbf{Q}}_l}(G), K_{2t+\epsilon}(\mathbf{Z}_l(\mu_N))).$$

Let $\mathbf{F}_{l^e}$ denote the residue field of the local ring $\mathbf{Z}_l(\mu_N)$. By [21] the reduction map

$$K_{2t+\epsilon}(\mathbf{Z}_l(\mu_N)) \longrightarrow K_{2t+\epsilon}(\mathbf{F}_{l^e})$$

yields an isomorphism on torsion of order prime to l

$$Tor(\mathbf{Q}/\mathbf{Z}[1/l], K_{2t+\epsilon}(\mathbf{Z}_l(\mu_N))) \stackrel{\cong}{\longrightarrow} K_{2t+\epsilon}(\mathbf{F}_{l^e})$$

where, by [13], $K_{2t+2}(\mathbf{F}_{l^e}) = 0$ and $K_{2t+1}(\mathbf{F}_{l^e}) \cong \mathbf{Z}/(l^{(t+1)e} - 1)$. In fact, [21] implies that there is a direct sum splitting of the form $K_{2t+\epsilon}(\mathbf{Z}_l(\mu_N)) \cong$

$$\frac{K_{2t+\epsilon}(\mathbf{Z}_l(\mu_N))}{Tor(\mathbf{Q}/\mathbf{Z}[1/l], K_{2t+\epsilon}(\mathbf{Z}_l(\mu_N)))} \oplus Tor(\mathbf{Q}/\mathbf{Z}[1/l], K_{2t+\epsilon}(\mathbf{Z}_l(\mu_N))).$$

Lemma 3.4.

In the situation of §3.3 $\rho_{2t+\epsilon,G}(V)(Z) \in K_{2t+\epsilon}(\mathbf{Z}_l(\mu_N))$ has a trivial component in the summand $Tor(\mathbf{Q}/\mathbf{Z}[1/l], K_{2t+\epsilon}(\mathbf{Z}_l(\mu_N)))$ for all V, Z.

Proof

We have to show that the element $\rho_{2t+\epsilon,G}(V)(Z) \in K_{2t+\epsilon}(\mathbf{Z}_l(\mu_N))$ maps to zero in $K_{2t+\epsilon}(\mathbf{F}_{l^e})$. Let $Frob_l$ denote the l-th power Frobenius automorphism on $\mathbf{F}_{l^e}$. From the formula

$$\rho_{2t+\epsilon,G}(V)(Z) = \sum_i n_i \cdot (\phi_i^l - l^{t+1}\phi_i)_*(Ind_{H_i}^G(Z))$$

we see that the image in $K_{2t+\epsilon}(\mathbf{F}_{l^e})$ of each term in the sum lies in the image of the endomorphism

$$Frob_{l_*} - l^{t+1} : K_{2t+\epsilon}(\mathbf{F}_{l^e}) \longrightarrow K_{2t+\epsilon}(\mathbf{F}_{l^e})$$

but, from the calculations of [13], this endomorphism is trivial. □

Remark 3.5. Regulators and congruences

Let L be an l-adic local field. An l-adic regulator is a natural homomorphism of the form $K_{2n+1}(L) \otimes \mathbf{Q}_l \longrightarrow L$. There are a number of such regulators to choose from and five of them are described in ([19] §5.1). Choosing such a regulator one may form the composite, denoted by $\hat{\rho}_{2n+1,G}$,

$$K_{2n+1}(\mathbf{Z}_l[G]) \stackrel{\rho_{2n+1,G}}{\longrightarrow} \mathrm{Hom}_{\Omega_{\mathbf{Q}_l}}(R_{\overline{\mathbf{Q}}_l}(G), K_{2n+1}(\mathbf{Z}_l(\mu_N))) \longrightarrow$$

$$\mathrm{Hom}_{\Omega_{\mathbf{Q}_l}}(R_{\overline{\mathbf{Q}}_l}(G), K_{2n+1}(\overline{\mathbf{Q}}_l)) \longrightarrow \mathrm{Hom}_{\Omega_{\mathbf{Q}_l}}(R_{\overline{\mathbf{Q}}_l}(G), \overline{\mathbf{Q}}_l) \cong K_1(\mathbf{Q}_l[G])$$

where the last isomorphism was proved in [12].

Let $\mathbf{Q}_l\{G\}$ denote the $\mathbf{Q}_l$-vector space whose basis consists of the conjugacy classes of elements of G ([18] Definition 4.5.4). By ([18] Proposition 4.5.14) there is an isomorphism of the form

$$\mathrm{Hom}_{\Omega_{\mathbf{Q}_l}}(R_{\overline{\mathbf{Q}}_l}(G), \overline{\mathbf{Q}}_l) \cong \mathbf{Q}_l\{G\}.$$

Let $\mathbf{Z}_l\{G\} \subset \mathbf{Q}_l\{G\}$ denote the free $\mathbf{Z}_l$-lattice whose basis consists of the conjugacy classes of elements of G.

By analogy with the case when $n = 0$ ([18] §4.3) one might ask whether there are higher-dimensional congruences of the following form. **Question:** Does $\hat{\rho}_{2n+1,G}(K_{2n+1}(\mathbf{Z}_l[G]))$ lie in $l^{n+1}\mathbf{Z}_l\{G\}$, at least when G is an l-group?

Example 3.6. Iwasawa theory – a profinite example

Let $\mathbf{Z}_l \cong \lim_{\overleftarrow{m}} \mathbf{Z}/l^m$ denote the l-adic integers. If G is replaced by the profinite group $\mathbf{Z}_l \times G$ with G finite but not necessarily abelian we may take a limit to obtain a homomorphism of the form

$$K_{2n+1}(\mathbf{Z}_l[[\mathbf{Z}_l \times G]]) \longrightarrow \mathbf{Q}_l[[T]]\{G\}.$$

In the arithmetic situation G is the Galois group of a number field extension. In this case, elements of $\mathbf{Q}_l[[T]]\{G\}$ appear in the work of Ritter and Weiss [15], Nguyen Quang Do [8] and Burns-Greither [4]. Particularly important are the Iwasawa power series arising from l-adic L-series by virtue of the Iwasawa Main Conjecture [27] (see also [20] Chapters 6 and 7).

Question: Does the Iwasawa power series lift to $K_{2n+1}(\mathbf{Z}_l[[\mathbf{Z}_l \times G]])$ in the arithmetical situation alluded to above?

4 Assembly maps

4.1. In this section l will be an odd prime and each spectrum $\underline{X}$ in the stable homotopy category will be l-adically completed, as in [6]. For example, $\underline{K(\mathbf{Z}_l[G])}$ will denote the l-adic completion of the algebraic K-theory spectrum of $\mathbf{Z}_l[G]$ and $\underline{bu}$ will denote l-adically completed connective topological unitary K-theory. We shall write $K_n(R; \mathbf{Z}_l)$ for the homotopy of the l-adically completed K-theory spectrum of the ring R, $\pi_n(\underline{K(R)})$. By Lemma 3.4 we have an induced homomorphism

$$\rho_{2t+\epsilon,G} : K_{2t+\epsilon}(\mathbf{Z}_l[G]; \mathbf{Z}_l) \longrightarrow \mathrm{Hom}_{\Omega_{\mathbf{Q}_l}}(R_{\overline{\mathbf{Q}}_l}(G), K_{2t+\epsilon}(\mathbf{Z}_l(\mu_N); \mathbf{Z}_l)).$$

Let BG_+ denote the disjoint union of the classifying space of G with a disjoint basepoint. It would be very interesting to evaluate the composite of $\rho_{2t+\epsilon,G}$ with the assembly map

$$\mu_{2t+\epsilon,G} : \pi_{2t+\epsilon}(\underline{K(\mathbf{Z}_l)} \wedge (BG_+)) \longrightarrow K_{2t+\epsilon}(\mathbf{Z}_l[G]; \mathbf{Z}_l).$$

From the Hesselholt-Madsen result mentioned in §2.1 we have a decomposition of l-adic spectra of the form

$$\underline{K(E)} \simeq \underline{F\Psi} \vee \Sigma\underline{F\Psi} \vee (\vee_{j=1}^{[E:\mathbf{Q}_l]} \Sigma\underline{bu})$$

where $\underline{F\Psi}$ is the spectrum associated to the summand $F\Psi^{g^{l^{a-1}d}}$ of the K-theory space for E in §2.1.

For the moment we shall content ourselves with studying the torsion-free part of $\rho_{2t+\epsilon,G} \cdot \mu_{2t+\epsilon,G}$. That is, we shall compose $\rho_{2t+\epsilon,G} \cdot \mu_{2t+\epsilon,G}$ with the projection

$$\mathrm{Hom}_{\Omega_{\mathbf{Q}_l}}(R_{\overline{\mathbf{Q}}_l}(G), K_{2t+\epsilon}(\mathbf{Z}_l(\mu_N); \mathbf{Z}_l))$$
$$\longrightarrow \mathrm{Hom}_{\Omega_{\mathbf{Q}_l}}(R_{\overline{\mathbf{Q}}_l}(G), \oplus_{j=1}^{[\mathbf{Q}_l(\mu_N):\mathbf{Q}_l]} \pi_{2t+\epsilon}(\Sigma \underline{bu})).$$

There is a Galois equivariant isomorphism of the form

$$\oplus_{j=1}^{[\mathbf{Q}_l(\mu_N):\mathbf{Q}_l]} \pi_{2t+\epsilon}(\Sigma \underline{bu}) \cong Ind_{\{1\}}^{\mathrm{Gal}(\mathbf{Q}_l(\mu_N)/\mathbf{Q}_l)}(\pi_{2t+\epsilon}(\Sigma \underline{bu}))$$

and therefore an isomorphism of the form

$$\mathrm{Hom}_{\Omega_{\mathbf{Q}_l}}(R_{\overline{\mathbf{Q}}_l}(G), \oplus_{j=1}^{[\mathbf{Q}_l(\mu_N):\mathbf{Q}_l]} \pi_{2t+\epsilon}(\Sigma \underline{bu})) \cong \mathrm{Hom}(R_{\overline{\mathbf{Q}}_l}(G), \pi_{2t+\epsilon}(\Sigma \underline{bu})).$$

Composing $\rho_{2t+\epsilon,G}$ with this projection yields a homomorphism, which is natural in G, of the form

$$\tilde{\rho}_{2t+\epsilon,G} : K_{2t+\epsilon}(\mathbf{Z}_l[G]; \mathbf{Z}_l) \longrightarrow \mathrm{Hom}_{\Omega_{\mathbf{Q}_l}}(R_{\overline{\mathbf{Q}}_l}(G), K_{2t+\epsilon}(\mathbf{Z}_l(\mu_N); \mathbf{Z}_l))$$
$$\longrightarrow \mathrm{Hom}(R_{\overline{\mathbf{Q}}_l}(G), \pi_{2t+\epsilon}(\Sigma \underline{bu})).$$

On the other hand we have a splitting of the form

$$\underline{K(\mathbf{Z}_l)} \wedge (BG_+) \simeq (\underline{F\psi} \vee \Sigma \underline{F\psi} \vee \Sigma \underline{bu}) \wedge (BG_+)$$

so that there is an isomorphism

$$\pi_{2t+\epsilon}(\underline{K(\mathbf{Z}_l)} \wedge (BG_+))$$
$$\cong \pi_{2t+\epsilon}(\underline{F\psi} \wedge (BG_+)) \oplus \pi_{2t+\epsilon}(\Sigma \underline{F\psi} \wedge (BG_+)) \oplus \pi_{2t+\epsilon}(\Sigma \underline{bu} \wedge (BG_+)).$$

Question: What is the composite of $\tilde{\rho}_{2t+\epsilon,G}$ with the restriction of the assembly map to the summand $\pi_{2t+\epsilon}(\Sigma \underline{bu} \wedge (BG_+))$

$$\pi_{2t+\epsilon}(\Sigma \underline{bu} \wedge (BG_+)) \stackrel{\mu_{2t+\epsilon,G}}{\longrightarrow} K_{2t+\epsilon}(\mathbf{Z}_l[G]; \mathbf{Z}_l)$$
$$\stackrel{\tilde{\rho}_{2t+\epsilon,G}}{\longrightarrow} \mathrm{Hom}(R_{\overline{\mathbf{Q}}_l}(G), \pi_{2t+\epsilon}(\Sigma \underline{bu}))?$$

There is another way to map between these groups. Recall that for each pointed space X we have the standard "universal coefficient" map ([1]; [16]) which takes the form

$$\pi_j(\underline{bu} \wedge X) = bu_j(X; \mathbf{Z}_l) \longrightarrow \mathrm{Hom}(bu^i(X), bu_{j-i}(*; \mathbf{Z}_l)).$$

When we set $X = (BG)_+$ and $i = 0$ we obtain

$$\pi_j(\underline{bu} \wedge (BG)_+) \longrightarrow \mathrm{Hom}(bu^0((BG)_+), \pi_j(\underline{bu})),$$

which is constructed from the product $bu \wedge \underline{bu} \longrightarrow \underline{bu}$. There is also a pairing $BU \wedge \underline{bu} \longrightarrow \underline{bu}$ which induces

$$\mathrm{adj} : \pi_j(\underline{bu} \wedge (BG)_+) \longrightarrow \mathrm{Hom}(KU^0((BG)_+), \pi_j(\underline{bu})).$$

Here "adj" stands for the adjoint of the slant product ([1] p.228). Furthermore, if we identify $R_{\overline{\mathbf{Q}}_l}(G)$ with the complex representation ring $R(G)$ of G then the associated vector bundle construction $R(G) \longrightarrow KU^0(BG_+)$ yields a homomorphism $R_{\overline{\mathbf{Q}}_l}(G) \longrightarrow KU^0(BG_+)$ which induces an isomorphism between $KU^0((BG)_+)$ and the completion of $R_{\overline{\mathbf{Q}}_l}(G)$ in the augmentation ideal topology [2]. In any case this homomorphism induces a map of the form

$$\mathrm{Hom}(KU^0((BG)_+), \pi_j(\underline{bu})) \longrightarrow \mathrm{Hom}(R_{\overline{\mathbf{Q}}_l}(G), \pi_j(\underline{bu})).$$

The following result relates the slant product to the torsion-free part of the regulator map when G is any finite group.

Theorem 4.2.

Let l be an odd prime and G a finite group. Then, in the notation of §4.1 when $t \geq 0$ and $\epsilon = 1, 2$, the composite

$$\pi_{2t+\epsilon}(\Sigma\underline{bu} \wedge (BG_+)) \stackrel{\mu_{2t+\epsilon,G}}{\longrightarrow} K_{2t+\epsilon}(\mathbf{Z}_l[G]; \mathbf{Z}_l) \stackrel{\tilde{\rho}_{2t+\epsilon,G}}{\longrightarrow} \mathrm{Hom}(R_{\overline{\mathbf{Q}}_l}(G), \pi_{2t+\epsilon}(\Sigma\underline{bu}))$$

is equal to the composite

$$\pi_{2t+\epsilon}(\Sigma\underline{bu} \wedge (BG)_+) \stackrel{\mathrm{adj:}}{\longrightarrow} Hom(KU^0((BG)_+), \pi_{2t+\epsilon}(\Sigma\underline{bu}))$$

$$\longrightarrow Hom(R_{\overline{\mathbf{Q}}_l}(G), \pi_{2t+\epsilon}(\Sigma\underline{bu})) \stackrel{(\psi^l - l^{s+1})^*}{\longrightarrow} Hom(R_{\overline{\mathbf{Q}}_l}(G), \pi_{2t+\epsilon}(\Sigma\underline{bu}))$$

where ψ^l denotes the l-th Adams operation on $R_{\overline{\mathbf{Q}}_l}(G)$.

Proof

Firstly, for a group G and a subgroup H of finite index, there is a commutative *projection formula* diagram [14] involving the transfer map in algebraic K-theory, the transfer map in stable homotopy and the assembly map.

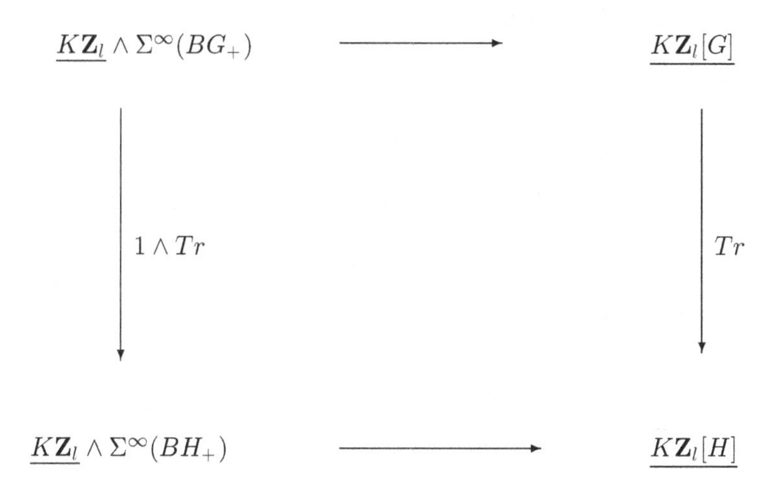

If E is a ring spectrum, another manifestation of the projection formula is a commutative diagram of the following form.

$$\begin{array}{ccc}
E^p(BH_+) \otimes E_q(BG_+) & \xrightarrow{1 \wedge Tr} & E^p(BH_+) \otimes E_q(BH_+) \\
\Big\downarrow {\scriptstyle Tr \wedge 1} & & \Big\downarrow {\scriptstyle (-\backslash -)} \\
E^p(BG_+) \otimes E_q(BG_+) & \xrightarrow{(-\backslash -)} & E^{p-q}(S^0)
\end{array}$$

where $(-\backslash -)$ is the slant product on E-theory defined in ([1] p.228). A similar diagram holds for the slant pairing induced by $BU \wedge \underline{bu} \longrightarrow \underline{bu}$.

Now I shall sketch the proof of the theorem. We may assume that $\epsilon = 1$ because both maps are zero where $\epsilon = 2$.

Suppose we have $\tilde{Z} \in \pi_{2s+1}(\underline{K\mathbf{Z}_l} \wedge (BG_+))$ and $\rho \in R_{\overline{\mathbf{Q}}_l}(G)$ with the image of $\tilde{Z}$ under the assembly map being $Z \in K_{2s+1}(\mathbf{Z}_p[G]; \mathbf{Z}_l)$. Then Z maps to the homomorphism sending ρ to

$$\rho'_*(Z) = \sum_i n_i \cdot (\phi_i^l - l^{s+1}\phi_i)_*(Ind_{H_i}^G(Z)) \in K_{2s+1}(\mathbf{Z}_l(\mu_N)).$$

With $H = H_i$ in the projection formula diagram

$$Ind_{H_i}^G(Z) = Tr_*(Z) = \mu_*(1 \wedge Tr)_*(\tilde{Z})$$

where μ denotes an assembly map. However

$$(\phi_i^l - l^{s+1}\phi_i)_* \mu_*(1 \wedge Tr)_*(\tilde{Z}) = (\phi_i^l - l^{s+1}\phi_i)\backslash(1 \wedge Tr)_*(\tilde{Z})$$
$$= Tr_*(\phi_i^l - l^{s+1}\phi_i)\backslash\tilde{Z},$$

by the slant product projection formula. Adding all the terms together and interpreting Tr_* as $Ind_{H_i}^G$ we obtain

$$\rho'_*(Z) = \sum_i n_i \cdot Ind_{H_i}^G(\phi_i^l - l^{s+1}\phi_i)\backslash\tilde{Z} = (\psi^l(\rho) - l^{s+1}\rho)\backslash\tilde{Z}.$$

The identification of the transfer $Tr_* : KU^0(BH_+) \longrightarrow KU^0(BG_+)$ with the completion of $Ind_H^G : R(H) \longrightarrow R(G)$ is a result of Mike Boardman (see [7]). Projecting to the $\underline{bu}$-summands completes the proof. □

Remark 4.3. At the moment I do not have enough expertise with the groups $K_{2t+\epsilon}(\mathbf{Z}_l(\mu_N); \mathbf{Z}_l)$ to examine the image of $\rho_{2t+\epsilon,G} \cdot \mu_{2t+\epsilon,G}$ in the other summands of $\mathrm{Hom}_{\Omega_{\mathbf{Q}_l}}(R_{\overline{\mathbf{Q}}_l}(G), K_{2t+\epsilon}(\mathbf{Z}_l(\mu_N); \mathbf{Z}_l))$. However, it would be much more interesting to know what is happening in the "torsion summands".

References

[1] J.F. Adams: *Stable Homotopy and Generalised Homology*; University of Chicago Press (1974).

[2] M.F. Atiyah: Characters and cohomology of finite groups; Pub. Math. no. 9 IHES (Paris) (1961) 23-64.

[3] R. Boltje: A canonical Brauer induction formula; Astérisque t.181-182 (1990) 31-59.

[4] D. Burns and C. Greither: Equivariant Weierstrass preparation and values of L-functions at negative integers; Documenta Math. (2003) Extra volume in honour of Kazuya Kato.

[5] D. Grayson: Higher algebraic K-theory II (after D.G. Quillen); Algebraic K-theory (1976) 217-240, Lecture Notes in Math. #551, Springer Verlag.

[6] L. Hesselholt and I. Madsen: On the K-theory of local fields; Annals of Math. (1) 158 (2003) 1-113.

[7] D.S. Kahn and S.B. Priddy: Applications of the transfer to stable homotopy; Bull. A.M.Soc. 78 (6) (1972) 981-987.

[8] T. Nguyen Quang Do: Quelques applications de la Conjecture Principale Équivariante; (handwritten notes for Journées Arithmétiques lecture in Lens; June 2002).

[9] R. Oliver: SK_1 for finite group rings II; Math. Scand. 47 (1980) 195-231.

[10] R. Oliver: Lower bounds for $K_2^{\mathrm{top}}(\hat{\mathbf{Z}}_p\pi)$ and $K_2(\mathbf{Z}\pi)$; J. Alg. (2) 94 (1985) 425-487.

[11] R. Oliver: K_2 of p-adic group-rings of abelian groups; Math. Zeit. (4) 195 (1987) 505-558.

[12] J. Queyrut: S-groupes des classes d'un ordre arithmétiques ; J.Alg. 76 (1982) 234-260.

[13] D.G. Quillen: On the cohomology and K-theory of the general linear groups over a finite field; Annals of Math. 96 (1972) 552-586.

[14] D.G. Quillen: High Algebraic K-theory I; Algebraic K-theory I (1973) 85-147, Lecture Notes in Math. # 341, Springer-Verlag.

[15] J. Ritter and A. Weiss: Some steps towards equivariant Iwasawa theory in the noncommutative situation; two lectures at the Fields Institute Conference on the Eqivariant Tamagawa Number Conjecture, McMaster University (September 2003).

[16] L. Smith: On the relation of connective K-theory to homology; Proc. Cambs. Philos. Soc. 68 (1970)

[17] V.P. Snaith: Explicit Brauer Induction; Inventiones Math. 94 (1988) 455-478.

[18] V.P. Snaith: *Explicit Brauer Induction (with applications to algebra and number theory)*; Cambridge studies in advanced mathematics #40, Cambridge University Press (1994).

[19] V.P. Snaith: Local fundamental classes derived from higher dimensional K-groups; Proc. Great Lakes K-theory Conf., Fields Institute Communications Series #16 (A.M.Soc. Publications) 285-324 (1997).

[20] V.P.Snaith: *Algebraic K-groups as Galois Modules*; Birkhäuser Progress in Mathematics series #206 (2002).

[21] A.A. Suslin: On the K-theory of local fields; J.P.A.Alg. (2-3) 34 (1984) 301-318.

[22] P. Symonds: A splitting principle for group representations; Comm. Math. Helv. 66 (1991) 169-184.

[23] M.J. Taylor: Locally free class-groups of groups of prime power order; J. Alg. 50 (1978) 463-487.

[24] M.J. Taylor: A logarithmic approach to class-groups of integral group-rings; J. Alg. 66 (1980) 321-353.

[25] M.J. Taylor: *Class Groups of Group Rings*, L.M.Soc. Lecture Notes Series #91, Cambridge University Press (1984).

[26] M.J. Taylor: On Fröhlich's conjecture for rings of integers of tame extensions; Inventiones Math. 63 (1981) 41-79.

[27] A. Wiles: The Iwasawa conjecture for totally real fields; Annals of Math. 131 (1990) 493-540.

On complete resolutions

Olympia Talelli

Abstract

This is a survey on complete resolutions for groups. We present criteria of existence of complete resolutions and examine relations between them. We also discuss the connection between complete resolutions for groups and free and proper actions of groups on homotopy spheres.

0 Introduction

A complete resolution of a finite group G is a doubly infinite acyclic complex $F = (F_i)_{i\in\mathbb{Z}}$ of projective $\mathbb{Z}G$-modules, together with a map $\varepsilon : F_0 \to \mathbb{Z}$ such that $(F_i)_{i\geq 0} \xrightarrow{\varepsilon} \mathbb{Z}$ is a resolution of G. Complete resolutions appeared first in the calculation of the Tate cohomology of finite groups. Farrell [6] constructed a cohomology theory for groups of finite virtual cohomological dimension, which generalizes the Tate cohomology theory of finite groups. For the calculation of this cohomology, complete resolutions were used, where here, a complete resolution of a group G is a doubly infinite acyclic complex $(F_i)_{i\in\mathbb{Z}}$ of projective $\mathbb{Z}G$-modules, which agrees with a projective resolution of G above a certain dimension. Ikenaga [9] generalized the Farrell–Tate cohomology to a class of groups which properly contains the groups of finite virtual cohomological dimension, namely, the class of groups which admit a complete resolution and every projective has finite injective dimension. It is worth noting that in [18] C.T.C Wall suggested generalizing Farrell theory to rings where projectives have finite injective dimension and injectives has finite projective dimension.

Tate cohomology has now been generalized to the class of all groups by Benson and Carlson [2], Mislin [12] and Vogel [8]. The generalized Tate cohomology of a group G, $\widehat{H}^i(G,-)$, $i \in \mathbb{Z}$, or complete cohomology of G, shares basic properties with ordinary cohomology but it also has some distinctive features:

(a) $\widehat{H}^i(G,P) = 0$ for every projective $\mathbb{Z}G$-module P and $i \in \mathbb{Z}$

Partly supported by Special Account for Research Grants (70/4/6411). The project is co-funded by the European Social Fund and National Resources EPEAK II–Pythagoras

(b) there is a canonical natural transformation

$$\tau : H^*(G,-) \longrightarrow \widehat{H}^*(G,-)$$

such that every natural transformation from ordinary cohomology to a cohomological functor which vanishes on projectives, factors uniquely through τ

(c) if there exists an $n \in \mathbb{Z}$ such that $H^i(G,P) = 0$ for all projective $\mathbb{Z}G$-modules P and all $i > n$ then $\tau : H^i(G,-) \cong \widehat{H}^i(G,-)$ for all $i > n$.

(d) For any group G, $cdG < \infty$ if and only if $\widehat{H}^0(G,\mathbb{Z}) = 0$.

The generalized Tate cohomology can not always be calculated via complete resolutions as they do not always exist.

However, there are many advantages in having the generalized Tate cohomology computed via complete resolutions. For example, in this case the Eckmann–Shapiro Lemma holds for the generalized Tate cohomology, spectral sequences can be constructed analogous to the Lyndon–Hochschild–Serre spectral sequence in ordinary cohomology, cup product is constructed which is compatible with that in ordinary cohomology.
This article is a survey on complete resolutions.
In §1 we recall the definitions of complete cohomology and complete resolutions and discuss their basic properties.
In §2 we present criteria of existence and methods of construction of complete resolutions and consider relations between the various criteria.
In §3 we discuss the state of a Conjecture, in which having the complete cohomology calculated via complete resolutions plays an important role. The Conjecture [15] states that periodicity in the cohomology of a group after some steps characterizes those groups which act freely and properly on a finite dimensional homotopy sphere.

1 Complete cohomology and Complete resolutions

In [12] generalized Tate cohomology or complete cohomology is defined for any group G, specializing to the ones defined by Farrell and Ikenaga, when the latter are defined. Let us recall briefly the definition.

A sequence of additive functors $T^* = \{T^n \mid n \in \mathbb{Z}\}$ is a $(-\infty,+\infty)$-cohomological functor from $\mathbb{Z}G$-modules to abelian groups if there is a natural long exact sequence

$$\cdots \longrightarrow T^n B' \longrightarrow T^n B \longrightarrow T^n B'' \longrightarrow T^{n+1} B' \longrightarrow \cdots$$

associated to any short exact sequence $0 \to B' \to B \to B'' \to 0$ of $\mathbb{Z}G$-modules.

Definition.

(i) A $(-\infty, +\infty)$-cohomological functor T^* is called P-complete if $T^n(P) = 0$ for every projective $\mathbb{Z}G$-module P and for every $n \in \mathbb{Z}$.

(ii) Let T^* be a cohomological functor from $\mathbb{Z}G$-modules to abelian groups. A P-completion of T^* consists of a cohomological functor $\widehat{T}^*$ from $\mathbb{Z}G$-modules to abelian groups, together with a morphism $T^* \to \widehat{T}^*$ of $(-\infty, +\infty)$-cohomological functors such that the following conditions hold:

(1) $\widehat{T}^*$ is P-complete

(2) If V^* is a P-complete cohomological functor then any morphism $T^* \to V^*$ of $(-\infty, +\infty)$ cohomological functors factors uniquely through the given morphism $T^* \to \widehat{T}^*$.

It is clear from standard arguments that the P-completion of T^*, if it exists, is uniquely determined up to natural isomorphism.
It was shown in [12] that

$$\widehat{H}^n(G, -) = \varinjlim_{j \geq 0} S^{-j} H^{n+j}(G, -), \qquad n \in \mathbb{Z}$$

with $S^{-j}H^{n+j}(G, -)$ denoting the jth left satellite of $H^{n+j}(G, -)$, is the P-completion of $H^n(G, -)$, $n \in \mathbb{Z}$. Note that $H^n(G, -)$ is regarded as a $(-\infty, +\infty)$-cohomological functor with the convention that $H^n(G, -) = 0$ for $n < 0$.

Different, but equivalent definitions of complete cohomology can be found in [2] and [8].

A striking property of the complete cohomology is the following [11]:

$$\widehat{H}^0(G, \mathbb{Z}) = 0 \text{ if and only if } cdG < \infty.$$

In general, $\widehat{H}^0(G, \mathbb{Z})$ is very hard to compute for groups of infinite cohomological dimension. Some calculations in this direction were done in [5].

Sometimes the complete cohomology can be calculated using complete resolutions. Let us recall the definition.

Definition. A complete resolution for a group G, $(\mathcal{F}, \mathcal{P}, n)$, consists of an acyclic complex $\mathcal{F} = \{(F_i, \partial_i) \mid i \in \mathbb{Z}\}$ of projective modules and a projective resolution $\mathcal{P} = \{(P_i, d_i) \mid i \geq 0\}$ of G such that $\mathcal{F}$ and $\mathcal{P}$ coincide in sufficiently high dimensions

$$\begin{array}{ccl} & & F_{n-1} \longrightarrow \cdots \longrightarrow F_0 \longrightarrow F_{-1} \to \cdots \\ & \nearrow & \\ \cdots \longrightarrow F_{n+1} \longrightarrow F_n & & \\ & \searrow & \\ & & P_{n-1} \longrightarrow \cdots \longrightarrow P_0 \longrightarrow \mathbb{Z} \longrightarrow 0 \end{array}$$

The number n is called the coincidence index of the complete resolution.

A map of complete resolutions $(\mathcal{F}, \mathcal{P}, n) \to (\mathcal{F}', \mathcal{P}', m)$ is a pair of chain maps $f : \mathcal{F} \to \mathcal{F}'$, $g : \mathcal{P} \to \mathcal{P}'$ such that $f_i = g_i$ for $i \geq \max(m, n)$.
Let silp$G$ denote the supremum of the injective lengths of the projective $\mathbb{Z}G$-modules [7]. The following proposition, which was shown in [9], says that if G admits a complete resolution and silp$G < \infty$ then any two complete resolutions of G are homotopy equivalent.

Proposition 1.1. [9] *Let G be a group which admits complete resolutions and assume that* silp$G < \infty$. *If* $(\mathcal{F}, \mathcal{P}, n), (\mathcal{F}', \mathcal{P}', m)$ *are complete resolutions of G then there are homotopy equivalences* $f : \mathcal{F} \simeq \mathcal{F}'$, $f' : \mathcal{P} \simeq \mathcal{P}'$ *which agree in sufficiently high dimensions.*

It follows now from [12, Lemma 2.4] that if G admits a complete resolution and silp$G < \infty$ then the complete cohomology of G is calculated via complete resolutions, i.e.

$$\widehat{H}^i(G, M) = H^i(Hom_{\mathbb{Z}G}(\mathcal{F}, M))$$

for any $\mathbb{Z}G$-module M and all $i \in \mathbb{Z}$.
Ikenaga didn't use the notion of silp$G$ but that of generalized cohomological dimension of G, $\underline{cd}G = \sup\{k : \mathrm{Ext}^k_{\mathbb{Z}G}(M, F) \neq 0,\ M\ \mathbb{Z}\text{-free},\ F\ \mathbb{Z}G\text{-free}\}$. Note however that $\underline{cd}G \leq$ silp$G \leq \underline{cd}G + 1$. He showed that $\underline{cd}G < \infty$ is equivalent to a certain extension condition; namely:

Proposition 1.2. [9] silp$G \leq m$ *is equivalent to the following extension condition:*
For every exact sequence

$$0 \longrightarrow \ker \partial_m \longrightarrow P_m \xrightarrow{\partial_m} P_{m-1} \longrightarrow \cdots \longrightarrow P_0 \longrightarrow M \longrightarrow 0$$

with P_i *projective* $\mathbb{Z}G$*-modules for* $0 \leq i \leq m$, *any map* $\ker \partial_m \to P$, P *projective, extends to a map* $P_m \to P$.

Moreover, if G admits a complete resolution and silp$G < \infty$ *then G admits a complete resolution of coincidence index* $\leq$ silp$G$.

The following result shows that not every group admits a complete resolution, so in particular the complete cohomology is not always calculated via complete resolutions.

Proposition 1.3. [17] *If a group G admits a complete resolution of coincidence index n, then*

(i) $H^i(G, P) \neq 0$ *for some projective* $\mathbb{Z}G$*-module P and some* $i \leq n$

(ii) *the finitistic dimension of* $\mathbb{Z}G$ *is finite and fin.dim of* $\mathbb{Z}G \leq n + 1$.

Note that $\mathbb{Z}G$ has finite finitistic dimension means that there is a bound on the projective dimensions of the modules of finite projective dimension.

Corollary 1.4. *If a group G contains a free abelian subgroup of infinite rank, then G does not admit a complete resolution.*

Proof. It follows from Prop. 1.3 that a free abelian group $\mathcal{A}$ of infinite rank does not admit a complete resolution since $H^i(\mathcal{A}, P) = 0$ for any projective $\mathbb{Z}\mathcal{A}$-module P and any i. The result now follows, since admitting a complete resolution is a subgroup closed property. □

1.5 Examples of complete resolutions

(a) G finite.

Since G is finite we have the following $\mathbb{Z}$-split $\mathbb{Z}G$-monomorphism

$$\begin{aligned} 0 \longrightarrow \mathbb{Z} &\xrightarrow{\rho} \mathbb{Z}G \\ 1 &\longrightarrow \sum_{g \in G} g \end{aligned} \tag{1}$$

If C is a $\mathbb{Z}$-free $\mathbb{Z}G$-module then we can embed C in a projective $\mathbb{Z}G$-module by tensoring (1) with C over $\mathbb{Z}$. Indeed $0 \to C \to \mathbb{Z}G \underset{\mathbb{Z}}{\otimes} C$ is $\mathbb{Z}G$-exact, where $\mathbb{Z}G \underset{\mathbb{Z}}{\otimes} C$ is a $\mathbb{Z}G$-module via the diagonal action. As $\overset{\searrow}{\mathbb{Z}G} \underset{\mathbb{Z}}{\otimes} \overset{\searrow}{C} \simeq \overset{\searrow}{\mathbb{Z}G} \underset{\mathbb{Z}}{\otimes} C$ it follows that $\overset{\searrow}{\mathbb{Z}G} \underset{\mathbb{Z}}{\otimes} \overset{\searrow}{C}$ is a projective $\mathbb{Z}G$-module. If $R_{-1} = coker \rho$ then R_{-1} is $\mathbb{Z}$-free and by using repeatedly the above embedding we obtain the following $\mathbb{Z}G$-exact sequence

$$0 \longrightarrow \mathbb{Z} \xrightarrow{\rho} \mathbb{Z}G \longrightarrow \mathbb{Z}G \underset{\mathbb{Z}}{\otimes} R_{-1} \longrightarrow \mathbb{Z}G \underset{\mathbb{Z}}{\otimes} R_{-1} \underset{\mathbb{Z}}{\otimes} R_{-1} \longrightarrow \cdots$$

(with the maps factoring through R_{-1} and $R_{-1} \underset{\mathbb{Z}}{\otimes} R_{-1}$ respectively)

with $\mathbb{Z}G$, $\mathbb{Z}G \underset{\mathbb{Z}}{\otimes} R_{-1}$, $\mathbb{Z}G \underset{\mathbb{Z}}{\otimes} R_{-1} \underset{\mathbb{Z}}{\otimes} R_{-1}$, ... projective $\mathbb{Z}G$-modules.

If now $\cdots \to P_n \to P_{n-1} \to \cdots \to P_0 \xrightarrow{\varepsilon} \mathbb{Z} \to 0$ is a projective resolution for G then

$$\cdots \longrightarrow P_n \longrightarrow P_{n-1} \cdots \longrightarrow P_0 \underset{\rho\varepsilon}{\longrightarrow} \mathbb{Z}G \longrightarrow \mathbb{Z}G \underset{\mathbb{Z}}{\otimes} R_{-1} \longrightarrow \mathbb{Z}G \underset{\mathbb{Z}}{\otimes} R_{-1} \underset{\mathbb{Z}}{\otimes} R_{-1} \longrightarrow \cdots$$

is a complete resolution of G of coincidence index 0.

(b) $vcdG < \infty$.

Let H be a torsion free subgroup of G of finite index with $cdH = n$. Consider $G = \bigcup_{t \in T} tH$ a coset decomposition. As $|T| < \infty$ we have the following $\mathbb{Z}$-split $\mathbb{Z}G$-monomorphism

$$\begin{aligned} 0 \longrightarrow \mathbb{Z} &\xrightarrow{\rho} \mathbb{Z}(G/H) \\ 1 &\longrightarrow \sum_{t \in T} tH \end{aligned} \tag{2}$$

If C is a $\mathbb{Z}G$-module which is projective as a $\mathbb{Z}H$-module then we can embed C in a projective $\mathbb{Z}G$-module by tensoring (2) with C over $\mathbb{Z}$.
Indeed $0 \to C \to \mathbb{Z}(G/H) \underset{\mathbb{Z}}{\otimes} C$ is $\mathbb{Z}G$-exact, where $\mathbb{Z}(G/H) \underset{\mathbb{Z}}{\otimes} C$ is a $\mathbb{Z}G$-module via the diagonal action. As $\mathbb{Z}(\overset{\searrow}{G}/H) \underset{\mathbb{Z}}{\otimes} \overset{\searrow}{C} \simeq \mathbb{Z}\overset{\searrow}{G} \underset{\mathbb{Z}H}{\otimes} C$ it follows that $\mathbb{Z}(G/\overset{\searrow}{H}) \underset{\mathbb{Z}}{\otimes} \overset{\searrow}{C}$ is a projective $\mathbb{Z}G$-module.
Now consider $\cdots \to P_i \xrightarrow{\partial_i} P_{i-1} \to \cdots \to P_0 \to \mathbb{Z} \to 0$ a projective resolution of G. Since $cdH = n$ we have that $R_n = \mathrm{im}\partial_n$ is a projective $\mathbb{Z}H$-module, hence if Γ is any $\mathbb{Z}$-free $\mathbb{Z}G$-module then $\overset{\searrow}{R_n} \otimes \overset{\searrow}{\Gamma}$ is a $\mathbb{Z}G$-module which is projective as a $\mathbb{Z}H$-module. If now $K = coker\rho$ then by using repeatedly the above embedding we obtain the following $\mathbb{Z}G$-exact sequence

$$0 \longrightarrow R_n \longrightarrow \mathbb{Z}(G/H) \underset{\mathbb{Z}}{\otimes} R_n \underset{K \underset{\mathbb{Z}}{\otimes} R_n}{\searrow\!\nearrow} \mathbb{Z}(G/H) \underset{\mathbb{Z}}{\otimes} K \underset{\mathbb{Z}}{\otimes} R_n \underset{K \underset{\mathbb{Z}}{\otimes} K \underset{\mathbb{Z}}{\otimes} R_n}{\searrow\!\nearrow} \mathbb{Z}G \otimes K \underset{\mathbb{Z}}{\otimes} K \underset{\mathbb{Z}}{\otimes} R_n$$

with $\mathbb{Z}(G/H) \underset{\mathbb{Z}}{\otimes} R_n$, $\mathbb{Z}(G/H) \underset{\mathbb{Z}}{\otimes} K \underset{\mathbb{Z}}{\otimes} R_n$, $\mathbb{Z}(G/H) \underset{\mathbb{Z}}{\otimes} K \underset{\mathbb{Z}}{\otimes} K \underset{\mathbb{Z}}{\otimes} R_n$,... projective $\mathbb{Z}G$-modules.
It follows now that

$$\cdots \longrightarrow P_n \begin{cases} \nearrow \mathbb{Z}(G/H) \underset{\mathbb{Z}}{\otimes} R_n \longrightarrow \mathbb{Z}(G/H) \underset{\mathbb{Z}}{\otimes} K \underset{\mathbb{Z}}{\otimes} R_n \longrightarrow \cdots \\ \searrow P_{n-1} \longrightarrow \cdots \longrightarrow P_0 \longrightarrow \mathbb{Z} \longrightarrow 0 \end{cases}$$

is a complete resolution of G of coincidence index n.

(c) Groups with periodic cohomology after k-steps.

If a group G has periodic cohomology with period q after k-steps, i.e. if the functors $H^i(G,-)$ and $H^{i+q}(G,-)$ are naturally equivalent for all $i > k$, then G has a projective resolution which is periodic after k-steps [14]. Hence there is a $\mathbb{Z}G$-exact sequence of the form

$$0 \longrightarrow R_{k+q} \longrightarrow P_{k+q-1} \longrightarrow \cdots \longrightarrow P_k \underset{R_k}{\searrow\!\nearrow} P_{k-1} \longrightarrow \cdots \longrightarrow P_0 \longrightarrow \mathbb{Z} \longrightarrow 0$$

with P_i projective $\mathbb{Z}G$-modules for all i and $R_{k+q} \simeq R_k$.
Clearly by splicing together copies of

$$0 \longrightarrow R_k \longrightarrow P_{k+q-1} \longrightarrow \cdots \longrightarrow P_k \longrightarrow R_k \longrightarrow 0$$

we obtain a complete resolution of G of coincidence index k.

It was shown in [17] that:

Proposition 1.6. *The complete cohomology of a group G can be calculated via complete resolutions if and only if G admits a complete resolution and* $\text{silp}G < \infty$.

If the complete cohomology of a group G can be calculated via complete resolutions, then the complete cohomology has many properties analogous to those of the Farrell–Tate cohomology theory, where the role of $vcdG$ is played by silp$G$, namely [9]:

(1) the natural map $H^i(G,-) \to \widehat{H}^i(G,-)$ is an isomorphism for $i > \text{silp}G$

(2) Eckmann–Shapiro's Lemma holds: if $H \leq G$ and M is a $\mathbb{Z}H$-module then
$$\widehat{H}^i(H,M) = \widehat{H}^i(G, Hom_{\mathbb{Z}H}(\mathbb{Z}G, M)), \quad i \in \mathbb{Z}$$

(3) $\widehat{H}^i(G,M) = 0$ if M is an injective or a projective $\mathbb{Z}G$-module, for all $i \in \mathbb{Z}$. Consequently, the functors $\widehat{H}^i(G,-)$ are effaceable and coeffaceable and thus we have dimension shifting in both directions.

(4) cup product is constructed in $\widehat{H}^*$ with the usual properties, which is compatible with that in ordinary cohomology, i.e. the diagram
$$\begin{array}{ccc} H^p(G,M) \otimes H^q(G,N) & \xrightarrow{\cup} & H^{p+q}(G, M \otimes N) \\ \downarrow & & \downarrow \\ \widehat{H}^p(G,M) \otimes \widehat{H}^q(G,N) & \xrightarrow{\cup} & \widehat{H}^{p+q}(G, M \otimes N) \end{array}$$
commutes for all p, q, and all $\mathbb{Z}G$-modules M, N.

2 silp$G$, spli$G$, fin. dim of $\mathbb{Z}G$ and Complete resolutions

It follows from the proof of [9, Thm. 2] that:

Proposition 2.1. *If a group G admits a $\mathbb{Z}G$-exacts sequence*
$$0 \longrightarrow \bigoplus_{i_n \in I_n} \mathbb{Z}(G/G_{i_n}) \longrightarrow \cdots \longrightarrow \bigoplus_{i_0 \in I_0} (G/G_{i_0}) \longrightarrow \mathbb{Z} \longrightarrow 0$$
such that G_{i_j} admits a complete resolution for all i_j and there is an integer m with $\text{silp}G_{i_j} \leq m$ *for all i_j then G admits a complete resolution and* $\text{silp}G < \infty$.

In particular, if G admits a finite dimensional contractible $G-CW$-complex X with finite cell stabilizers then G admits a complete resolution and $\text{silp}G < \infty$, since the augmented cellular chain complex of X satisfies the hypotheses of Prop. 2.1.

Let spli$G$ denote the supremum of the projective lengths of the injective $\mathbb{Z}G$-modules [7]. It was shown in [7] that:

Proposition 2.2. *If* spli$G < \infty$ *then G admits a complete resolution and* silp$G < \infty$.

It is implicit in Cornick's and Kropholler's results in [4] that:

Proposition 2.3. *If there is a $\mathbb{Z}$-split $\mathbb{Z}G$-exact sequence $0 \to \mathbb{Z} \to A$ with A $\mathbb{Z}$-free and $\mathrm{pd}_{\mathbb{Z}G} A < \infty$ then G admits a complete resolution and the complete cohomology of G can be calculated via any complete resolution.*

It turns out that the conditions in Proposition 2.2 and 2.3 are equivalent, namely:

Theorem 2.4. [17] *The following conditions are equivalent for a group G*

(1) *G admits a complete resolution and* silp$G < \infty$

(2) *There is a $\mathbb{Z}$-split $\mathbb{Z}G$-exact sequence $0 \to \mathbb{Z} \to A$ with A $\mathbb{Z}$-free and $\mathrm{pd}_{\mathbb{Z}G} A < \infty$*

(3) spli$G < \infty$.

Proof. (1)⇒(2) Consider a complete resolution for the group G

$$\longrightarrow F_{n+1} \xrightarrow{\partial_{n+1}} F_n \begin{cases} \nearrow & F_{n-1} \xrightarrow{\partial_{n-1}} \cdots \longrightarrow F_0 \longrightarrow F_{-1} \to \cdots \\ \searrow & P_{n-1} \longrightarrow \cdots \longrightarrow P_0 \longrightarrow \mathbb{Z} \longrightarrow 0 \end{cases} .$$

Let $R_n = \mathrm{im}\partial_n$, $n \in \mathbb{Z}$. If $\lambda : R_n \to P$ is a $\mathbb{Z}G$-homomorphism with P a projective $\mathbb{Z}G$-module, then since silp $G < \infty$ there is a positive integer m_0 and an integer m so that λ represents the zero element of $\mathrm{Ext}^{m_0}_{\mathbb{Z}G}(R_m, P)$. Hence we obtain the following commutative diagram

$$\begin{array}{ccccccccccccc} \longrightarrow F_{n+1} & \xrightarrow{\partial_{n+1}} & F_n & \xrightarrow{\partial_n} & F_{n-1} \longrightarrow \cdots \longrightarrow & F_1 & \xrightarrow{\partial_1} & F_0 & \xrightarrow{\partial_0} & F_{-1} & \xrightarrow{\partial_{-1}} \\ & & & & & & & \searrow R_0 \nearrow & & & \\ \| & & \| & & \downarrow f_{n-1} & \downarrow f_1 & & \downarrow f_0 & \downarrow f & & \\ \longrightarrow F_{n+1} & \xrightarrow[d_{n+1}]{} & F_n & \xrightarrow[d_n]{} & P_{n-1} \longrightarrow \cdots \longrightarrow & P_1 & \xrightarrow[d_1]{} & P_0 & \to & \mathbb{Z} \longrightarrow 0 & \end{array}$$

where $R_0 = \mathrm{im}\partial_0$.
Clearly $[f] \in \mathrm{Ext}^1_{\mathbb{Z}G}(R_{-1}, \mathbb{Z})$ and Yoneda product with $[f]$ induces an isomorphism: $\mathrm{Ext}^i_{\mathbb{Z}G}(\mathbb{Z}, -) \to \mathrm{Ext}^{i+1}_{\mathbb{Z}G}(R_{-1}, -)$ for all $i > n$.

That implies ([14], [18]) that $[f]$ is represented by an extension $0 \to \mathbb{Z} \to A \to R_{-1} \to 0$ with $pd_{\mathbb{Z}G} A < \infty$. The result now follows.
$(2) \Rightarrow (3)$
Let $0 \to \mathbb{Z} \to A \to B \to 0$ be a $\mathbb{Z}$-split, $\mathbb{Z}G$-exact sequence with A $\mathbb{Z}$-free and $pd_{\mathbb{Z}G} A < \infty$. If I is an injective $\mathbb{Z}G$-module, then $0 \to I \to I \otimes A \to I \otimes B \to 0$ is an exact sequence of $\mathbb{Z}G$-modules and to show that $pd_{\mathbb{Z}G} I < \infty$ it is enough to show that $pd_{\mathbb{Z}G} A \otimes I < \infty$. Now consider a $\mathbb{Z}G$-exact sequence $0 \to K \to P \to I \to 0$ with P a projective $\mathbb{Z}G$-module. Since A is $\mathbb{Z}$-free the sequence $0 \to K \otimes A \to P \otimes A \to I \otimes A \to 0$ is exact. Since $pd_{\mathbb{Z}G} A < \infty$ and K is $\mathbb{Z}$-free it follows that $pd_{\mathbb{Z}G} K \otimes A < \infty$. Hence $pd_{\mathbb{Z}G} I \otimes A < \infty$ and thus $pd_{\mathbb{Z}G} I < \infty$. By standard arguments, it follows that if every injective $\mathbb{Z}G$-module has finite projective dimension then spli $G < \infty$.
$(3) \Rightarrow (1)$ ([7])
We shall first show that spli$G < \infty \Rightarrow$ silp$G < \infty$.
Let P be a projective $\mathbb{Z}G$-module. The short exact $\mathbb{Z}$-sequence $0 \to \mathbb{Z} \to Q \to Q/\mathbb{Z} \to 0$ gives rise to the short exact $\mathbb{Z}G$-sequence $0 \to P \to P \underset{\mathbb{Z}}{\otimes} Q \to P \underset{\mathbb{Z}}{\otimes} Q/\mathbb{Z} \to 0$. So to show that P has finite injective dimension is equivalent to showing that $P \underset{\mathbb{Z}}{\otimes} D$ has finite injective dimension where D is a divisible abelian group. Let us denote $P \underset{\mathbb{Z}}{\otimes} D$ with $\tilde{P}$.
Consider the $\mathbb{Z}$-split, $\mathbb{Z}G$-exact sequence $0 \to IG \to \mathbb{Z} \xrightarrow{\varepsilon} \mathbb{Z} \to 0$ where ε is the augementation map and IG the augmentation ideal. This gives rise to the following $\mathbb{Z}$-split, $\mathbb{Z}G$-exact sequence sequence (diagonal action)

$$0 \longrightarrow \mathbb{Z} \longrightarrow Hom_{\mathbb{Z}}(\mathbb{Z}G, \mathbb{Z}) \longrightarrow Hom_{\mathbb{Z}}(IG, \mathbb{Z}) \longrightarrow 0$$

which, in its turn, gives rise to the following $\mathbb{Z}$-split $\mathbb{Z}G$-exact sequence

$$0 \longrightarrow Hom_{\mathbb{Z}}(Hom_{\mathbb{Z}}(IG, \mathbb{Z}), \tilde{P}) \longrightarrow Hom_{\mathbb{Z}}(Hom_{\mathbb{Z}}(\mathbb{Z}G, \mathbb{Z}), \tilde{P}) \longrightarrow \tilde{P} \longrightarrow 0.$$

Since $\tilde{P}$ is a direct summand of an induced module, it is relative projective hence $\tilde{P}$ is a direct summand of $Hom_{\mathbb{Z}}(Hom_{\mathbb{Z}}(\mathbb{Z}G, \mathbb{Z}), \tilde{P})$. Thus to show that $\tilde{P}$ has finite injective dimension it is enough to show that $Hom_{\mathbb{Z}}(\mathbb{Z}G, \mathbb{Z})$ has finite projective dimension. For the last conclusion we use the well known result that if $pd_{\mathbb{Z}G} B < \infty$ and Γ is a $\mathbb{Z}$-injective $\mathbb{Z}G$-module then $id_{\mathbb{Z}G} Hom_{\mathbb{Z}}(\overset{\searrow}{B}, \overset{\searrow}{\Gamma}) < \infty$. So it remains to show that $pd_{\mathbb{Z}G} Hom_{\mathbb{Z}}(\mathbb{Z}G, \mathbb{Z})$ is finite. Indeed from $0 \to \mathbb{Z} \to \mathbb{Q} \to \mathbb{Q}/\mathbb{Z} \to 0$ we obtain the short exact $\mathbb{Z}G$-sequence

$$0 \longrightarrow Hom_{\mathbb{Z}}(\mathbb{Z}G, \mathbb{Z}) \longrightarrow Hom_{\mathbb{Z}}(\mathbb{Z}G, Q) \longrightarrow Hom_{\mathbb{Z}}(\mathbb{Z}G, \mathbb{Q}/\mathbb{Z}) \longrightarrow 0$$

and since $Hom_{\mathbb{Z}}(\mathbb{Z}G, \mathbb{Q})$ and $Hom_{\mathbb{Z}}(\mathbb{Z}G, \mathbb{Q}/\mathbb{Z})$ are injective $\mathbb{Z}G$-modules and spli$G < \infty$ it follows that $pd_{\mathbb{Z}G} Hom_{\mathbb{Z}}(\mathbb{Z}G, \mathbb{Z}) < \infty$. Thus we proved that any

projective has finite injective dimension. By standard arguments now follows that silp$G < \infty$.
It remains to show that spli$G < \infty$ implies the existence of a complete resolution of G.
Consider an injective resolution of G

$$0 \longrightarrow \mathbb{Z} \longrightarrow I_0 \longrightarrow I_1 \longrightarrow \cdots \longrightarrow I_m \longrightarrow I_{m+1} \longrightarrow \cdots .$$

Then there are projective resolutions $(\mathcal{P}, \partial) \to \mathbb{Z}$, $(\mathcal{P}^i, \partial^i) \longrightarrow I_i$, $i = 0, 1, 2, \ldots$ such that the following diagram is commutative

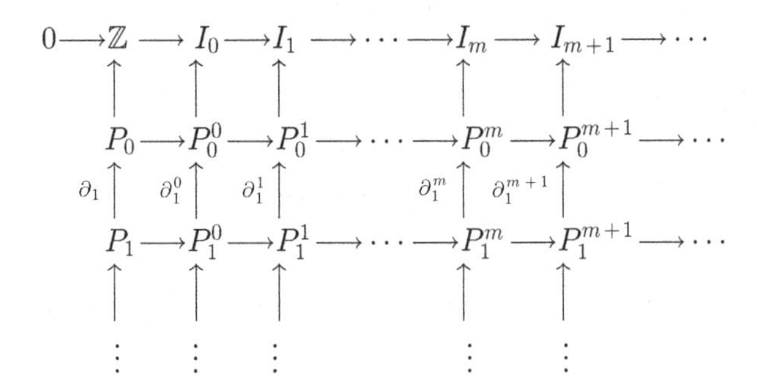

If spli$G = n$ then im$\partial_n^i = R_n^i$ is a projective $\mathbb{Z}G$-module for all i. Hence there is the following $\mathbb{Z}G$-exact sequence

$$0 \longrightarrow R_n \longrightarrow R_n^0 \longrightarrow R_n^1 \longrightarrow R_n^2 \longrightarrow \cdots$$

where $R_n = \mathrm{im}\partial_n$ and R_n^i are projective $\mathbb{Z}G$-modules. It is clear now that

$$\longrightarrow P_{n+1} \longrightarrow P_n \begin{array}{l} \nearrow R_n^0 \longrightarrow R_n^1 \longrightarrow R_n^2 \longrightarrow \cdots \\ \searrow P_{n-1} \longrightarrow \cdots \longrightarrow P_0 \longrightarrow \mathbb{Z} \longrightarrow 0 \end{array} .$$

is a complete resolution of G of coincidence index n. □

Remarks 2.5.

a) It follows from Prop. 1.6 and Thm. 2.4 that spli$G < \infty$ if and only if the complete cohomology of G is calculated via complete resolutions

b) If there is a $\mathbb{Z}$-split $\mathbb{Z}G$-exact sequence $0 \to \mathbb{Z} \xrightarrow{\rho} A$ with A $\mathbb{Z}$-free and $pd_{\mathbb{Z}G} A < \infty$ then one can construct a complete resolution for G as follows [4]:

Assume that $pd_{\mathbb{Z}G} A = n$ and let

$$\longrightarrow P_i \xrightarrow{\partial_i} P_{i-1} \longrightarrow \cdots \longrightarrow P_0 \longrightarrow \mathbb{Z} \longrightarrow 0$$

be a projective resolution of $\mathbb{Z}$. Then

$$\cdots \longrightarrow P_i \underset{\mathbb{Z}}{\otimes} A \xrightarrow{\partial_i \otimes 1_A} P_{i-1} \underset{\mathbb{Z}}{\otimes} A \longrightarrow \cdots \longrightarrow P_0 \underset{\mathbb{Z}}{\otimes} A \longrightarrow A \longrightarrow 0$$

is a projective resolution for A, hence if $R_n = \mathrm{im}\partial_n$ then $R_n \underset{\mathbb{Z}}{\otimes} A$ is a projective $\mathbb{Z}G$-module.

If follows now from $0 \to \mathbb{Z} \xrightarrow{\rho} A$ that any module of the form $R_n \underset{\mathbb{Z}}{\otimes} B$, where B is a $\mathbb{Z}$-free $\mathbb{Z}G$-module, can be embedded in a projective module. Indeed, if $K = coker\rho$ then $0 \to R_n \underset{\mathbb{Z}}{\otimes} B \to R_n \underset{\mathbb{Z}}{\otimes} B \underset{\mathbb{Z}}{\otimes} A \to R_n \underset{\mathbb{Z}}{\otimes} B \underset{\mathbb{Z}}{\otimes} K \to 0$ is a $\mathbb{Z}G$-exact sequence with $R_n \underset{\mathbb{Z}}{\otimes} B \underset{\mathbb{Z}}{\otimes} A$ a projective $\mathbb{Z}G$-module. By using now repeatedly this embedding we obtain the following $\mathbb{Z}G$-exact sequence

$$0 \longrightarrow R_n \longrightarrow R_n \underset{\mathbb{Z}}{\otimes} A \longrightarrow R_n \underset{\mathbb{Z}}{\otimes} K \underset{\mathbb{Z}}{\otimes} A \longrightarrow R_n \underset{\mathbb{Z}}{\otimes} K \underset{\mathbb{Z}}{\otimes} K \underset{\mathbb{Z}}{\otimes} A \longrightarrow \cdots$$

(the first map between tensor terms factoring through $R_n \underset{\mathbb{Z}}{\otimes} K$, the second through $R_n \underset{\mathbb{Z}}{\otimes} K \underset{\mathbb{Z}}{\otimes} K$)

with $R_n \underset{\mathbb{Z}}{\otimes} A$, $R_n \underset{\mathbb{Z}}{\otimes} K \underset{\mathbb{Z}}{\otimes} A$, $R_n \underset{\mathbb{Z}}{\otimes} K \underset{\mathbb{Z}}{\otimes} K \underset{\mathbb{Z}}{\otimes} A$, ... projective $\mathbb{Z}G$-modules.

Hence we have that

$$\cdots \longrightarrow P_{n+1} \longrightarrow P_n \begin{cases} \nearrow R_n \underset{\mathbb{Z}}{\otimes} A \longrightarrow R_n \underset{\mathbb{Z}}{\otimes} K \underset{\mathbb{Z}}{\otimes} A \longrightarrow \cdots \\ \searrow P_{n-1} \longrightarrow \cdots \longrightarrow P_0 \longrightarrow \mathbb{Z} \longrightarrow 0 \end{cases}$$

is a complete resolution of G of coincidence index n.

The complete resolutions in Examples (a), (b) of 1.5 are special cases of this construction.

c) As we saw in the proof of (3) $\Rightarrow$ (1), spli$G < \infty \Rightarrow pd_{\mathbb{Z}G} Hom_{\mathbb{Z}}(\mathbb{Z}G, \mathbb{Z}) < \infty$. Actually it is not difficult to show that spli$G < \infty$ if and only if $pd_{\mathbb{Z}G} Hom_{\mathbb{Z}}(\mathbb{Z}G, \mathbb{Z}) < \infty$.

Definition. Let C_s be the class of groups G whose complete cohomology is calculated via complete resolutions.

Note that by Thm. 2.4 and Remarks 2.5 a), C_s coincides with the class of groups of finite spli and the class of groups which admit a complete resolution and for which silp$G$ is finite.

We saw that groups which admit a finite dimensional contractible CW-complex with finite cell stabilizers are in C_s. Moreover, if a group G admits a finite dimensional free $G-CW$-complex homotopy equivalent to a sphere then it follows from results in [13] and Thm. 2.4 that G is in C_s.

Proposition 2.6. [17] *The class C_s is subgroup closed, extension closed and moreover if G acts cocompactly on a G-CW-complex with stabilizers in C_s, then G is in C_s.*

Proof. It was shown in [7] that the class of groups G with spli$G < \infty$ is subgroup and extension closed. Now if G acts cocompactly on a G-CW-complex with stabilizers in C_s, then there is an exact $\mathbb{Z}G$-sequence

$$0 \longrightarrow \underset{i_1 \in I_1}{\oplus} \mathbb{Z}(G/G_{i_1}) \longrightarrow \underset{i_0 \in I_0}{\oplus} \mathbb{Z}(G/G_{i_0}) \longrightarrow \mathbb{Z} \longrightarrow 0$$

with spli$G_{i_j} < \infty$ for all i_j and $| I_0 | < \infty$, $| I_1 | < \infty$.
The result now follows from Prop. 2.1.

As we have seen if spli$G < \infty$ then silp$G < \infty$. It is not known whether spli$G < \infty$ if and only if sipl$G < \infty$. However, it is easy to see that if spli$G < \infty$ then spli$G =$ silp$G =$ fin. dim of $\mathbb{Z}G$ ([13]). □

We expect that

CONJECTURE I The following are equivalent for a group G

(1) G admits a complete resolution

(2) fin.dim $\mathbb{Z}G < \infty$

(3) silp$G < \infty$

(4) spli$G < \infty$

Consider the class $H\mathcal{F}$, of hierarchically decomposable groups which was introduced by Kropholler [10] and is defined as follows. Let $H_0\mathcal{F}$ be the class of finite groups. Now define $\mathcal{H}_\alpha\mathcal{F}$ for each ordinal α inductively: if α is a successor ordinal then $\mathcal{H}_\alpha\mathcal{F}$ is the class of groups G which admit a finite dimensional contractible $G-CW$-complex with cell stabilizers in $\mathcal{H}_{\alpha-1}\mathcal{F}$, and if α is a limit ordinal then $\mathcal{H}_\alpha\mathcal{F} = \bigcup_{\beta<\alpha} \mathcal{H}_\beta\mathcal{F}$. A group belongs to $\mathcal{H}\mathcal{F}$ if it belongs to $\mathcal{H}_\alpha\mathcal{F}$ for some $\mathcal{A}$.

Note that $\mathcal{H}\mathcal{F}$ contains among others all groups of finite virtual cohomological dimension and all countable linear groups of arbitrary characteristic. Moreover, it is extension closed, subgroup closed, closed under countable directed unions and closed under amalgamated free products and HNN-extensions.

Proposition 2.7. [17] *If G is in $H\mathcal{F}$ then Conj.I is true.*

Proof. Note that for any group G, (1) $\Rightarrow$ (2) and (4) $\Rightarrow$ (1). It was shown in [4] that if G is in $H\mathcal{F}$ then (2) $\Leftrightarrow$ (3) $\Leftrightarrow$ (4). The proof now follows. □

3 Complete resolutions and free actions on finite dimensional homotopy spheres

We saw in 1.5 (c). that if a group G has periodic cohomology with period q after k-steps then G admits a complete resolution of coincidence index k. We believe that the periodicity isomorphisms are always induced by cup product with an element in $H^q(G, \mathbb{Z})$ and we showed in [17] that:

Theorem 3.1. *Let G be a group with periodic cohomology of period q after k steps. Then the following are equivalent*

(i) *the periodicity isomorphisms are induced by cup product with an element in $H^q(G, \mathbb{Z})$*

(ii) silp$G < \infty$

(iii) spli$G < \infty$

(iv) *the complete cohomology of G is calculated via complete resolutions.*

Corollary 3.2. [17] *If G is in $H\mathcal{F}$ and has periodic cohomology with period q after k-steps, then the periodicity isomorphisms are induced by cup product with an element in $H^q(G, Z)$.*

Proof. It is immediate from Prop. 2.7 and Thm. 3.1. □

It was conjectured in [16] that a countable group G has periodic cohomology after some steps if and only if G acts freely and properly on $\mathbb{R}^n \times S^m$, for some n and m. Note that if G acts freely and properly on $\mathbb{R}^n \times S^m$ then G must be countable, since $\mathbb{R}^n \times S^m$ is a separable metric space. A generalization of this conjecture was stated in [13] namely, Conj. $\mathcal{A}$: Periodicity in cohomology after some steps is the algebraic characterization of those groups G which admit a finite dimensional free $G - CW$-complex homotopy equivalent to a sphere. Conj. $\mathcal{A}$ was proved in [13] for the class $H\mathcal{F}_b$, i.e. the subclass of $H\mathcal{F}$ which consists of those groups in $H\mathcal{F}$ for which there is a bound on the orders of their finite subgroups.

Both the above conjectures have their roots in a conjecture of C.T.C. Wall [18] which says that a countable group G of finite virtual cohomological dimension acts freely and properly on $\mathbb{R}^n \times S^m$, for some n and m if and only if G has periodic Farrell–Tate cohomology. C.T.C. Wall's conjecture was proved in [3].

Conj. $\mathcal{A}$ was proved by Adem and Smith in [1] under the additional hypothesis that the periodicity isomorphisms are induced by cup product with an element in $H^q(G,\mathbb{Z})$. It follows now from their result and Corollary 3.2 that

Theorem 3.3. [17]. *If G is in $H\mathcal{F}$ then Conj. $\mathcal{A}$ holds for G.*

We believe that admitting a complete resolution is a type of finiteness condition stemming from the finite subgroups of the group and we propose:

CONJECTURE II A group G admits a complete resolution if and only if G admits a $\mathbb{Z}G$-exact sequence

$$0 \longrightarrow E_n \longrightarrow E_{n-1} \longrightarrow \cdots \longrightarrow E_0 \longrightarrow \mathbb{Z} \longrightarrow 0$$

where each E_i is a direct summand of a module of the form $\bigoplus_{j_i \in J_i} \mathbb{Z}(G/G_{j_i})$, with G_{j_i} finite subgroups of G for all j_i.

This in particular would imply:

CONJECTURE III If G has periodic cohomology after some steps and G is torsion free then $cdG < \infty$.

Conj. III reflects our belief that genuine periodicity in the cohomology of a group G after some steps comes form the finite subgroups of the group. Note that this is not a necessary condition, since there is a group G which acts on a tree with cyclic stabilizers, hence every finite subgroup of G has periodic cohomology with period 2, and yet G does not have periodic cohomology after any finite number of steps [15].

References

[1] A. Adem and J. Smith, *Periodic complexes and group actions.* Annals of Math. (2) (154) (2001).

[2] D.J. Benson and J. Carlson, *Products in negative cohomology,* J. Pure Appl. Algebra 82 (1992), 107–130.

[3] F. Connolly and S. Prassidis, *On groups which act freely on* $\mathbb{R}^m \times S^{n-1}$, Topology 28 (1989), 133–148.

[4] J. Cornick and P.H. Kropholler: *On complete resolutions,* Topology and its Applications 78 (1997), 235–250.

[5] F. Dembegioti, *On the zeroeth complete cohomology,* to appear in J. of Pure and Applied Algebra.

[6] F.T. Farrell, *An extension of Tate cohomology to a class of infinite groups,* J. Pure Appl. Algebra 10 (1977), 153–161.

[7] T.V. Gedrich and K.W. Gruenberg, *Complete cohomological functors on groups*, Topology and its Application 25 (1987), 203–223.

[8] F. Goichot, *Homologie de Tate-Vogel equivariant*, J. Pure Appl. Algebra 82 (1992), 39–64.

[9] B.M. Ikenaga, *Homological dimension and Farrell cohomology*, Journal of Algebra 87 (1984), 422–457.

[10] P.H. Kropholler, *On groups of type FP_∞*, J. Pure Appl. Algebra 90 (1993), 55–67.

[11] P.H. Kropholler, *Hierarchical decompositions, generalized Tate cohomology and groups of type FP_∞*, in: Combinatorial and Geometric Group Theory, London Math. Society Lecture Notes 204, Cambridge Univ. Press (1995).

[12] G. Mislin, *Tate cohomology for arbitrary groups via satelites*, Topology and its applications 56 (1994), 293–300.

[13] G. Mislin and O. Talelli, *On groups which act frealy and properly on finite dimensional homotopy spheres* Computational and Geometric Aspects of Modern Algebra, London Math. Society Lecture Notes 275, Cambridge Univ. Press (2000).

[14] O. Talelli, *On cohomological periodicity for infinite groups*, Comment. Math. Helv. 55 (1980), 178–192.

[15] O. Talelli, *On groups with periodic cohomology after 1-step*, J. Pure Appl. Algebra 30 (1983), 85–93.

[16] O. Talelli, *Periodicity in cohomology and free and proper actions on $\mathbb{R}^n \times S^m$*, to appear in the *Proceedings of Groups at St. Andrews 1997: Vol. II*, Cambridge Univ. Press.

[17] O. Talelli, *Periodicity in group cohomology and complete resolutions* to appear in the Bulletin of LMS.

[18] C.T.C. Wall, *Periodic projective resolutions*, Proc. London Math. Soc. 39 (1979), 509–533.

Structure theory for branch groups

John S. Wilson

University College,
Oxford

1 Introduction

Branch groups are certain groups of automorphisms of rooted trees, whose definition, formulated in 1997 by R. I. Grigorchuk, will be recalled below. The class of branch groups provides a convenient setting in which to study many important examples of groups, with relevance to topics in analysis, geometry, combinatorics, probability and computer science as well as group theory. We shall recall a few of these examples below; a fuller introducton to examples and to the contexts in which branch groups arise can be found in Grigorchuk's survey article [3]. Of further importance, and particularly relevant here, is the fact that the class of branch groups arises naturally in the study of just infinite groups (see [14], [15] and below). We recall that if $\mathcal{X}$ is a property of groups then a group G is called a *just non-$\mathcal{X}$ group* if each proper quotient of G has the property $\mathcal{X}$ but G itself does not have the property.

Let $(m_n)_{n\geqslant 0}$ be a sequence of integers with $m_n \geqslant 2$ for each n. The (*spherically homogeneous*) *rooted tree of type* (m_n) is a tree T with a vertex v_0 (called the *root vertex*) of valency m_0, such that every vertex at a distance $n \geqslant 1$ from v_0 has valency m_n+1. The distance of a vertex v from v_0 is called the *level* of v, and the set of vertices of level n is called the nth *layer* of T. We picture T with the root at the top and with m_n edges descending from each vertex of level n. Each vertex v of level r is the root of a rooted subtree T_v of type $(m_n)_{n\geqslant r}$. We may regard the set of vertices of T as an ordered set by writing $u \leqslant v$ if and only if u is a vertex of T_v.

Now let G be a group that acts faithfully as a group of automorphisms of T fixing the root vertex. For each vertex v we write $\mathrm{rist}_G(v)$ for the subgroup consisting of those elements of G that fix all vertices outside T_v, and for each $n \geqslant 0$ we write $\mathrm{rist}_G(n)$ for the group generated by all subgroups $\mathrm{rist}_G(v)$ with v of level n. The group G *acts as a branch group* on T if G satisfies the following two conditions for each $n \geqslant 0$:

(i) G acts transitively on the set of vertices of level n;

(ii) $|G : \mathrm{rist}_G(n)|$ is finite.

More generally, we say that G is a *branch group* if G acts faithfully as a branch group on some rooted tree.

Groups G of automorphisms of rooted trees satisfying (i) are subject to strong restrictions: it is proved in [3, Theorem 4] that each non-trivial normal subgroup of G contains the derived group of $\mathrm{rist}_G(n)$ for some n, and the proof of [5, Lemma 2] shows that if G has a non-trivial virtually abelian normal subgroup then $\mathrm{rist}_G(n)$ is abelian for some n. Addition of the condition (ii) excludes the possibility that some subgroup $\mathrm{rist}_G(n)$ is abelian, and so each branch group G has the following property:

(B1) G is just non-(virtually abelian) and G has no non-trivial virtually abelian normal subgroups.

Suppose that G acts faithfully as a branch group on a tree T. If v is a vertex then clearly $(\mathrm{rist}_G(v))^g = \mathrm{rist}_G(vg)$ for each $g \in G$, so that the normalizer $N_G(\mathrm{rist}_G(v))$ of $\mathrm{rist}_G(v)$ is just the stabilizer of v in G. Therefore $\bigcap_{v \in T} N_G(\mathrm{rist}_G(v)) = 1$. Restricted stabilizers are examples of basal subgroups (whose definition will be given in Section 3) and so every branch group has the following property:

(B2) $\bigcap(N_G(B) \mid B \text{ a basal subgroup}) = 1$.

The properties (B1) and (B2) concern the purely internal group-theoretic structure of G and make no reference to actions of G on geometric objects. It turns out that branch groups can be characterized completely by (B1), (B2) and another more technical property (B3), which again relates only to the structure of G as a group. Indeed, graphs having large rooted subtrees can be found within the structure of groups G satisfying (B1) and (B2), and in certain cases the graphs are themselves trees. Our primary object here is to describe, without proofs, the genesis of these graphs, their properties and their role in the characterization of branch groups. Before doing this, we recall some examples and give further group-theoretic motivation for studying branch groups.

2 Examples and motivation

In order to describe and work with examples of groups acting on rooted trees it is helpful to have a more explicit notation for vertices of trees.

For each $n \in \mathbb{N}$ let X_n be a finite set with $|X_n| \geqslant 2$. We consider the set T of finite strings $x_1 \ldots x_n$ with $n \in \mathbb{N} \cup \{0\}$ and $x_i \in X_i$ for each i. We regard T as a rooted tree, with edges linking all pairs of the form $(x_1 \ldots x_n, x_1 \ldots x_n x_{n+1})$, and with root vertex the empty string $\emptyset$. In this way we obtain an explicit representation of the rooted tree of type (m_n) where $m_n = |X_n|$. The nth layer L_n of this tree consists of all strings of length n, and for each $v \in T$, the set of strings in T beginning with v is the subtree T_v with root vertex v.

Example 1. The group $A = \mathrm{Aut}(T)$ is itself a branch group: it acts transitively with kernel $\mathrm{rist}_A(n)$ on L_n for each n. Since A is the inverse limit of its images in the groups $\mathrm{Sym}(L_n)$, it is a profinite group and its action on the set of vertices of T (regarded as a discrete space) is continuous.

For each integer $r > 0$ write D_r for the group of automorphisms of T that affect only the first r entries of strings. Thus D_r is a complement to $\mathrm{rist}_A(r)$ in A and it permutes the subtrees with roots in L_r. Each element of $\mathrm{rist}_A(r)$ is determined by its actions on these subtrees, which are isomorphic to the tree $T^{(r)}$ with vertices labelled by strings whose nth entry is in X_{r+n}. Given a family (α_l) of automorphisms of $T^{(r)}$ indexed by the elements $l \in L_r$ we define an automorphism of T, also denoted by (α_l), which acts on T_l as α_l acts on $T^{(r)}$ for each $l \in T^{(r)}$. Each automorphism of T can be written uniquely as a product $\sigma.(\alpha_l)$ with $\sigma \in D_r$ and each α_l in $\mathrm{Aut}(T^{(r)})$. As this notation suggests, the group $A = \mathrm{Aut}(T)$ is isomorphic to the permutational wreath product $\mathrm{Aut}(T^{(r)}) \,\mathrm{wr}\, D_r$. We note also that D_r is isomorphic to the permutational wreath product

$$\mathrm{Sym}(X_r) \,\mathrm{wr}\, \mathrm{Sym}(X_{r-1}) \,\mathrm{wr}\, \cdots \,\mathrm{wr}\, \mathrm{Sym}(X_1).$$

Example 2. Our next examples are certain wreath products with infinitely many factors.

Fix a 'trivial' element $1_n \in X_n$ for each n and write $1^{(n)}$ for the string of length n with all entries trivial. Suppose that H_n is a transitive permutation group on X_n for each $n \in \mathbb{N}$. We define a faithful action of H_n on T as follows. Let $\xi \in H_n$ and $v = x_1 \dots x_r \in T$. If $r \geqslant n$ and $x_i = 1_i$ for $i < n$ we let $v\xi$ be the string with ith entry x_i for $i \neq n$ and nth entry $x_n\xi$; otherwise we set $v\xi = v$. In other words, ξ is an element of $\mathrm{rist}_A(1^{(n-1)})$ and it permutes in the obvious way the rooted trees having roots $1^{(n-1)}x_n$ with $x_n \in X_n$. We identify each H_n with its image in $\mathrm{Aut}(T)$ and we write

$$W = \langle H_n \mid n \in \mathbb{N} \rangle.$$

Evidently W acts transitively on each L_n, and the kernel of the action on L_n is $\mathrm{rist}_W(n)$, so that W is a branch group.

Groups of this type (and more general ones with $\mathbb{N}$ replaced by arbitrary linearly ordered sets) were first discussed in Hall [8]. The isomorphisms described in Example 1 above induce isomorphisms

$$W \cong \langle H_n \mid n > r \rangle \,\mathrm{wr}\, H_r \,\mathrm{wr}\, H_{r-1} \,\mathrm{wr}\, \cdots \,\mathrm{wr}\, H_1$$

for each r; the group $\langle H_n \mid n > r \rangle$ acts naturally on a tree isomorphic to the trees with root in L_r.

It is not hard to prove that if the groups H_n are non-abelian simple groups then the only proper non-trivial normal subgroups are the groups $\mathrm{rist}_W(n)$. Therefore, if in addition the groups H_n are isomorphic permutation groups, then every non-trivial normal subgroup of W is isomorphic to a direct product of finitely many copies of W.

Example 3. The groups in Example 2 are locally finite. Finitely generated groups with similar wreath product decompositions were first found by P. M. Neumann [10]. For simplicity of exposition we fix $m \geqslant 5$ and consider only the tree $T(m)$ whose vertices are finite strings with entries from $\{1, 2, \ldots, m\}$. The discussion of Example 1 shows that each element g of $A = \mathrm{Aut}(T(m))$ is uniquely of the form $\sigma.(g_1, \ldots, g_m)$, with σ in the group $\Sigma_m = \mathrm{Sym}\{1, \ldots, m\}$ and $g_1, \ldots, g_m \in A$, and that this establishes an isomorphism $A \cong A \,\mathrm{wr}\, \Sigma_m$.

For $g \in A$ define $\bar{g} \in A$ by $\bar{g} = (g, \bar{g}, 1, \ldots, 1)$. This is a recursive definition: if the action of $\bar{g}$ on the nth layer L_n is known, it gives the action on L_{n+1}. For h, $k \in A$ we have

$$\overline{(hk)}^{-1}\bar{h}\bar{k} = ((hk)^{-1}hk, \overline{hk}^{-1}\bar{h}\bar{k}, 1, \ldots, 1) = (1, \overline{(hk)}^{-1}\bar{h}\bar{k}, 1, \ldots, 1).$$

We deduce successively that $\overline{(hk)}^{-1}\bar{h}\bar{k}$ acts trivially on L_n for $n = 1, 2, \ldots,$ and hence that $\overline{(hk)} = \bar{h}\bar{k}$. Therefore $g \mapsto \bar{g}$ is a homomorphism $A \to A$.

We regard the alternating group A_m as embedded in the obvious way in the subgroup D_1 of A (defined in Example 1) and we fix a finitely generated perfect subgroup P of A with $A_m \leqslant P$. Let $G = \langle h, \bar{h} \mid h \in P \rangle$. So G is finitely generated and perfect. We prove that $G \cong G \,\mathrm{wr}\, A_m$.

Let h, $k \in P$. Since $(2, 3, 4) \in G$, the subgroup G contains

$$[\bar{h}, \bar{k}^{(2,3,4)}] = [(h, \bar{h}, 1, \ldots, 1), (k, 1, \bar{k}, 1, \ldots, 1)] = ([h, k], 1, \ldots, 1).$$

Therefore $(h, 1, \ldots, 1) \in G$ for all $h \in P$ as P is perfect. Hence G contains

$$(h, 1, \ldots, 1)^{-1}\bar{h} = (1, \bar{h}, 1, \ldots, 1) \quad \text{and} \quad (1, \bar{h}, 1, \ldots, 1)^{(1,3,2)} = (\bar{h}, 1, \ldots, 1).$$

It follows that G contains $\{(g, 1, \ldots, 1) \mid g \in G\}$ as well as A_m. Thus

$$G \geqslant \{\sigma.(g_1, \ldots, g_m) \mid \sigma \in A_m, g_i \in G\}.$$

The reverse inequality is clear, and so the isomorphism $A \to A \,\mathrm{wr}\, \Sigma_m$ induces an isomorphism $G \to G \,\mathrm{wr}\, A_m$.

Write K_n for the kernel of the action of G on L_n for each n. Clearly $\mathrm{rist}_G(1)$ maps to the base group B of the wreath product under the isomorphism, and $\mathrm{rist}_G(1) = K_1$. Thus if $\mathrm{rist}_G(n) = K_n$ for some n then $\mathrm{rist}_B(n+1)$ is the kernel of the action of B on L_{n+1}, and hence $\mathrm{rist}_G(n+1) = K_{n+1}$. This and a similar induction show that $\mathrm{rist}_G(n) = K_n$ for each n, that G acts transitively on each L_n, and therefore that G is a branch group. In fact the argument of

[10, Lemma 5.1] shows that the subgroups K_r are the only proper non-trivial normal subgroups of G.

Examples like this (but with different permutation groups acting on the sets X_n) were used by Neumann [10] to answer questions of Pride concerning largeness of groups, and by Segal [12] to answer questions about subgroup growth of groups. In [17], [18] it was shown that if $m \geqslant 29$ then the group G constructed above does not have uniformly exponential growth; this leads to examples that answer a question of Gromov [6] on word growth of groups.

Example 4. The next example, the Grigorchuk group, has rather different properties from those considered so far.

Take T to be the tree $T(2)$ with vertices the finite strings with entries from $\{1,2\}$. Let a be the automorphism that changes the first entry of each non-empty string; thus a interchanges the two trees T_1, T_2 with roots the strings 1, 2.

We define b, c, d recursively as the maps (a,c), (a,d), $(1,b)$; thus b acts on T_1 as a acts on T and on T_2 as c acts on T, and similarly for c, d.

The (first) Grigorchuk group is the group $\Gamma = \langle a,b,c,d\rangle$. Grigorchuk proved that it has the following properties:

- Γ is a just infinite 2-group;
- Γ has solvable word and conjugacy problems;
- Γ has intermediate word growth and so is amenable.

More recently, Pervova [11] proved that the maximal subgroups of Γ have index 2; though this property is automatic for (locally) finite 2-groups, there is no reason to expect it to hold for arbitrary 2-groups. Grigorchuk and the author [4] have proved that Γ has the following additional properties:

- any two infinite finitely generated subgroups have isomorphic subgroups of finite index;
- the finitely generated subgroups of Γ are closed in the profinite topology on Γ, and Γ has solvable generalized word problem.

Grigorchuk [2] has introduced and studied many groups sharing some of the properties of Γ. Further important just infinite branch groups of a rather similar character are provided by the Gupta–Sidki p-groups [7]. Let p be an odd prime. The Gupta–Sidki p-group is the subgroup of $\mathrm{Aut}(T(p))$ generated by a, b, where a is the automorphism that acts as the p-cycle $(1,2,\dots,p)$ on the first entries of non-empty strings and b is the automorphism defined recursively by the formula $b = (a, a^{-1}, 1, \dots, 1, b)$. The Grigorchuk group and the Gupta–Sidki groups are perhaps the most easily described examples of finitely generated infinite torsion groups.

Example 5. Our final example is not a branch group but it satisfies conditions (B1), (B2). There are many such groups; the following particularly simple and elegant description of one was shown to me by Said Sidki. The group is the subgroup of the automorphism group of $T = T(6)$ generated by two automorphisms a, b. Let s, t be the automorphisms that permute the first entries of non-empty strings as the permutations $(1,2,3)(4,6,5)$ and $(1,4)(2,5)(3,6)$. The automorphisms a, b are defined recursively as follows:

$$a = (a, 1, \dots, 1)s, \quad b = (b, 1, \dots, 1)t.$$

Further motivation

The examples in (2), (3), (4) above are just infinite. In [14], [15] it was shown that if G is any just infinite group then exactly one of the following holds:

(a) G is virtually abelian;

(b) G embeds with finite index in some group $H \operatorname{wr} F$, where F is a finite transitive permutation group, H is not virtually cyclic and all subgroups of finite index in H are just infinite;

(c) G acts faithfully as a branch group on some tree.

Therefore just infinite groups fall naturally into three disjoint classes, one consisting of branch groups, and so branch groups arise inevitably in the study of the structure of just infinite groups.

3 The structure graph

Our goal now is to explain why conditions (B1), (B2) of the Introduction characterize a class of groups acting on trees somewhat wider than the class of branch groups, and then to formulate a third condition that distinguishes the branch groups. The immediate aim of the section is to use ideas from the study of just infinite groups in [14], [15] to extract from the structure of a group G satisfying (B1), (B2) a graph, called the *structure graph* of G. The arguments justifying the assertions that we make are harder than for just infinite groups since the powerful finiteness properties of just infinite groups are no longer available.

We noted in Example 1 that $\operatorname{Aut}(T)$ is a profinite group which acts continuously on (the set of vertices of) T. Given a profinite group G, it is therefore reasonable to ask when it has a *continuous* action as a branch group on a rooted tree. It turns out that the answer is exactly the same as for abstract groups, but with the notions of subgroup, epimorphic image etc. interpreted appropriately in the category of profinite groups. The proofs are easy modifications of those for the abstract case, for reasons partially explained in [15, Section 2].

3.1 The structure lattice

Let G be an abstract or profinite group. We write $\overline{L} = \overline{L}(G)$ for the set of subnormal subgroups of G having only finitely many conjugates. By a result of Wielandt ([15, Lemma 2.2]), the set $\overline{L}$ is closed for intersections and joins of pairs of subgroups, and so it is a sublattice of the lattice of subgroups of G.

Though we shall ultimately be interested in groups that are just non-(virtually abelian), we can start in greater generality. Let $\mathcal{X}$ be a property of groups satisfying the following conditions:

- $\mathcal{X}$ is inherited by subgroups, quotients, finite extensions and direct products with finitely many factors;
- every non-trivial $\mathcal{X}$-group has a non-trivial virtually abelian characteristic subgroup.

Examples of such properties of interest in this context are the properties of being finite, virtually abelian, virtually metabelian, virtually nilpotent, or virtually soluble. (An example studied in [1] corresponds to the case when $\mathcal{X}$ is the class of virtually metabelian groups.)

For the rest of this section, G will be a group having the following property:

(B1$_\mathcal{X}$) G is a just non-$\mathcal{X}$ group and G has no non-trivial virtually abelian normal subgroups.

It follows easily from (B1$_\mathcal{X}$) that $\overline{L}$ contains no non-trivial virtually abelian subgroups. Moreover if H, $K \in \overline{L}$ then $[H, K] = 1$ if and only if $H \cap K = 1$.

For subgroups H, K of $\overline{L}$ we write $H \sim K$ if $C_G(K) = C_G(H)$. One can show that the relation $\sim$ is a congruence on the lattice $\overline{L}$; that is to say, it is an equivalence relation, and if subgroups $H_1, H_2, K_1, K_2 \in \overline{L}$ satisfy $H_1 \sim H_2$, $K_1 \sim K_2$ then

$$H_1 \cap K_1 \sim H_2 \cap K_2 \quad \text{and} \quad \langle H_1, K_1 \rangle \sim \langle H_2, K_2 \rangle.$$

Thus the quotient set $\mathcal{L} = \overline{L}/{\sim}$ is a lattice, with the join and meet of elements $[H]$, $[K]$ defined by

$$[H] \vee [K] = [H \vee K] \quad \text{and} \quad [H] \wedge [K] = [H \wedge K].$$

We call $\mathcal{L}$ the *structure lattice* of G. It has greatest and least elements $[G]$ and $\{1\}$. It turns out that $\mathcal{L}$ is a Boolean lattice; that is, it satisfies the distributive laws and each element has a complement. Conjugation in G induces an action of G on the lattice $\mathcal{L}$, defined by $[H]^g = [H^g]$.

For each normal subgroup N of finite index in G let $\mathcal{L}_N$ be the set of fixed points of $\mathcal{L}$ under the action of N on $\mathcal{L}$. Clearly $\mathcal{L}_N$ is a Boolean sublattice of $\mathcal{L}$, and we have $\mathcal{L}_N = \{[H] \mid H \triangleleft N\}$. Clearly too $\mathcal{L} = \bigcup_N \mathcal{L}_N$. Each sublattice $\mathcal{L}_N$ is finite; this is a consequence of the following result whose proof is similar to the proof of [16, (12.6.7)].

Proposition 3.1. *Let F be a finite group that acts as a group of automorphisms of a modular lattice L. Then the set L^F of elements fixed by F is a sublattice. If L^F satisfies the maximal condition then so does L.*

Being a finite Boolean lattice, $\mathcal{L}_N$ is isomorphic to the lattice of subsets of a finite set. Indeed, if b is an atom of $\mathcal{L}_N$ and Y denotes the orbit of b under the action of G on $\mathcal{L}$, then the map $\theta : \{b_1, \ldots, b_r\} \mapsto b_1 \vee \cdots \vee b_r$ is an isomorphism from the lattice of subsets of Y to $\mathcal{L}_N$. Translated back to a statement about subgroups of G, the above statement becomes the following:

Proposition 3.2. *Let N be a normal subgroup of finite index in G. Then N has a normal subgroup B whose conjugates in G generate their direct product, and such that the elements of $\mathcal{L}_N$ are precisely the 2^m equivalence classes containing products of conjugates of B, where $m = |G : N_G(B)|$.*

We illustrate this result by considering permutational wreath products. Let F be a finite transitive permutation group on a set Ω, let H be a group satisfying $(\mathrm{B}1_{\mathcal{X}})$ and let G be the permutational wreath product $H \operatorname{wr} F$, with base group N. Then G also satisfies $(\mathrm{B}1_{\mathcal{X}})$, and $\mathcal{L} = \mathcal{L}(G)$ can be identified with the set of maps from Ω to $\mathcal{L}(H)$. Moreover $\mathcal{L}_N$ can be identified with the set of subsets of Ω, since each subgroup in $\overline{L}(G)$ normalized by the base group is equivalent in the relation $\sim$ to the direct product of some of the canonical direct factors of the base group. In particular, if $\mathcal{L}(H)$ has only the two elements $\{1\}$, $[H]$, then $\mathcal{L}(G)$ is equal to $\mathcal{L}_N$ and is isomorphic to the lattice of subsets of Ω.

This process can be reversed: if G is an arbitrary group satisfying $(\mathrm{B}1_{\mathcal{X}})$ and if N is a normal subgroup of finite index in G then G is isomorphic to a large subgroup of a wreath product associated with N. Taking B as in Proposition 3.2, we let K be the kernel of the action of G by conjugation on the set $Y = \{B^g \mid g \in G\}$ and let $H = K/C_K(B)$. Then H has property $(\mathrm{B}1_{\mathcal{X}})$ and there is a natural embedding of G with large image in $H \operatorname{wr} (G/K)$, where we regard $F = G/K$ as a permutation group via its action on Y.

3.2 Basal subgroups and the graph $\mathcal{B}$

The subgroup B in Proposition 3.2 is a basal subgroup of G. A *basal* subgroup of G is a subgroup with only finitely many distinct conjugates, such that the conjugates generate their direct product. Hence basal subgroups are normal in their normal closures and belong to $\overline{L}$. If $B \in \overline{L}$ then B is basal if and only if any two distinct conjugates of B intersect trivially.

If G acts faithfully as a group of automorphisms of a rooted tree then the rigid stabilizer of each vertex v is a basal subgroup: we have $(\mathrm{rist}_G(v))^g = \mathrm{rist}_G(vg)$ for each $g \in G$, so that $\mathrm{rist}_G(v)$ has only finitely many conjugates, which evidently generate their direct product. This is the motivation for our interest in basal subgroups.

We list some straightforward properties of basal subgroups.

Lemma 3.3. *Suppose that A, B are basal subgroups of G.*

(a) *$A \cap B$ is basal.*
(b) *If $1 \neq A \leqslant B$ then $N_G(A) \leqslant N_G(B)$.*
(c) *If $A \sim B$ then $N_G(A) = N_G(B)$.*
(d) *If $N_G(A) \leqslant L \leqslant G$ then $\langle A^L \rangle$ is basal and $N_G(\langle A^L \rangle) = L$.*
(e) $\mathrm{stab}_G([A]) = N_G(A)$.

Lemma 3.4. *If $H \in \overline{L} \setminus \{1\}$ then H contains a basal subgroup B such that $H \sim \langle B^g \mid B^g \leqslant H \rangle$.*

At this point we need to make some restriction on the class $\mathcal{X}$. For simplicity we shall assume from now on that $\mathcal{X}$ is the class of virtually abelian groups, so that our group G satisfies condition (B1).

One can show that if G is residually finite then G has descending chains $(K_i)_{i \in \mathbb{N}}$ of normal subgroups of finite index with $\bigcap K_i = 1$, and moreover that if $(K_i)_{i \in \mathbb{N}}$ is any such chain then each basal subgroup B of G is normalized by some subgroup K_i. It follows that $\mathcal{L}$ is the union of countably many finite sublattices $\mathcal{L}_N$ and so is countable. If $\mathcal{L}$ is infinite then it is isomorphic to the lattice of closed and open subsets of Cantor's ternary set, and since any countable residually finite group can act faithfully with all orbits finite on this lattice, it is clear that $\mathcal{L}$ itself reflects little of the structure of G. To obtain useful actions we have to pass to subsets of $\mathcal{L}$.

Write

$$\mathcal{B} = \mathcal{B}(G) = \{[B] \mid B \neq 1,\ B\,\text{basal}\}.$$

The action of G on $\mathcal{L}$ induces an action on $\mathcal{B}$. We recall the condition (B2) stated in the Introduction:

(B2) $\bigcap (N_G(B) \mid B \text{ a basal subgroup}) = 1$.

Since each subgroup $N_G(B)$ has finite index, (B2) implies that G is residually finite, and moreover (B2) is precisely the condition that G acts faithfully on $\mathcal{B}$.

Therefore from now on we assume that G satisfies both (B1) and (B2). The set $\mathcal{B}$ is infinite; regarded as a partially ordered set it satisfies the maximal condition and it has no minimal elements, but it is not generally a rooted tree. We regard $\mathcal{B} = \mathcal{B}(G)$ as a graph by linking with an edge each pair (a, b) with b maximal in $\{c \in \mathcal{B} \mid c < a\}$. We call $\mathcal{B}$ the *structure graph* of G. It turns out that $\mathcal{B}$ is a connected graph, and of course G acts as a group of graph automorphisms of $\mathcal{B}$.

To illustrate the utility of the structure graph we make some remarks about automorphisms of G. One can prove that G has a chain of characteristic subgroups of finite index. From this fact alone it follows that (1) $\mathrm{Aut}(G)$ is residually finite and (2) $\mathrm{Aut}(G)$ is profinite if G is profinite. However more information is available, since there is a canonical faithful representation of $\mathrm{Aut}(G)$ in a profinite subgroup of the group $A = \mathrm{Aut}(\mathcal{B})$.

Let A be the automorphism group of $\mathcal{B}$, and let $\bar{G}$ be the image of G in A. It is easy to show that $C_A(\bar{G}) = 1$. For each $h \in N_A(\bar{G})$, let $h^\star$ be the automorphism $g \mapsto g^h$ of $\bar{G}$. Then the map $h \mapsto h^\star$ is an isomorphism from $N_A(\bar{G})$ to the full group of abstract automorphisms of $\bar{G}$. It follows that $\mathrm{Aut}(G) \cong N_A(\bar{G})$.

Let $\tilde{A}$ be the subgroup of A consisting of the automorphisms that induce bijections of $\mathcal{B} \cap \mathcal{L}_N$ for each characteristic subgroup N of finite index in G. Then $\tilde{A}$ is the inverse limit of its images in the groups $\mathrm{Sym}(\mathcal{B} \cap \mathcal{L}_N)$, and so $\tilde{A}$ is a profinite group canonically associated with G. It is clear that $N_A(\bar{G}) \leqslant \tilde{A}$. Assertions (1), (2) above follow, and we also see that if G is profinite then every automorphism of G as an abstract group is continuous.

4 Structure trees

Let G be a group that satisfies conditions (B1) and (B2). We shall show that the structure graph

$$\mathcal{B} = \mathcal{B}(G) = \{[B] \mid B \neq 1,\ B \text{ basal}\}$$

contains G-invariant subsets which have the structure of rooted trees on whose layers G acts transitively. The proofs of most assertions of this section are simple modifications of those in [15, Section 7].

We fix an infinite strictly descending chain $\mathcal{F} = (G_i)_{i \geqslant 0}$ of normal subgroups of finite index with $G_0 = G$ and $\bigcap G_i = 1$. An element b of $\mathcal{B}$ is called a *vertex* if there is an integer i such that b is an atom in $\mathcal{L}_{G_i}$; the least such i is the *depth* of b. We write $\mathcal{T} = \mathcal{T}(\mathcal{F})$ for the set of all vertices; since each $\mathcal{L}_{G_i}$ is finite, so is the set of vertices of depth i for each i.

Clearly $[G]$ is a vertex, being an atom in $\mathcal{L}_{G_0}$. If a, b are vertices and $b < a$ then the definition of vertices implies that b has greater depth than a.

We regard $\mathcal{T}$ as a graph by linking with an edge each pair (a, b) with $a, b \in \mathcal{T}$ and b maximal in $\{v \in \mathcal{T} \mid v < a\}$. The action of G on $\mathcal{L}$ and $\mathcal{B}$ induces an action on $\mathcal{T}$, and the sets of vertices of fixed depth are unions of orbits for this action. In fact, the orbits of G in its action on $\mathcal{T}$ are the non-empty sets of elements of equal depth. For each $b \in \mathcal{B}$ there is a vertex t with $t \leqslant b$, and it follows that $\mathcal{T}$ has no minimal vertices. Hence $\mathcal{T}$ is a rooted tree on each of whose layers G acts transitively. We call $\mathcal{T}$ the *structure tree* of G with respect to $\mathcal{F}$. The action of G on $\mathcal{T}$ is faithful, and if G is profinite then the map from G to $\mathrm{Aut}(\mathcal{T})$ is continuous, and so is an embedding of profinite groups.

The following result illustrates the intimate relationship between the subgroup structure of G and the properties of the tree $\mathcal{T}$.

Proposition 4.1. *Let a, b be vertices of $\mathcal{T}$ of depths i, j with $i \leqslant j$ and write $a = [A], b = [B]$ with A, B basal. The following are equivalent:*

(a) *$b \leqslant a$ in $\mathcal{L}$;*

(b) *the path in $\mathcal{T}$ from $[G]$ to b passes through a;*

(c) *A and B do not centralize each other.*

The existence of the tree $\mathcal{T}$ is the crucial ingredient in the characterization, in terms of their actions on rooted trees, of the groups with the properties (B1), (B2).

Let H act faithfully as a group of automorphisms of a tree S. We say that H acts as a *generalized branch group* on S if H acts transitively on each layer of S and for all n the following holds: $\mathrm{rist}_H(n)$ is non-abelian and $H/(\mathrm{rist}_H(n))'$ is virtually abelian.

Theorem 4.2. *Let G be an abstract or profinite group. Then G can act faithfully as a generalized branch group on a rooted tree if and only if G satisfies conditions* (B1), (B2).

In order to characterize branch groups we need an additional condition forcing rigid level stabilizers to have finite index.

(B3) For each non-trivial basal subgroup A, the normal closure in G of the subgroup $\bigcap(N_G(B) \mid B\,\text{basal}, A \cap B = 1)$ has finite index in G.

Condition (B3) can be stated more succinctly using the notion of a rigid normalizer. The *rigid normalizer* $R_G(A)$ of a non-trivial basal subgroup A of G is defined by

$$R_G(A) = \bigcap(N_G(B) \mid B\,\text{basal},\ A \cap B = 1).$$

The subgroup $R_G(A)$ is the unique maximal element of $[A]$ and it is a basal subgroup. If G acts faithfully on a rooted tree T and if the rigid stabilizer of each vertex of T is non-trivial, then $R_G(\mathrm{rist}_G(v)) = \mathrm{rist}_G(v)$ for each vertex v of T. Condition (B3) is the requirement that the normal closure of the rigid normalizer of each non-trivial basal subgroup has finite index in G. The role of rigid normalizers in the following characterization was first fully elucidated by my former student Philip Hardy in his Ph.D. thesis [9].

Theorem 4.3. *Let G be an abstract or profinite group. Then G is a branch group if and only if it satisfies conditions* (B1), (B2) *and* (B3).

5 The relation between the graph $\mathcal{B}$ and trees

Let G be an abstract or profinite group G satisfying (B1), (B2), and fix a rooted tree T on which G has a generalized branch action. The structure graph can be reconstructed from T:

Lemma 5.1. (a) $\mathcal{B} = \{[\langle \mathrm{rist}_G(v)^L \rangle] \mid v \in T, \ \mathrm{stab}_G(v) \leqslant L \leqslant G\}$.

(b) *Suppose that* $v \in T$ *and* $L, M \leqslant G$ *with* $L, M \geqslant \mathrm{stab}_G(v)$. *Then* $\langle \mathrm{rist}_G(v)^L \rangle \sim \langle \mathrm{rist}_G(v)^M \rangle$ *if and only if* $L = M$.

(c) *If* $v \leqslant w$ *in* T *then* $\mathrm{rist}_G(w) \sim \langle \mathrm{rist}_G(v)^{\mathrm{stab}_G(w)} \rangle$.

(d) *Let* $b \in \mathcal{B} \setminus \{[G]\}$. *The elements* a *of* $\mathcal{B}$ *with* $a > b$ *such that* a, b *are joined by an edge are in bijective correspondence with the subgroups* L *of* G *that contain* $\mathrm{stab}_G(b)$ *as a maximal subgroup.*

We see that the subgroup structure of G above each of its vertex stabilizers determines both the vertices and edges of $\mathcal{B}$.

There is a canonical map $\phi_T : T \to \mathcal{B}$ defined by

$$\phi_T(v) = [\mathrm{rist}_G(v)] \quad \text{for each } v \in T.$$

This map is injective and preserves the action of G. Philip Hardy [9] has proved that ϕ_T is surjective if and only if the action of G on T satisfies the following condition:

(S) for each vertex v of T and each subgroup L of G such that $L \geqslant \mathrm{stab}_G(v)$, there is a vertex $w \geqslant v$ such that $L = \mathrm{stab}_G(w)$.

Let B be a non-trivial basal subgroup of G and u a vertex of T. We call B a *vertex subgroup* of G with respect to u if there is an integer $m \geqslant 0$ such that $\prod_v \mathrm{rist}_G(v)' \leqslant B \lhd \mathrm{rist}_G(u)$, where the product is over all vertices of level m in T_u. If G acts as a branch group on T, then B is a vertex subgroup with respect to u if and only if $B \sim \mathrm{rist}_G(u)$; thus in this case u is uniquely determined.

We can now state various conditions under which $\mathcal{B}$ is a tree.

Theorem 5.2. *Suppose that* G *is an abstract or profinite group satisfying* (B1) *and* (B2). *The following are equivalent:*

(a) $\mathcal{B}$ *is a rooted tree;*

(b) *there exists a rooted tree on which* G *has a faithful generalized branch action with condition* (S);

(c) *for each normalizer* M *of a non-normal basal subgroup of* G *there is a unique subgroup* L_M *of* G *which contains* M *as a maximal subgroup.*

If G also satisfies (B3), *these are equivalent to*

(d) *G acts faithfully as a branch group on some rooted tree such that each non-trivial basal subgroup is a vertex subgroup.*

Suppose now that $\mathcal{B}$ is a tree. Then $\mathcal{B}$ is a rooted tree, and G acts on it with a generalized branch action satisfying condition (S). Furthermore $\mathrm{rist}_G([A]) = R_G(A)$ for each non-trivial basal subgroup A. Any tree T on which G acts as a generalized branch group can be embedded canonically (by the map ϕ_T) in $\mathcal{B}$. The trees in $\mathcal{B}$ that arise are obtained using the notion of *deletion of layers* introduced in [5]. Given an infinite set Ω of non-negative integers containing 0 we form a new tree T_Ω as follows: its vertices are the vertices of $\mathcal{B}$ whose levels are in Ω, and vertices u, v of T_Ω with $v < u$ are linked by an edge if and only if v is maximal in $\{w \in T_\Omega \mid w < u\}$, where the partial order on T_Ω is the one inherited from $\mathcal{B}$. Clearly G also acts as a generalized branch group on T_Ω (respecting topologies in the profinite case), and if G acts as a branch group on $\mathcal{B}$ then it acts as a branch group on T_Ω.

In [5] the implications were studied of two properties of branch actions of a group G on a tree T that hold in a number of cases of interest:

($*$) for each vertex u of T the stabilizer of u in G acts as a (transitive) cyclic group of prime order on the edges descending from u;

($**$) whenever u, u' are incomparable vertices and $v < u$, there is an element $g \in G$ with $u'^g = u'$ and $v^g \neq v$.

It was shown directly in [5] that the structure graphs of branch groups satisfying ($*$) and ($**$) are trees, and that each of these branch groups acts on a maximal tree that is essentially unique. Philip Hardy has proved that a generalized branch group satisfying ($*$) and ($**$) must also satisfy (S), and this allows these results to be extended to generalized branch groups. The conditions ($*$) and ($**$) can often be verified directly for particular examples. It was shown in [5] that they hold for a number of familiar examples (including the Grigorchuk group and the Gupta–Sidki groups). Further examples of branch groups whose structure graph is a tree are Examples 1, 2, 3 described in Section 2, with each H_n primitive in Example 2. The structure graph of the group of Example 2 fails to be a tree if at least one of the permutation groups H_n has point stabilizer contained in two incomparable subgroups of H_n.

References

[1] A. M. Brunner, S. Sidki and A. C. Vieira. A just non-solvable torsion-free group defined on the binary tree. *J. Algebra* **211** (1998), 99–114.

[2] R. I. Grigorchuk. The growth degrees of finitely generated groups and the theory of invariant means. *Izv. Akad Nauk. SSSR Ser. Mat.* **48** (1984), 939–985. (English transl. *Math. USSR Izv.* **25** (1985), 259–300.)

[3] R. I. Grigorchuk. Just infinite branch groups. In *New horizons in pro-p groups* (Birkhäuser, Boston, 2000), pp. 121–179.

[4] R. I. Grigorchuk and J. S. Wilson. A structural property concerning abstract commensurability of subgroups. *J. London Math. Soc.* (2) **68** (2003), 671–682.

[5] R. I. Grigorchuk and J. S. Wilson. The uniqueness of the actions of certain branch groups on rooted trees. *Geom. Dedicata* **100** (2003), 103–116.

[6] M. Gromov. Structures métriques pour les variétés riemanniennes. CEDIC, Paris, 1981.

[7] N. Gupta and S. Sidki. On the Burnside problem for periodic groups. *Math. Z.* **182** (1983), 385–388.

[8] P. Hall. Wreath powers and characteristically simple groups. *Proc. Cambridge Philos. Soc.* **58** (1962), 170–184.

[9] P. D. Hardy. Aspects of abstract and profinite group theory. Ph.D. thesis, University of Birmingham (2002).

[10] P. M. Neumann. Some questions of Edjvet and Pride about infinite groups. *Illinois J. Math.* **30** (1986), 301–316.

[11] E. L. Pervova. Everywhere dense subgroups of one group of tree automorphisms. *Trudy Mat. Inst. Steklova* **231** (2000), 356–367. (English transl. *Proc. Steklov Inst. Math.* **231** (2000), 139–150.)

[12] D. Segal. The finite images of finitely generated groups. *Proc. London Math. Soc.* (3) **82** (2001), 597–613.

[13] J. S. Wilson. Some properties of groups inherited by subgroups of finite index. *Math. Z.* **114** (1970), 19–21.

[14] J. S. Wilson. Groups with every proper quotient finite. *Proc. Cambridge Philos. Soc.* **69** (1971), 373–391.

[15] J. S. Wilson. On just infinite abstract and profinite groups. In *New horizons in pro-p groups* (Birkhäuser, Boston, 2000), pp. 181–203.

[16] J. S. Wilson. *Profinite groups* (Clarendon Press, Oxford, 1998).

[17] J. S. Wilson. On exponential growth and uniformly exponential growth for groups. *Invent. Math.* **155** (2004), 287–303

[18] J. S. Wilson. Further groups that do not have exponential growth. *J. Algebra* **279** (2004), 292–301

Printed in the United States
by Baker & Taylor Publisher Services